THE PENTAGON, CLIMATE CHANGE, AND WAR

THE PENTAGON, CLIMATE CHANGE, AND WAR

**Charting the Rise and Fall
of U.S. Military Emissions**

NETA C. CRAWFORD

The MIT Press
Cambridge, Massachusetts
London, England

The MIT Press would like to thank the anonymous peer reviewers who provided comments on drafts of this book. The generous work of academic experts is essential for establishing the authority and quality of our publications. We acknowledge with gratitude the contributions of these otherwise uncredited readers.

This book was set in Adobe Garamond and Berthold Akzidenz Grotesk by Jen Jackowitz. Printed and bound in the United States of America.

Library of Congress Cataloging-in-Publication Data is available.

ISBN: 978-0-262-04748-7

10   9   8   7   6   5   4   3   2   1

Contents

IV THE WAY AHEAD

INTRODUCTION

"Every tonne of CO_2 emissions adds to global warming."[1] There are many ways human activity has put global warming gases—carbon dioxide, methane, nitrous oxide, fluorinated gases, and water vapor—into the atmosphere. The United Nations Intergovernmental Panel on Climate Change (IPCC) assessment reports have consistently warned of potential ecological and social catastrophe if humans fail to reduce their greenhouse gas emissions. For many, the catastrophe has arrived; the world's forests are on fire, coral reefs are dying, and humans are on the move, fleeing excessive heat and escaping vast zones of too much or too little rainfall.

For more than a decade, the United States national security establishment has been a leader within the federal government in putting aside questions about whether climate change is real to focus on understanding the strategic implications of climate change; policy analysis and doctrine take for granted that global warming is already occurring, that it has and will affect the U.S. military, and that climate change may spark war. Current and former military and intelligence officials, and several think tanks, including the RAND Corporation, the Center for Naval Analysis, and the much newer Center for Climate & Security have consistently raised the alarm about climate change. In 2019, a group of senior military and security experts announced in the Netherlands their formation of the International Military Council on Climate and Security (IMCCS) and began in 2020 to produce annual assessments, called the *World Climate and Security Report*.[2] Members of the national security community have become more urgent in asserting that climate change may cause armed conflicts.[3] In 2013, United States Navy

Admiral Samuel J. Locklear II, then chief of United States Pacific Command (renamed Indo-Pacific Command in 2021), stated that instability caused by climate change "is probably the most likely thing that is going to happen . . . that will cripple the security environment, probably more likely than the other scenarios we all often talk about."[4] Or as one U.S. military officer said to me recently, "People will start fighting over access to water pretty quickly." Books such as Gwynne Dyer's *Climate Wars: The Fight for Survival as the World Overheats* and Harald Welzer's *Climate Wars: What People Will Be Killed for in the 21st Century* have contributed to a sense of alarm.[5]

The Biden administration put climate change at the heart of its national security and stressed the potential for climate change caused conflicts. President Biden said, "We will prioritize defense investments in climate resiliency and clean energy."[6] Yet, unlike the scientific consensus that greenhouse gas emissions cause global warming and that emissions must be reduced, scholars are divided over the relationship between climate change and war. Adding the mission of responding to potential climate change caused conflicts to the DOD's already extensive and expensive list of missions will likely increase the military budget and their greenhouse gas emissions.

The greening of the military is important, but little of the analysis has focused on the fact that the U.S. military has historically been a great driver of fossil fuel use and continues to be the largest emitter in the federal government. In fact, although the U.S. military funded some of the first scientific research on global warming and was an early advocate of anticipating and reacting to the threats and challenges posed by climate change, the Pentagon resisted including military emissions during the negotiations that led to the 1997 Kyoto Protocol. For the first time, the Fiscal Year 2021 National Defense Authorization Act required the DOD to produce a report on its greenhouse gas emissions over the previous ten years, and the 2022 National Defense Authorization Act required the administration to produce a plan to reduce DOD emissions.[7]

Since many who work in my field (international security) are not yet well versed in the language of greenhouse gas emissions, it is important to clarify a few terms of art. The international standard for calculating the greenhouse gas (GHG) emissions of institutions includes counting emissions through the product cycle of inputs, production, and outputs. Figure 0.1 illustrates the

Overview of GHG Protocol scopes and emissions across the value chain

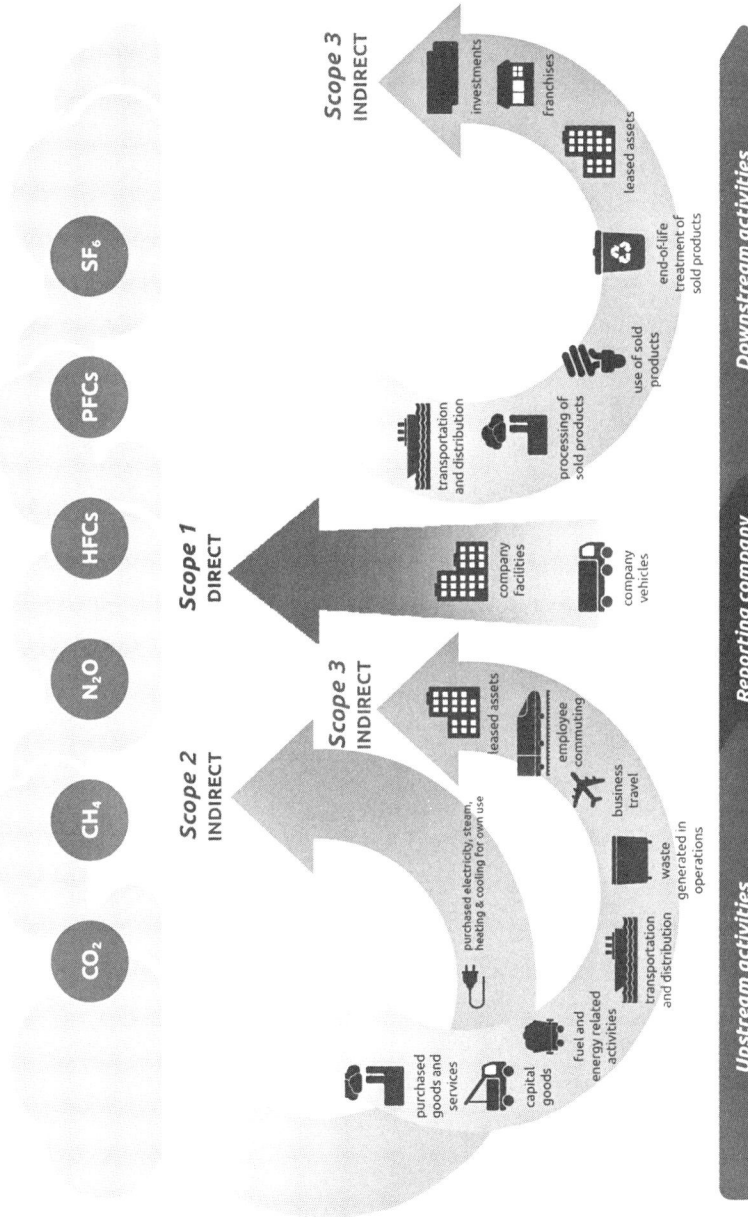

Figure 0.1

Overview of GHG protocol scopes and emissions across the value chain.

Source: United States Environmental Protection Agency, https://www.epa.gov/climateleadership/scope-3-inventory-guidance.

kinds of emissions that are counted for a typical business. In the clouds of this illustration, in order, are the chemical formulas for greenhouse gases—carbon dioxide (CO_2); methane (CH_4); nitrous oxide (N_2O); hydrofluorocarbons (HFCs); perfluorochemicals (PFCs); and sulfur hexafluoride (SF_6). CO_2 equivalent, or CO_2e, is a measure of all greenhouse gases that takes into account the different warming potentials of different greenhouse gases. Although it is not in the illustration, water vapor is also a greenhouse gas. Scope 1 emissions are those directly produced by the institution; scope 2 emissions are those that occur from the production of the electricity and steam power that the institution purchases; and scope 3 emissions are those that occur upstream and downstream of what an organization does—its value chain.

The U.S. military, like all armed forces, emits greenhouse gases directly (scope 1 emissions) when it operates vehicles and consumes fuel at its bases and installations. Emissions also occur when the DOD purchases energy (scope 2 emissions), such as electricity and steam. For a military organization, scope 3 include the goods and services it acquires, employee commuting, employee business travel, and the emissions of leased assets. Downstream emissions include those from waste disposal and the use of its products. A comprehensive understanding of military downstream emissions would include the emissions released as a consequence the use of force, such as when a city or oil infrastructure is bombed, and the emissions associated with reconstruction.

The U.S. Department of Energy (DOE) and DOD have reported that U.S. Department of Defense scope 1 and 2 emissions were 54.8 million metric tons of CO_2e (MMTCO$_2$e) in fiscal year 2019. Table 0.1 reproduces the DOD's accounting of its total scope 1 and 2 emissions for 2010–2019, which track surges in the wars in Afghanistan, Iraq, and Syria. The next year, fiscal year 2020, DOD emissions were lower, 51.5 MMTCO$_2$e.[8] This continues a recent trend in reductions in U.S. military greenhouse gas emissions. At the time of this writing, U.S. military greenhouse gas emissions are the largest share—currently about 75 percent—of all U.S. government GHG emissions.

The military's scope 1 and 2 emissions in any one year are just a snapshot of a slice of a larger system. The Department of Energy reported some of the scope 3 emissions for the Department of Defense for a few years—for fiscal

Table 0.1

DOD scope 1 and 2 emissions, CO_2e, for fiscal years 2010–2019

DoD Agency Wide Emission (MTCO2e)	2010	2011	2012	2013	2014	2015	2016	2017	2018	2019
Installation	27,014,937	25,680,151	24,387,027	24,098,546	23,820,379	23,927,275	22,068,190	21,320,275	20,872,157	20,683,925
Operational	49,508,306	48,753,207	44,945,277	39,499,301	38,056,442	38,671,866	37,239,955	37,073,325	34,534,487	34,088,337
Total Emissions	76,523,243	74,433,358	69,332,305	63,597,848	61,876,821	62,599,140	59,308,145	58,393,599	55,406,644	54,772,262

Source: Office of the Assistant Secretary of Defense for Sustainment, "Report on Greenhouse Gas Emission Levels" (Department of Defense, August 2021).

years 2008 and 2010–2016. Included in this accounting of scope 3 emissions were transportation and distribution losses, employee air and ground business travel, employee commuting, and contracted wastewater treatment and municipal solid waste disposal. For the years the DOE reported those emissions, they averaged 7.4 $MMTCO_2e$ annually.[9] The DOE accounting of Department of Defense scope 3 emissions did *not* include military-industrial emissions or the consequences of military operations abroad.

A comprehensive account of the military's greenhouse gas emissions (its carbon "footprint" or "bootprint") would also include all scope 3 emissions, which encompasses the emissions created by the military industries production of equipment and war material.[10] There are dozens of weapons manufacturers in the United States, and many more companies that are not considered part of the defense industry that make components that become part of the military-industrial process.[11] The emissions (scope 1 and 2) of the top twelve largest military-industrial companies in the United States totaled to about 5.2 $MMTCO_2e$ in 2019. These weapons are made up of components produced by other suppliers; I estimate that the entire supply chain for the twelve largest military industries contributed an additional 45.7 $MMTCO_2e$ in 2019 and so the total emissions of the top twelve U.S. military manufacturers, including their supply chain, were about 50.9 $MMTCO_2e$, or the size of the emissions produced by another country with a medium-sized economy. Thus, U.S. military and military industry emissions for the top twelve U.S. military manufacturers totaled 105.7 $MMTCO_2e$ in 2019. If the annual emissions of the United States military and military industry are compared to the emissions of a country, they would be about the same as those of Kuwait or the Czech Republic. They are about equal to the combined emissions of the world's forty-five smallest emitting countries; or the combined annual emissions of Sweden, Denmark, and Norway; or the annual emissions of the states of Virginia or Wisconsin; or the combined emissions of Maine, New Hampshire, Vermont, and Massachusetts.

This accounting of total military and military-industrial emissions is conservative. It does not count the Department of Energy emissions associated with the production of nuclear weapons. Nor does it include the standard and nonstandard scope 1 and 2 emissions of the Coast Guard, a military service and law enforcement organization that has been part of the U.S. Department of Homeland Security (DHS) since March 2003. Nor does my estimate include all U.S. military-industrial emissions, but only the top twelve companies that produce military equipment, and an estimate of their supply chains. My estimate also does not include the emissions of the defense contractors who perform an increasing share of service work for the U.S. military. Nor does it include the emissions caused by the destruction that is characteristic of war, including of oil wells and oil storage facilities. Nor have I estimated the emissions associated with reconstruction after war—such as the emissions from the production of steel and concrete for new buildings.

How should we think about the scale of U.S. military and military-industrial emissions and whether they are significant enough that policy makers and the public ought to be concerned about reducing them? The answer depends on one's perspective. On the one hand, the United States economy, at the moment the world's largest, is so large that the U.S. military contributes a small portion of the nation's entire greenhouse gas emissions. In 2019 total gross U.S. greenhouse gas emissions were 6,558 million metric tons of CO_2e.[12] Other sources and sectors produce large quantities of greenhouse gas emissions. Power plants and transportation are large producers of carbon dioxide; ruminant farm animals and natural gas systems produce large quantities of methane; agriculture and wastewater treatment are large sources of nitrous oxide.[13] United States military emissions, at about 55 $MMTCO_2e$ in 2019 were responsible for less than 1 percent of total U.S. emissions. When military-industrial emissions are included, total military and military-industrial emissions are just under 2 percent of total U.S. emissions, still not a large portion. The conclusion, from this perspective, might be that we don't need to worry too much about the scale of U.S. military greenhouse gas emissions, especially since they seem to be going down rather steadily.

On the other hand, as the U.S. Department of Defense says about itself, it is the single largest energy consumer in the United States.[14] The DOD is also the largest fossil fuel user in the federal government, and consequently, its largest greenhouse gas emitter. In 2020, the DOD accounted for 76 percent of all federal energy consumption and 75 percent of all federal greenhouse gas emissions.[15] Indeed, the U.S. military is the single largest institutional fossil fuel user in the world and thus the world's single largest greenhouse gas emitter. As I have noted, annual DOD greenhouse gas emissions are comparable in scale to the annual greenhouse gas emissions of entire countries.[16] In fact, because many countries' emissions are comparatively small, DOD annual emissions are larger than the combined emissions of many small countries. The following table 0.2, a snapshot of U.S. military and military-industrial emissions in a comparative perspective, also shows that U.S. military emissions

Table 0.2

Comparison of CO_2e emissions of the U.S. military–related emissions with selected country CO_2 emissions in 2019

	Total emissions of MMTCO$_2$ in 2019	Population in millions	Emissions per capita of MMTCO$_2$ in 2019
China	11,535.2	1,420	8.1
United States	5,107.3	329.1	15.5
Canada	584.8	37.3	15.7
United Kingdom	364.9	66.9	5.4
Peru	56.3	32.9	1.7
U.S. military emissions (CO_2e) per active duty, National Guard, and reserve forces*	54.8	2.2	24.9
U.S. military emissions (CO_2e) per active duty, National Guard, reserve, and DOD civilian personnel*	54.8	2.9	18.6
Total U.S. military emissions and estimated emissions of the top 12 defense contractors (CO_2e), per active duty, reserve, and civilian personnel**	117.5	2.9	39.9
Hungary	53.1	9.6	5.5
Sweden	44.7	10.1	4.5
Switzerland and Lichtenstein	39.4	8.6	4.6
Denmark	31.1	5.8	5.4

Sources: Country data, rounded to the nearest tenth: M. Crippa et al., *Fossil CO₂ Emissions of All World Countries: 2020 Report* (Joint Research Centre, European Commission, 2020), https://publications.jrc.ec.europa.eu/repository/handle/JRC121460. See also DOD, DOD Personnel, Workforce Reports & Publications, https://dwp.dmdc.osd.mil/dwp/app/dod-data-reports/workforce-reports.

*Military emissions scope 1 and 2 military emissions.

**Military emissions scope 1 and 2; military industry scope 1 and 2, and estimated supply chain emissions.

are more intensive than average national emissions. With 2.9 million people employed in active duty, guard, reserve, and civilian capacities, the United States military and military-industrial emissions per capita are greater than the CO_2 emissions per capita of many countries.

U.S. military emissions are also large in comparison to the military emissions of other countries. For example, Stuart Parkinson estimates that the direct military emissions of the UK in 2017–2018 were 3.03 million metric tons (MMTCO$_2$e). Parkinson and Linsey Cottrell estimate total military greenhouse gas emissions for the European Union's twenty-seven members in 2019 were 24.83 MMTCO$_2$e, not including the UK. EU military emissions were about half the amount of U.S. military emissions in the same year. Parkinson and Cottrell's research estimate the military-industrial emissions of the EU and UK as, respectively, 1.7 to 2.3 MMTCO$_2$e for Europe in 2019 and 1.23 MMTCO$_2$e for the UK in 2017–2018.[17] There are, as far as I know, no estimates of Chinese military or military-industrial emissions, though I expect that given China's reliance on coal-fired industry, its military and military-industrial emissions are substantial. While military emissions in any one year may be enormous, most greenhouse gas emissions persist in the environment for decades, sometimes centuries. Their impact is thus cumulative.

THE DEEP CYCLE AND CLIMATE CHANGE AS AN EXISTENTIAL THREAT

In 2007, several top U.S. national security experts compared the risk of climate change and the potential consequences of nuclear war. These former officials, who had served presidents of both parties, concluded: "The collapse and chaos associated with extreme climate change futures would destabilize virtually every aspect of modern life. The only comparable experience for many in the group was considering what the aftermath of a U.S.-Soviet nuclear exchange might have entailed during the height of the Cold War."[18] Indeed, thousands of nuclear weapons remain in the hands the United States, Russia, the UK, France, China, India, Pakistan, Israel, and North Korea. Global warming and nuclear weapons are the most serious threats to security and survival that humans and other species have ever faced.

Nuclear war and global warming seem quite opposite in some respects—the potential instantaneous fire of nuclear explosions and the fear of a subsequent "nuclear winter" versus the slow but certain melting of glaciers and sea level rise, punctuated by wildfire or violent storms. However, the existential threat posed by nuclear war and military greenhouse gas emissions are both manifestations of the theory that military force is necessary to secure us from some potential danger and to ensure a way of life. According to nuclear deterrence strategy, nuclear weapons are meant to deter dangerous adversaries. They also allow great powers to determine the shape of world order. Further, the militaries that are symbolic of sovereign states are there to protect the people inside them from invasion or domestic unrest. While armed forces are sometimes used in wars of aggression, they are first of all, intended to provide security. An individual's consumption of fossil fuels and the deforestation that fueled economic growth were meant to secure us from poverty, hunger, and cold. Fossil fuel consumption has also been part of the industrial growth cycle, raising standards of living, and making some people enormously wealthy.

How did U.S. military emissions grow to be so large? How did our military forces become part of the problem, adding to the threat that climate change poses to our survival? Why did the DOD resist counting all military emissions in the 1990s? What caused the military to start attending to climate change? Why have U.S. military emissions declined in recent years? Are the predictions of "climate wars" realistic or alarmist? Is the U.S. national security strategy appropriately working to avert the likelihood and risk of climate wars? If climate change could cause war, how does the United States need to be prepared for those wars and for all other potential military contingencies?

The book tracks the formation of a three-stranded braid and offers a way to reconceptualize and change the seemingly inevitable and tightly woven relationship between fossil fuel use and military, industrial, and strategic institutions.

The first strand, the growth of U.S. military emissions since the nineteenth century, is rooted in the way U.S. military and foreign policy decision makers have thought about the relationship between war and fossil fuels. I argue that the U.S. economy and military have, for more than 170 years, been on a path that has become the *deep cycle*: a long-term cycle of economic growth, fossil fuel use, and dependency. The dominant narrative describing the rise of greenhouse gases, which stresses increasing human population, does not mention the role of militarization and war. For example, the National Oceanic and Atmospheric Administration said in 2016: "For most of human evolution, CO_2 levels hovered around 278 ppm [parts per million], helping to maintain the global climate in a relatively stable state conducive to agriculture and the growth of human populations. That all changed starting in the 1850s with massive deforestation around the world." The cause of deforestation in the 1850s was increasing agricultural production to meet the needs of a growing human population. "Then in the 1950s, a dramatic increase in the burning of fossil fuels—coal to make electricity and steel, oil for vehicles and manufacturing—vastly accelerated the rate of CO_2 being pumped into the atmosphere."[19] The demand for manufactured goods accelerated in line with population growth.

Yet, war and mobilization for it has been a spark for deforestation, innovation, industrialization, and increasing fossil fuel use in the United States and elsewhere. Specifically, the military's role in the search for markets in Asia, the Civil War, the wars of westward expansion, the colonization of the Philippines, and World War I and World War II industrialization prompted innovations in transportation that accelerated fossil fuel use and necessitated the acquisition of bases for refueling military and commercial vehicles. In this, I agree with emerging literature in climate change and transportation research that shows that military demand for fossil fuel has been and continues to be a key driver for the adoption of fossil fuels.[20] Of course, the perceived necessity for access to fossil fuels was also driven by the logic of trade/market globalization and consumer capitalism and it is difficult to disentangle the role of military and commercial drivers of increased emissions. As the U.S. military grew increasingly dependent on coal, and then oil, in the nineteenth and twentieth centuries, it focused on developing the tactics, strategies, and military bases and equipment necessary to ensure easy access to a secure supply of fossil fuels. Indeed the U.S. armed forces and other militaries have, at times, shown flashes of tactical brilliance in ensuring that they would have a steady supply of fuel. This was the institutionalization of fossil fuel demand

and use, which then resulted in past and current military greenhouse gas emissions. Part of this strand is how, in the last several decades, the U.S. military has also become increasingly interested in fuel efficiency as a way to increase its mobility and decrease vulnerability to fuel shortages and adversaries who could attack fuel in transit.

The second strand consists of the ways the military's dependence on fossil fuels for war have shaped the world beyond the military itself. Institutions were constructed over the last two centuries to realize decision makers' beliefs about the role of fossil fuels in war—weapons, political alliances, bases, military doctrine, infrastructure, and industries that have come to shape our world and the choices we believe are inevitable. In other words, beliefs about fossil fuels became military doctrines that were self-evident to military leaders who stressed the necessity of acquiring or protecting access to fossil fuels and refueling stations, and denying those fuels to enemies. Further, military industries were, in times of war, disproportionately supported by government demand and subsidies and this in turn shaped and stimulated the larger industrial economy, further cementing the importance of fossil fuels and the necessity of protecting access to them. I show that, at least in the past two centuries, beliefs about the role of fossil fuels and the institutional structures that realized those beliefs have led to or intensified some regional rivalries and armed conflicts. And so, as the military protected its capacity to go anywhere and do anything policy makers should want, they spent resources—including fossil fuel—to ensure access to oil. The desire to protect access to oil for both war and industry have at times led to war, and so this is also a tale of strategic blindness and inflexibility.

The third strand is the military and international security community's dawning realization of the causes of global warming and its members' far-sighted understanding and beliefs about global warming's daunting consequences, including the link between climate change and social and political stress. Humans are increasingly conscious of and reacting to the effects of climate change on their local environments: the increase and intensification of flooding, storms, fires, drought, and famine that have killed people and led to migration as people move from areas where they cannot grow food, or where life has become untenable. The fact that the atmosphere has been altered by fossil fuel use, and this in turn has stressed human institutions, caused some military leaders, academics, and strategists to warn that global warming could to lead to increased risk of armed conflict—that climate change is a "threat multiplier." They even warn of potential "climate wars." But the fear about climate war has been contested by others who argue that even as governments are strained to meet the demands caused by climate disasters, even as some resources such as water and arable land become scarce, armed conflict is not inevitable. While human security—public health, food security, and the means to make a living—will be impacted by climate change, there is concern that "securitizing" climate change risks militarizing our response, which itself could be wasteful and ineffective, or at worst, counterproductive, exacerbating global warming.[21]

The three strands—beliefs about war and fossil fuel, the institutions constructed over hundreds of years to make war possible and successful, and the environmental and social consequences of those beliefs and actions, including the potential for human insecurity and war brought about by climate change—are woven tightly and seemingly inextricably together. Any one country's military and military-industrial emissions are a consequence of its larger understanding of its strategic context—its grand ambitions, its view of the threats it faces at the moment, and its fears about the future. In other words, policy makers' beliefs about the world—embedded in military doctrines, organizational interests, the physical infrastructure of its domestic and overseas bases, and the scale of its military industry—leave an atmospheric trace. To the extent that countries enter into and continue militarized strategic rivalries, they can extend and deepen the cycle of military fossil fuel demand, alliances, acquisition of overseas bases, and military industrialization. And a military's emissions are also a consequence of the political, strategic, and economic choices that have over several decades developed into institutions—or in the case of the United States, over about two hundred years. In sum, through their belief systems, their accounts of cause and effect, decision makers construct a world of institutions and practices that have both intended and unintended consequences. Thus, reducing military emissions will require rethinking strategic assumptions and breaking a deep cycle of economic growth, expansion, military mobilization and military-industrial production that is reliant on fossil fuels.

OUTLINE OF THE BOOK

The book has four parts. Part I, "The Deep Cycle," focuses the history of the Pentagon's fossil fuel use, focusing on the beliefs that shaped U.S. military doctrine and force structure and how fossil fuels shaped war, U.S. alliances, and bases. The belief held among U.S. foreign policy decision makers that fossil fuels are more than the means to greater convenience and productivity, but a vital strategic necessity—worth making extraordinary efforts to gain access to and forming unsavory contracts and alliances, and perhaps even killing for—grew out of the exigencies of the wars of the nineteenth and twentieth centuries and all the beliefs about what is necessary and legitimate to do in war.

Chapter 1, "So Hungry, So Thirsty: Coal, Oil, and War" shows that from the point when the U.S. Navy supplemented sail with steam engines fueled by coal to propel its ships, the U.S. military has been increasingly dependent on fossil fuels to perform what it considers essential missions. Steam engines, initially powered by wood and coal also powered U.S. weapons manufacturing in the nineteenth century. Since then, the military has been focused on fossil fuel logistics—finding and protecting reserves of coal and oil, securing access to fuel at the lowest price and sometimes any cost, and denying its adversaries access to fuel. When the search for markets led the United States to extend its power to Asia, coaling stations were necessary, as were ports, most notably in Hawaii. The coaling stations in the Pacific enabled commerce, the U.S. war in the Philippines, and America's long occupation there. The nineteenth century concerns about coal prefigured the twentieth century obsession with petroleum. Coal and oil powered the U.S. Navy in World War I, and U.S. fuel exports enabled British and French airpower. This deep cycle has reinforced and motivated geographic expansion to protect access to fuel, increasing military power and military spending. By the start of World War II, oil was central to military strategy, and both coal and oil powered U.S. industrialization through that war. Petroleum products were an essential element in U.S. and Allied mobility in World War II, as well as powering the factories that produced war material and explosives. Coal was the dominant source of U.S. energy until 1949, when it was surpassed by petroleum consumption.

The necessity for refueling also determined the location of U.S. bases after World War II. Further, as the United States became more dependent on Persian Gulf oil, it became increasingly concerned in the 1970s and 1980s about access to oil and the potential use of the "oil weapon." All the while that the United States fought cold and hot wars, U.S. military industry was powered by fossil fuels. Alliances with oil providers, including sometimes unsavory regimes, were believed to be necessary to protect access to the oil that powered U.S. industry.

Chapter 2, "The Life Blood and the Deep Cycle: Oil and U.S. Military Doctrine since World War II" describes the second and third strands: the world that foreign policy decision makers have made through our dependence on fossil fuels, and the ways the military is adapting to climate change and preparing for its likely and potential effects. Chapter 2 tracks how the military defined security in terms of access to fossil fuels and in particular how U.S. military doctrine and force structure over the past fifty years were driven by the mission to protect access to Persian Gulf oil. While U.S. military spending during the Cold War was largely driven by the competition with the Soviet Union, and in a larger sense the ideological confrontation with communism, it was also motivated by a concern to protect access to oil, in part to make sure that oil was available for war, but also because oil was increasingly seen as vital to the U.S. economy and way of life. Concern about access to fuel turned to anxiety about being cut off from it after the 1973 oil embargo and the Soviet Union's invasion of Afghanistan in 1979 and the Iranian Revolution in that same year. The seeming reasonableness of this anxiety was reaffirmed after the 1990 invasion of Kuwait by Iraq. However, even as U.S. dependence on Persian Gulf oil has declined in the last twenty years, the force posture associated with "defending the Persian Gulf" has remained. I show that taken-for-granted assumptions and scarcely challenged conceptions of U.S. interests and threats to those interests have yielded a legacy of bases, equipment, doctrine, and alliances that infuse current

U.S. military doctrine. The perceived need for fuel still shapes elements of U.S. grand strategy, military doctrine, and foreign policy in the Middle East and is also driving some Chinese military policy.

In part II, "The U.S. Military and Climate Change," I turn to the links between the climate change science, U.S. military fuel use, and emissions. Chapter 3, "Climate Change Science and the Politics of Counting Military Emissions," describes how the warming effects of the greenhouse gases carbon dioxide and water vapor were first discovered and the link between climate change science and national security that some scientists made in the 1950s. The U.S. Navy began to discuss the implications of climate change for their operations in the 1990s, and in the early 2000s the DOD began to be more concerned with fuel efficiency. Despite the navy's early recognition of the reality of climate change, the U.S. military wanted to maintain military preeminence—which its leaders believed required the capacity to have unconstrained greenhouse gas emissions. Thus, during the December 1997 Kyoto Protocol negotiations the DOD and U.S. negotiators worked to make sure the treaty would not constrain the U.S. military. Although this has largely been forgotten, the U.S. military resisted counting greenhouse gas emissions in the 1990s because it wanted to preserve American military supremacy. Thus, the Kyoto Protocol exempted counting some military emissions in national emissions reporting, ensuring that there would be no full and transparent accounting of the United States'—or for that matter, *any* government's—military greenhouse gas emissions. At the Paris Agreement in 2015, this exemption was modified, but full disclosure and military emissions cuts are optional, left to the discretion of the individual country. In practical terms, however, the United States has not, at the time of this writing, changed the form of its submissions regarding military emissions.[22]

Chapter 4, "A Guide to U.S. Military and Military-Industrial Emissions since 1975," uses publicly available data to estimate U.S. military greenhouse gas emissions. Only recently has the DOE published greenhouse gas emissions data for the DOD, for 2008 and continuously since 2010. Yet because greenhouse gases remain in the atmosphere for many years, it is also important to understand past emissions. Unsurprisingly, U.S. military emissions have tracked U.S. hot wars, the Cold War, and the overall size of the Pentagon.

Importantly, U.S. military emissions have declined in recent years—even while the United States was at war. I describe the causes of those emissions reductions, which include greater efficiency, switching fuels, and base closures. I also use publicly available data to estimate the scale of military-industrial greenhouse gas emissions for the top U.S. weapons manufacturers. It turns out that military-industrial emissions are roughly comparable or even larger than military emissions.

Part III, "U.S. Military Doctrine and National Security Strategy," describes the U.S. military and the larger national security community's understanding of the effects of climate change and the role of armed forces in dealing with global warming. The national security community has focused on three factors: their own vulnerability to the effects of climate change, how climate change may affect the capacities of other states and increase conflict, and more recently, their strategies for reducing their own emissions.

The military has become progressively concerned about climate change as both a complicating factor in operations and war and as a potential spur to war. Chapter 5, "Energy and Climate Security: Finding and Fixing Vulnerabilities," outlines how the U.S. military has become steadily aware of and concerned about the threats that global warming pose to its operations and infrastructure and has been leading a charge to reduce its vulnerabilities. Specifically, the U.S. armed forces worry about how global warming will alter the environment at installations and bases, complicate training and change the operational environment; bases are vulnerable to too much heat and too much and too little water, and operations will be complicated by changes in the physical environment in which they occur. Vulnerability to the effects of global warming is a powerful motivator. John Conger, a former assistant secretary of defense for energy, installations, and the environment during the Obama administration found that his colleagues at the DOD were not too concerned about emissions, making comments like, "OK, we'll lower emissions, but only in the context of helping our [military] missions." But Conger found that the DOD was more receptive to climate change directives when they thought about how rising seas affected naval installations—"how it's more about protecting ourselves from the environment than protecting the environment."[23]

Chapter 6, "Climate Change as 'Threat Multiplier,'" tracks the U.S. military and intelligence services' growing concern that climate change will cause regional and perhaps global instability, including war. Climate change has indeed moved from a factor that might exacerbate conflicts, to a near certain cause for conflict. Lt. General Laura Potter, deputy Army chief of staff for intelligence told the Association of the United States Army in October 2021 that, "The inescapable fact is that climate change leads to political instability. Through cascading cause-and-effect dynamics, the effects of global warming can compound, causing pressure on vulnerable governments."[24]

In a sense, it is natural that those tasked with providing for security would be attentive to the potential for climate change to cause war. Indeed, the military's authority and institutional power rests on its traditional role in providing military force as the solution to situations that are or may become military threats or as the tool for achieving U.S. ambitions. Further, many current and former officials say that the world has already seen the first climate wars—citing the Syrian civil war that ultimately spawned ISIS as an example. If those who predict climate wars are right, we are headed to a terrible future.

Yet, the scholars who study the links between climate and conflict believe that the wars some say were already driven by global warming had complex causes, and climate change may not have been a significant factor. Further, conflict and war need not be an inevitable outcome of the stresses caused by global warming. Nevertheless, out of an abundance of caution and perhaps a desire to keep its budget share (about 50 percent of all discretionary spending), the military leadership of the United States may be highlighting the risks of climate war and over-insuring against threats that are less likely than the much more likely direct threats posed by climate change. But preparing for climate war may be counterproductive if it keeps military spending and military emissions high.

Part IV, "The Way Ahead," begins from the assumption that it is not too late to forestall or avoid some of the direct threats that global warming has in store for us if the Pentagon continues business as usual or does not reduce emissions as rapidly as possible. What is the U.S. military doing, and what else could it do to reduce its emissions and reduce the likelihood of war? This raises the questions of whether a necessary step is downsizing the military and if it is possible to defend the United States with a smaller military.

Chapter 7, "A Lean Green Fighting Machine? Mitigation versus Adaptation," focuses on how the U.S. military is already responding to climate change. The armed forces have, for nearly two decades, been concerned with increasing fuel efficiency—largely as part of a push to increase range and to reduce fuel consumption in order to decrease the military's vulnerability to attacks on fuel in transit. Emissions cuts have been the side effect of attempting to reach other goals, for example, extending the range of aircraft or closing unneeded bases. It is also the case that for more than a decade, the military has, in response to government-wide mandates, worked to increase energy efficiency at buildings and on bases, where renewable energy is now in use. The military has also been an innovator in green technologies.

But the Pentagon seems to be assuming that the worst effects of climate change are unavoidable, and the DOD therefore is focused on adaptation. Further, even in December 2021, when the Biden administration ordered the federal government to act, through a "whole-of-government approach," to move to net-zero emissions by 2050, the DOD and other agencies could be exempted "when it is in the interest of national security."[25] What this means, in practice, is that while it reduces emission on the one hand, it simultaneously works to keep its core missions and functions intact, and perpetually rebuilds bases damaged by storms and floods—even as reconstruction and adaptation impede making more significant cuts in emissions that would help reduce the likelihood of the worst-case scenarios that the DOD leadership fears.

In chapter 8, "The Path to Climate Security," I argue that the United States is at a critical inflection point, where it could choose between the modest military emissions reductions and more significant reductions. I return to the concept of a deep cycle of war, militarization, and increased emissions and argue that war has fostered increased greenhouse gas emissions in both the military-industrial and civilian sectors. The U.S. can and should break out of that cycle. Although the United States' diminished dependence on Persian Gulf oil, and the DOD's belief that climate change is a looming challenge and may be a "threat multiplier" should have caused the U.S. military to reevaluate its doctrine and force structure, U.S strategy still has not been significantly revised. Rather, the DOD and national security elites appear be fighting the last wars, with doctrine much the same as it has been since the mid-nineteenth century—what I describe as an oil consumption, militarization, and conflict cycle. Specifically, even as the Pentagon assumes that global warming is a fait accompli, the strategy is essentially the same as it was in 1990, which is essentially the same as it was in 1890; it works to assure that access to fuel is unimpeded. In other words, the DOD has prioritized maintaining and expanding military capability that it believes is essential to preserving U.S. superiority. I argue that the U.S. can and should rethink its security posture and make reductions in military emissions that are much deeper than the U.S. political and military leadership currently contemplates.

Military supremacy—the capacity to control outcomes anywhere in the world—is not the same as self-defense. The concern that counting and limiting greenhouse gas emissions would limit the United States' capacity to defend itself was ill placed and, in any case, the United States faces more risks and certain consequences from climate change than from loss of access to Persian Gulf oil or from most potential military conflicts.

Some believe that the military can continue to modernize its weapons and all the while be a leader in technological innovations that will help us reduce emissions and respond to climate change. The DOD has already done so. But I think it is a mistake to invest this way; counting on the military to innovate the United States out of climate change crises is inefficient at best and will not likely lead to reductions in U.S military emissions on the scale and timeline that are required and that could be accomplished by reducing the size of the military. In addition, while the potential for climate change to spark climate wars is conceivable, militarizing the U.S. response to climate change raises the prospect of wasting resources or even potentially increasing the risk of war. The human misery caused by climate change is not best avoided or diminished by military means. While the Pentagon is moving in the right direction by gradually reducing military emissions, I show that more significant military emissions cuts are possible. Indeed, the

U.S. military can reduce emissions, restructure its forces, and still defend the United States from *likely* military threats.

At the UN Climate Conferences in Paris in December 2015 and in Glasgow in November 2021, the world's leaders emphasized the necessity of reducing greenhouse gas emissions. The Biden administration committed the United States to dramatically reducing greenhouse gas emissions, including U.S. government emissions, by 2030.[26] This a now familiar and urgent refrain. However, if someone is searching for a magic bullet—the single most significant greenhouse gas contributor or activity whose emissions, if curbed, would end the climate crisis—military emissions are not it. There is no *single* action that will end the climate crisis; if the world is to avoid the worst consequences of global warming, many sources of greenhouse gas emissions must be reduced or eliminated, and the resources must be found to convert to lower or zero-emissions energy sources and prepare for the climate changes already baked into the system. Everything must be on the table for emissions reductions, including activities that Americans don't often question, such as military emissions and military doctrines and operations that ensure access to oil, and war-related emissions. Emissions reductions will be difficult achieve unless the United States reckons with the scale of military emissions, rethinks national security strategy, and reduces military spending, which for the last several years has consistently been more than half of all U.S. government discretionary spending. Further, just as the preparation and conduct of war have increased emissions throughout economies, reducing military emissions, and avoiding war, could have a positive ripple effect in economies, including in military industry, that would foster larger emissions reductions.

I THE DEEP CYCLE

1 SO HUNGRY, SO THIRSTY: COAL, OIL, AND WAR

Coal lay in ledges under the ground since the Flood, until a laborer with pick and windless brings it to the surface. We may well call it black diamonds. Every basket is power and civilization. For coal is a portable climate. It carries the heat of the tropics to Labrador and the polar circle: and it is the means of transporting itself whithersoever it is wanted.
—Ralph Waldo Emerson, 1860[1]

Fuel stands first in importance of the resources necessary to a Fleet. Without ammunition, a ship may run away, hoping to fight another day, but without fuel she can neither run, nor reach her station, nor remain on it, if remote, nor fight.
—Alfred T. Mahan, April 21, 1911[2]

How did the United States military come to consume so much fossil fuel? In one sense, the answer is simple. Energy—to manufacture weapons and equipment, to transport people to a defensive position or an offensive battle, and to provide mobility once engaged—makes war possible. But all militaries need energy. Although the focus here is the United States military's use of fossil fuels from the nineteenth century to the end of World War II—I discuss the post–World War II era in chapter 2—these dynamics were not unique to the United States. Understanding how the U.S. military in particular became dependent on fossil fuel requires that we track three interrelated processes.

First, the ever-expanding uses and utility of fossil fuels, coal and then oil, for all elements of war—to enable industrial production, mobility, and explosive power—created an enormous appetite for fossil fuel. The military's dependence on fossil fuels began in the nineteenth century when the U.S. Navy's desire for mobility led it to invest in coal-burning steamships, and then the U.S. Army and later the U.S. Army Air Forces invested in liquid fuels for tanks, trucks, and aircraft and the navy switched to oil-powered surface ships and submarines. The military's desire for fuel sometimes prompted efficiency and innovation, but it frequently made *more* fuel appear to be an obvious necessity. The utilization of fossil fuels for war accelerated and shaped energy transitions in the broader economy, which in turn made fossil fuels ever more valuable and worth protecting by any means necessary. Beginning in the nineteenth century, wars accelerated the resources governments made available—initially wood and then fossil fuel to power innovations and military industries—and accelerated the development of means of transportation and infrastructure that then diffused into the larger economy and outlasted the wars. And as wars fundamentally boosted and sped up fossil fuel demand, shaping industries and economic demand, this in turn bolstered the sense that fossil fuel was an essential commodity that had to be protected at all costs. The U.S. military and political leadership's decision to add the protection of access to fuel and the infrastructure for refueling to its conception of U.S. vital interests became a circular logic. War required fuel, the economies that made total war possible required fuel, and the reason for war increasingly was to protect those economic interests and that fuel.

Second, the requirement to ensure access to fuel shaped foreign policy and military strategy. The enlarging reach of the United States from a continental to a global economic and military power prompted the U.S. armed forces to value different and sometimes distant territory, whether to acquire bases for refueling or to find and secure the now seemingly indispensable coal and petroleum resources that would power mobility. This was a self-reinforcing dynamic, where the need for refueling to expand and protect U.S. interests required bases over ever-larger portions of the globe, while the bases and the fuel became themselves strategic interests. This logic of ever-expanding commercial interests, alliances, and military infrastructure was similar for the United Kingdom whose power also first rested on access to coal and then oil.

Third, strategies to deny others access to fuel were the other side of the strategic coin. Denying an adversary access to fuel could be accomplished through formal and informal alliances with governments and corporations that controlled access to fuel and commercial arrangements. An alliance with the leaders of an oil-controlling state could and sometimes did lead to supporting undemocratic and in some cases unsavory political regimes—such as the U.S. alliance with Iran in the 1970s and with Iraq in the 1980s, and arguably with Saudi Arabia. If an alliance partner failed, supported an adversary, or became an enemy, control of their oil resources could be accomplished through occupation, sabotage, blockades, or bombing. And thus, the desire to keep fuel out of enemy hands meant that the same resource would frequently be destroyed.[3]

In retrospect, it all seemed inevitable and necessary, the consequence of the basic human drives for physical security from the elements and each other, enacting the logic of competition in an anarchic world where each state must provide for its own security. The logic of each choice unfolded as if we were traveling on a rocket whose coordinates had been set by engineers long before we got on board, with the choice to get on board in the first place having been seen as imperative. The events described here were part of a deep cycle of fossil fuel dependency—first coal and then oil—that begins with the idea that fuel is essential, the "life blood" of the military and the larger economy. Anxiety about fuel, namely having it or preventing others from access, military expansion to guarantee access to fuel, and increased fuel use in the military-industrial sector went hand in hand. Hard as it is to imagine, humanity and U.S. military leaders could have made other choices. It all seemed so logical and necessary at the time. It still does. But, as I will show, while each step seemed to inexorably lock the United States on a path to greater fuel dependency, increased military industrialization, and tighter dependence on oil producers, none of this was "inevitable."

IN THE BEGINNING

For centuries, wind and muscle power were the primary means to move armed forces around, while woodfires and strong arms shaped forged weapons. Little was different when the first English settlers arrived in North America. For example, air and sea currents enabled the Pilgrims' arrival in North America in late 1620 after sixty-six days aboard the three-masted *Mayflower*, even as those same forces kept the Pilgrims from Virginia, their original destination. The *Mayflower* was armed with cannon and carried at least one experienced English soldier, Myles Standish, the man in charge of the colony's defenses. Since the ship was not maneuverable in shallow water, the Pilgrims had brought with them a disassembled "shallop" on the Mayflower's deck. When reassembled, the thirty-foot-long single-masted vessel with oars was a kind of mini-galley ship, with a shallow draft capable of landing on beaches. This was how Standish and the other armed men of the *Mayflower* first made their way ashore to explore the coast of Cape Cod and take from Native American food caches, near what is now called Provincetown, and how they escaped their armed clash with the Nauset Wampanoag people at what is now First Encounter Beach, at Eastham on Cape Cod.

Little altered the reliance on arms, legs, wind, water, and wood in North America through the Revolutionary War and the War of 1812. On the water, sail and strong arms worked with or against the tides. Wind, waves, and the shape of a harbor could and did constrain the where and the when of sea maneuvers. The rate of land travel was limited by the pace of foot soldiers, and the comparatively few cavalry; although sometimes supplemented by plunder, food, fodder, ammunition, and cannon were constrained by what could be carried in packs or on carts by beasts that themselves would need to be fed. Rivers and canals could and did ease the load and increase mobility until travelers came to the end of their reach, or drought made navigation impossible. And while the size of naval guns grew, it was a different situation on land where "anything too heavy for horses to pull across open country was ruled out for field artillery."[4]

Prior to the fossil fuel revolution, weapons were generally made one by one, forged by skilled artisans. The Springfield Armory—which provided most U.S. firearms until the middle of the twentieth century, and manufactured fewer than five thousand muskets per year before 1807—relied on the Mill River for power, but the river's "limited drainage basin meant that it often experienced both floods and droughts."[5] Although waterpower was not the only limiting factor, it was so important that the U.S. government eventually built dams on the Mill and Connecticut rivers to power the Springfield Armory and other weapons manufacturers there.[6]

Inventions powered by wood and then fossil fuels gradually changed all that in the nineteenth century. The desire to pump water out of coal mines inspired the successive invention and modification of the steam engine by Thomas Savory, Thomas Newcomen, and James Watt. In retrospect, though militaries were often slow to adopt new technologies that might require rethinking cherished doctrines, it is difficult to overstate the revolutionary effects of the innovations spurred and necessitated by the adoption of fossil fuel over the last two centuries. Coal-powered steam engines, first installed at the Springfield Armory in the 1830s, transformed the productive capacity of military industry, powering most military-industrial factories by the end of the century, enabling the industrial-scale manufacture of handguns, rifles, and cannons. As steam engines powered navies and eventually railroads, the mobility, speed, and the capacity for bringing firepower to bear against an adversary grew on the sea and on land. And so also the invention of the internal combustion engine in the late nineteenth century and the fabrication of hundreds of uses of petroleum for war would transform war in the twentieth century. And all of that required the logistics and infrastructure to protect and transport coal and petroleum from the source to make it available for military industry and war.

FROM THE WAR OF 1812 THROUGH THE CIVIL WAR

The *North River Steamboat* launched in 1807 was the first paddle boat designed by an American, Robert Fulton. Fueled by wood, the steamer (as steamships were often called) had an average speed of almost five miles per hour. Fulton built the U.S. Navy's first armed steamer, *Demologos* during the War of 1812 for the defense of New York Harbor, but the ship was not ready before the end of the war and in any case, the twin-masted ship had limited capacity to store coal. At her top speed of six knots *Demologos* was not as fast as the frigate *Constitution*, also known as Old Ironsides, whose best speed was 13.5 knots.[7] The first steam engines in the United States were powered by wood—then in plentiful supply. Indeed, apart from draft animals, wood was the dominant form of energy consumption in the United States until about 1880.[8] The U.S. Navy was also slow to adopt the steamship "tea kettles" because their paddles were vulnerable to attack, and the inefficiency of the early steam engines meant they were limited in range and tethered to nearby coaling stations. Increases in efficiency eventually allowed commercial steamers to travel much further. The two-masted paddle-wheeled passenger coal-powered steamer *Sirius*, the first to cross the Atlantic, traveled from Ireland to New York in 1837, cutting the then average travel time between continents by more than half, to eighteen days. While the *Sirius* showed the potential for steamers to ply the oceans, its coal bunkers were nearly empty when it pulled into New York, and the inefficiency of steam engines limited steamers' military use at sea and largely confined them to rivers, lakes, and coastal duty.[9]

The U.S. Navy began a dedicated program of wooden paddle-wheeled steamship construction and launched the more efficient and maneuverable propeller-driven steamers, starting with the three-masted *Princeton*, commissioned in 1843. At this point both paddleboats and propeller steamers were powered by a combination of coal and sail; sail could be used when there was wind, saving the steam engines to be engaged for maneuver later, or they could be used simultaneously, achieving slightly better speeds than by steam power alone. Thus, the lack of suitable wind would no longer becalm a fleet. In the early 1840s, the navy estimated that six steamers would consume around fourteen thousand tons of coal in a year.[10] To fuel these ships, in 1842 the secretary of the navy called for installations along the coasts of North America and South America and as far west as Hawaii.[11] Secretary of State Daniel Webster drafted the Tyler Doctrine proclaiming that Hawaii, called the Sandwich Islands by the British explorer Captain Cook, was within the U.S. sphere of influence. Among other things, the United States saw the islands as a desirable coaling station and port for whalers.[12]

Steamships played a crucial role in the U.S. invasion of Mexico in 1846 as they moved troops and supplies to Texas and eventually navigated the Rio Grande, carrying some of Zachary Taylor's force into Mexico. The steam-powered paddle boats reduced the army's reliance on plunder for food and fodder. Ocean-going steamers also participated in the U.S. blockade of the

Mexican coast, the conquest of California, and Winfield Scott's bombardment of Vera Cruz (now known as Veracruz) and its nearby fort in March 1847. More often remembered as the first large-scale amphibious assault in U.S. military history, where men rowed ashore in surfboats specially constructed for the landing, at Vera Cruz the steamers participated in getting the landing boats close to the shore. Afterward, the bombardment of the fort and the city were conducted from batteries on shore, and from offshore by the three-masted paddle steamer the *Mississippi*, and the *Princeton*. Also joining the bombardment were the newly constructed steamers *Vixen* and *Spitfire*. Those vessels, under construction in New York for the Mexican Navy prior to the war, "were offered" by the manufacturers to the U.S. Navy and purchased in New York by the United States Navy soon after the outbreak of the war.[13]

In the decades following the Mexican–American War, the United States government became increasingly concerned with the availability and quality of coal and the characteristics of steam engines. The U.S. government subsidized the development of fossil fuel technology by paying for tests on the qualities of different types of coal, and sponsoring engineers to develop innovations in steam engines and boat design. Secretary of the Navy James Dobbin reported some of the expenditures and the results of the previous nine years of experiments and tests to Congress in 1854, illustrating the government's concern to resolve a host of technical questions, including the efficiency of steam engines. Dobbin's report also included tables comparing the speed and fuel use of the steamships then in the U.S. arsenal. Table 1.1 illustrates the characteristics of some of those ships. But the "logs show that the engines were seldom worked up to their full power, and that when sails were used, the consumption of coal was invariably lessened—economy of fuel, and not a high rate of speed, being the great desideratum."[14]

In the commercial sector, mechanical energy was still largely supplied by wind and water power: in 1850 and 1860, wind and water were providing 64 percent and 58 percent of the power for mechanical work in the United States. Most trains were still powered by wood. Thus, the U.S. Navy was leading the way in adopting coal. In 1855, the annual report of the secretary of the navy stated simply that the "increase in the number of steam-ships will make further purchase of coal necessary, and require depots for the purpose, both

Table 1.1

Characteristics of mid-nineteenth-century U.S. naval steamers

Name and number of guns	Fulton 6	Mississippi 4	Princeton 7–9	Vixen 1
Year commissioned	1837	1841	1843	1846
Type	Side wheel and sail	Side wheel and sail	Propeller and sail	Side wheel and sail
Coal capacity	275 tons, 14 days	620 tons, 20.5 days	176 tons, 11 days	80 tons
Average speed steam, knots/hour; coal consumed in pounds/hour	8.90 1,515	7.32 3,401	6.87 1,452	6.19 585
Average speed steam and sail, knots/hour; coal consumed in pounds/hour	10.01 1,818	7.55 2,666	8.08 1,577	6.71 434
Maximum speed steam, knots/hour; coal consumed in pounds/hour	10.50 1,820	10.00 4,650	10.00 2,627	9.00 900
Maximum speed steam and sail, knots/hour; coal consumed in pounds/hour	11.75 2,185	11.00 3,024	10.50 750	9.00 720

Source: James C. Dobbin, "Steam Navy of the United States: Letter from the Secretary of the Navy Transmitting Papers Giving Information in Reference to the Steam Navy of the United States," 33d Congress, 1st Session, 1854, 3 and 12.

on the home and foreign stations."[15] The heavy rate of coal consumption for even the most efficient steamships meant that they would need frequent coaling, and thus, in the minds of those who wanted a more expansive U.S. military presence, coaling stations had to extend as far as U.S. interests. In a second term as secretary of state, this time for President Willard Fillmore, Daniel Webster set U.S. sights on "opening" Japan, which had largely isolated itself from commerce and diplomatic relations with the West since 1638. The Japanese islands were rumored to have coal, and Webster believed this to be true. Webster hoped to establish "a line of steamers from California to China," and said that it was essential for the United States to secure Japanese coal, "that great necessity of commerce" the "gift of Providence, deposited by the Creator of all things in the depths of the Japanese Islands, for the benefit of all the human family."[16] Webster said that "we should obtain from the Emperor of Japan permission, to purchase from his subjects the necessary supplies of coal, which our steamers on their out- and inward voyages may require."[17]

Commodore Matthew Perry, who had captained the *Mississippi*, the *Vixen*, and other steamships and participated in the bombardment of Vera Cruz during the Mexican–American War was chosen to lead the U.S. expedition to Japan and China in 1852. Perry saw expansion in the Pacific as a race against the British. The British Navy, which dominated the world's oceans, had enormous supplies of coal from mines in the United Kingdom and an extensive infrastructure of coal refueling stations throughout the globe. Perry's chief objectives were to find and secure supplies of coal for the United States, acquire "ports of refuge," and create commercial and diplomatic ties with Japan and so facilitate American expansion.[18] Perry left Hampton Roads, Virginia, in November 1852 on the *Mississippi*, his old ship, which had been specially reconfigured to carry 600 tons of coal over its original design capacity of 450 tons. Realizing the entire voyage would require substantial deliveries of both U.S. and British coal for the journey, Perry contracted to have a private firm supply the fuel. Thus, Perry proved that the United States could arrange in advance for coal to be delivered to ports for his ship and the four other steamers that joined him, but the squadron still needed to purchase coal *en route*. The questions were, how much would the United States pay? And would force be necessary to come to an agreement with Japan?

Coal was selling for $60 a ton in Shanghai, but Perry's expedition ultimately found coal he could purchase for $3 a ton on Formosa. However, the search for local supplies of coal continued. After disbelieving the local's insistence that there was no coal to be found on one of the islands, members of the crew of the *Macedonian*, including a civilian geologist, literally followed a trail of coal from a village on Keelung further inland to a mine. Perry signed the *Convention of Kanagawa*, a treaty guaranteeing U.S. access to coaling at two ports and promising to only purchase coal from Japanese government officials.[19] Upon his return to the United States Perry extolled the virtues of commerce with China and Japan in a paper presented to the American Geographical and Statistical Society in March 1856. In his presentation, Perry underscored that Japan, Borneo, Formosa, and China all had "coal, the most valuable to commerce of all the minerals since the introduction of steam in aid of navigation."[20] Perry, a believer in his country's "manifest destiny" advocated for the United States to colonize the uninhabited Bonin Islands located South of Tokyo, because they would be useful for those trading with Japan and China. He wrote:

> It requires no sage to predict events so strongly foreshadowed to us all; still "Westward" will "the course of empire take its way." But the last act of the drama is yet to be unfolded; and notwithstanding the reasoning of political empirics, Westward, Northward, and Southward, to me it seems that the people of America will, in some form or other, extend their dominion and their power, until they shall have brought with their mighty embrace multitudes of the islands of the great Pacific, and placed the Saxon race upon the eastern shores of Asia.[21]

Perry predicted that the two "exponents of freedom and absolutism"—the United States and Russia—would inevitably meet in competition for the region. He wrote, "I think I see in the distance the giants that are growing up for that fierce and final encounter."[22] Just five years later, the United States was at war with itself.

Sometimes called the world's first industrial war, wood and coal burning played an important role during the U.S. Civil War. The North maintained a blockade of the southern coast with a combination of sail and steamships, and the Confederacy ran the blockade with sail and steamers, although the bulk of the blockade running was done by the faster steamers.[23]

In addition to its role in storing carbon, wood was also an important source of power during the Civil War. Throughout U.S. history until the early 1900s, forests supplied wood for home heating, transportation, and industrial power. Until about 1870, U.S. railroads ran primarily on wood-fueled steam engines.[24] Both sides cut down forests to make obstacles, build bridges, repair railways, construct "corduroy" roads of logs set perpendicular to the direction of travel over otherwise impassable sand and mud. Matthew Carr notes that during the Civil War, "Forests were stripped and cut down to make breastworks, trench fortifications, and chevaux-de-frise [sharpened stick obstacles] or set on fire in the course of battles."[25] The Confederacy cut down enormous trees along the Yazoo Pass to render the river unusable by the Union as it tried to move troops toward Vicksburg in early 1863.[26] The forests were also used for firewood and to support tents; as one Union soldier, Henry Hitchcock, remarked "It is bad for the live oaks and cedars that so many soldiers are camped round here in cold weather."[27] And to make it easier for artillery to see who was approaching, acres of trees were cleared in front of battle lines. In other instances, forest fires that were sparked by accident, such as at the Battle of the Wilderness in early 1864, or perhaps deliberately set as a way to destroy Confederate crops, such as in the Shenandoah Valley in September 1864, sometimes roared out of control. Megan Kate Nelson notes that "some areas of the eastern theater were entirely cleared of trees by the end of the war" and she estimates that two million "southern oaks, pines maples, and cedars were converted" into roads, bridges, fortifications, and obstacles, and "that another 25,000 ultimately died from their war wounds."[28]

The U.S. acquisition of coaling depots in the Pacific continued during the Civil War including an agreement with the government of the Kingdom of Hawaii to lease property in Honolulu to erect a coaling station, and the acquisition of a lease in 1864 for a coaling depot in Yokohama.[29] The United States established other coaling stations during the Civil War, including in Havana and Guantanamo, Cuba; Cape Haitian and Nicholas Mole, Haiti; Rio de Janeiro, Brazil; and Lisbon, Portugal.[30] Rear Admiral Robley Evans argued that the North's use of anthracite coal was a significant advantage in the Civil War.

> Only by the use of that fuel was the Federal fleet enabled to maintain the greatest blockade the world has ever known, on thousands of miles of coastline, from the Virginia Capes to the Mexican boundary on the Gulf of Mexico. The blockade-runners were obliged to use soft coal, and that was their undoing in most cases. Some got through the lines in fogs and bad weather; but, for the most part, they were detected by their trails of smoke and flame long before they could espy the blockading craft burning smokeless fuel, and either driven away or captured. It has been said that the Confederacy was "starved to death"; maybe this was one of the factors that has been overlooked by the historian.[31]

Steam-powered railroad trains, mainly fueled by wood, made the transport of food, cannon, ammunition, and soldiers possible at a huge scale and with speed, although "in spite of the railroads, the armies were still dependent on horse-drawn transport (over abominable roads) for resupply in the field."[32] The railways were also an essential element in moving coal to the seaboard for the Union steamships that were maintaining the blockade of the South and for the armed steamboats that plied the rivers.[33]

The dependence on rail transport also exposed a vulnerability of that system to attack. Indeed, southern railways used by the Confederacy were targeted by the Union Army and the bombardment of Atlanta and Rome, Georgia; Columbia and Charleston, South Carolina; and Fayetteville, North Carolina was justified in part because of their role in the rail system. The North added 4,000 miles of railway and rationalized their system.[34] On the other hand, if the Confederates attacked the Northern railway system at the beginning of the war, by the end, the Confederate military often destroyed their own railway, reasoning that it was better to eliminate the rails than allow the North to use the system.[35]

POST–CIVIL WAR EXPANSION

After the Civil War, regional railroad lines and later the transcontinental railroad, increasingly powered by coal, would both require soldiers to defend the trains from Native Americans, and would enable the U.S. Army to move the troops necessary to promote expansion and defend the railroad and the settlers in the west. "The railroads endowed the armies with mobility in the power of prompt reinforcement and re-supply; but in so doing it exacted . . . the paradoxical price which has always been the cost of mechanical advance in war. In freeing the armies they also enslaved them—to their rail lines."[36] Of course the vast tracts of land granted to the railroads had been expropriated from the Native Americans who already lived there, and thus, the army was sometimes used to ensure that the tracks would be built. In sum, as it would in the future, innovations came with the need for more infrastructure, including fuel and armed forces to support and defend it.

But of the two services, the navy had the greatest dependency on fossil fuel in the nineteenth century. By the 1880s, the view that naval power was crucial for any country to be a great power, but that the United States had underinvested in it, was widespread in U.S. government circles. The role of sea power was underscored when Rear Admiral Stephen B. Luce, who led the Naval Academy at Annapolis after the Civil War, was charged with starting a Naval War College in Newport, Rhode Island, in 1884. Luce tapped Captain Alfred T. Mahan to be a lecturer on naval history. When he received Luce's letter inviting him to the post, Mahan was on duty off the coast of South America on a coal steamer, the *Wachusett*. A graduate of Annapolis, Mahan had served as one of the captains maintaining the blockade of the South during the Civil War. Although his father, Dennis Hart Mahan, taught at West Point and had authored important books on tactics, Alfred T. Mahan's chief qualifications were perhaps that he was known to Luce from a stint teaching at the Naval Academy and for writing a short book on Civil War naval operations. Mahan said yes in September 1884, and after more time at sea, and before he joined the War College faculty, he spent the winter of 1885–1886 in New York studying the history of naval warfare in preparation for his lectures.

By the time Mahan arrived in Newport, Luce had been assigned to command the North Atlantic Squadron, and Mahan was designated president of the Naval War College. Mahan's lecture notes were turned into his next, and perhaps most influential book, *The Influence of Sea Power upon History, 1660–1783*, published in 1890. It brought Mahan wide acclaim, including honorary degrees at Cambridge, Oxford, and Harvard Universities. Mahan's arguments were in turn used by Theodore Roosevelt, albeit an already committed navalist, who would in a few years be undersecretary of the navy.[37] *The Influence of Sea Power upon History* largely focused on Britain's naval strategy well before the rise of steam-powered warships. But Mahan speculated on the similarities and differences between the new steam-powered ships and the age of galleons and sail. Mahan argued that steamships had some advantages, although the "necessity of renewing coal makes the cruiser of the present day even more dependent than of old on his port."[38]

But this was obvious to anyone who had studied Commodore Perry's trip to Japan. If the United States wanted to expand its commerce, it needed a merchant marine to carry it and a navy to protect it, Commodore Robert Shufeldt told Congress in 1878. "In *no other* way can our commerce be reestablished or our prestige restored upon the ocean. In no other way can the country be relieved of its surplus products or an additional impetus be given to its industries."[39] President Chester Arthur began a significant naval modernization effort that included the purchase of new coal-powered steel-hulled battleships in the early 1880s. In 1887, the United States acquired exclusive access to Pearl Harbor and the right to establish a coaling station at Pago Pago Harbor in Samoa.[40] President Benjamin Harrison's only mention of the navy in his March 1889 Inaugural Address included the statement, "The necessities of our Navy require convenient coaling stations and dock and harbor privileges."[41] Indeed, Peter Shulman observed in *Coal and Empire* that "most of the overseas locations Americans sought after the Civil War . . . were desired and justified at least in part for their value as strategic coaling depots for the navy."[42]

Nevertheless, Mahan argued that the United States lacked exactly the kind of infrastructure that was characteristic of great maritime powers like Britain: "Having therefore no foreign establishments, either colonial or military, the ships of war of the United States, in war, will be like land birds, unable to fly far from their own shores. To provide resting-places for them, where they can coal and repair, would be one of the first duties of a government proposing to itself the development of the power of the nation at sea."[43] For Mahan, the essentials of communication, what we today call logistics, were "first, fuel; second, ammunition; last of all, food."[44] In *The Influence of Sea Power upon History*, Mahan, like Perry, argued that forward bases were essential for resupply. "The renewal of coal is a want more frequent, more urgent, more peremptory, than any known to the sailing-ship. It is vain to look for energetic naval operations distant from coal stations. It is equally vain to acquire distant coaling stations without maintaining a powerful navy; they will but fall into the hands of the enemy. But the vainest of all delusions is the expectation of bringing down an enemy by commerce-destroying alone, with no coaling stations outside the national boundaries."[45] Mahan was certainly not alone among those who recognized the importance of coaling, and advocated expansion. But he was perhaps, besides Roosevelt and Senator Cabot Lodge, its most far-sighted advocate at the time in terms of understanding the role of refueling in projecting American military power and in arguing that it was better to keep one's adversaries away by denying them coaling stations.

During the 1890s the United States continued to modernize the navy at the urging of Teddy Roosevelt, Mahan, and others. The United States was thus prepared for war in 1898 against Spain with a modernized navy, including its first armored battleships, the USS *Texas* and USS *Maine*, both coal-powered, twin-masted ships. However, in the months prior to the war, the U.S. Navy had no collier ships to carry the coal the warships would need. Undersecretary of the Navy Roosevelt organized the purchase or lease of colliers and by the time the navy began its blockade of Cuba, the United States had six colliers available. The plan was that the navy would purchase more colliers and coal as needed and as they could find it at neutral ports. During the fiscal year ending on June 30, 1898, the United States purchased 452,551 tons of coal.[46]

Of the two types of coal abundant in the United States, the U.S. Navy preferred the dirtier bituminous variety from the mountains of West Virginia at the turn of the century. As noted earlier, anthracite coal would burn cleaner and longer, and was considered "smokeless." Indeed, during the Civil War, the navy purchased anthracite coal in large quantities, but, as Rear Admiral Evans noted, bituminous coal became the favored fuel in subsequent decades.[47] Although it was more volatile, prone to spontaneous combustion, and more visible to enemies—its black plumes of smoke were visible as far away as seventy miles—bituminous coal "burned hotter, generating faster speeds."[48] Spontaneous combustion in the hold of the battleship USS *Maine*, a twin-masted steamer, then at port in Havana in February 1898, may have been the cause of the explosion that sunk the ship, although at the time, the public assumed it was the Spanish who had blown it up.

Coal shaped both the Spanish and the American operations in the war. "Under modern conditions coal is the very life of the ship, and without it, no matter what her power or efficiency in other respects, she is utterly helpless."[49] The Spanish had much farther to travel to supply their forces in Cuba and Puerto Rico than the United States did, which based its fleet for the war in the Caribbean in Key West and Tampa, Florida.[50] Both countries would also have to travel thousands of miles to reach the Philippines. One of the first orders of business for the United States in the war in the Caribbean was to establish a coaling base in Guantanamo Bay, Cuba, which the U.S. Navy did in June 1898. The United States had already annexed Midway Island in 1867 and leased land for a naval base in Pearl Harbor, Hawaii, in 1887, but shortly after the declaration of war against Spain in March 1898, President William McKinley proposed that the United States annex Hawaii, where the United States already had a coal depot, in Honolulu. The resolution annexing the Hawaiian Islands passed Congress in July 1898, and the annexation of Wake Island and American Samoa followed in 1899 and 1900. But no matter if the coal was shipped from Hampton Roads, Virginia, or Wales, it had to travel 11,000 to 14,000 miles to fuel U.S. operations in Asia.

The British coal from Newcastle and Cardiff, Wales, was also highly sought after, not least because Britain had the largest network of commercial and naval coaling stations located all over the globe. On the way to the Philippines, the U.S. fleet refueled in Hawaii, and took the Spanish-controlled island of Guam, just 1,500 miles from the Philippines without a fight when the Spanish surrendered. Admiral George Dewey's forces, already based in Hong Kong, needed to refuel and it was British coal, bought in Hong Kong, that allowed the United States to swiftly take Manilla in 1898.[51] As the U.S. Navy writes in its own current analysis of the Spanish–American War, "A significant factor driving American expansionism in the Caribbean and Pacific was not only the proposed isthmian canal and increase of commerce, but the coal needs of the United States Navy to support such activity."[52]

At the conclusion of the war with Spain, the U.S. Navy continued to increase its capacity for coal refueling even as its coal purchases declined in 1899 to 281,169 tons.[53] In March 1899, the U.S. Navy took land, including the "entire island of Manglar," to establish a base and coaling facilities. The United States established a contract with a private U.S.-based company to construct a coal dock and other amenities for a coaling station in Pago Pago, Samoa, in September 1898, which, when complete, would hold 5,000 tons of coal. When the United States annexed the Hawaiian Islands, the president set aside more land in November 1898 in Honolulu to erect coal sheds capable of holding 20,000 tons of coal. With the purchase of Guam from Spain, the entire island was designated a U.S. naval station in 1899 and plans were made to construct facilities sufficient to store 10,000 tons of coal. Further, the U.S. Navy fleet based at Manilla required 4,000 tons of coal per month for its operations at Cavite Naval Station in the Philippines.[54] The navy also decided that "the only method of absolutely insuring rapid coaling is for the Department to have on hand a certain number of barges loaded with coal ready for immediate use."[55]

In the 1899 annual report, the secretary of the navy was clear that given U.S. ambitions, the country would require even more coaling capacity. "As a matter of fact it may be stated, without fear of contradiction, that at present it would be impossible for a United States fleet to carry on active operations during a war anywhere about the coast of Europe, Africa, a large portion of Asia, and South America, for want of coal." The argument was, in part, that the United States needed to use its military capacity to protect its neighbors from potential aggression by European powers: "This country has assumed a position of great responsibility in connection with the maintenance of the sovereignty of the territory of the South American Republics against any aggression on the part of a European nation, yet there is not a single port where coal or supplies may be obtained in time of war by a United States ship anywhere on the eastern or western coast of South America. In the West Indies and the Pacific Ocean we are now somewhat better off." But the United States needed even more capacity. "It is of paramount importance, however, to establish a coal depot on each side of the Isthmus of Panama near the termini of the Isthmian canal."[56]

In 1900, Captain Asa Walker put the reliance on coal in more romantic terms: "The modern man of war presents no canvas to the winds; within her bowels is an insatiable monster whose demand is ever for coal and still more coal. Every cubic inch of available space is filled with fuel, and when this is consumed the vast machine becomes an inert mass. Coal then may be considered as the lifeblood of the man of war, and upon its supply depends her existence as a living factor in the battle equation."[57] As figure 1.1 illustrates, the "insatiable" navy's bases thus came to span the globe.

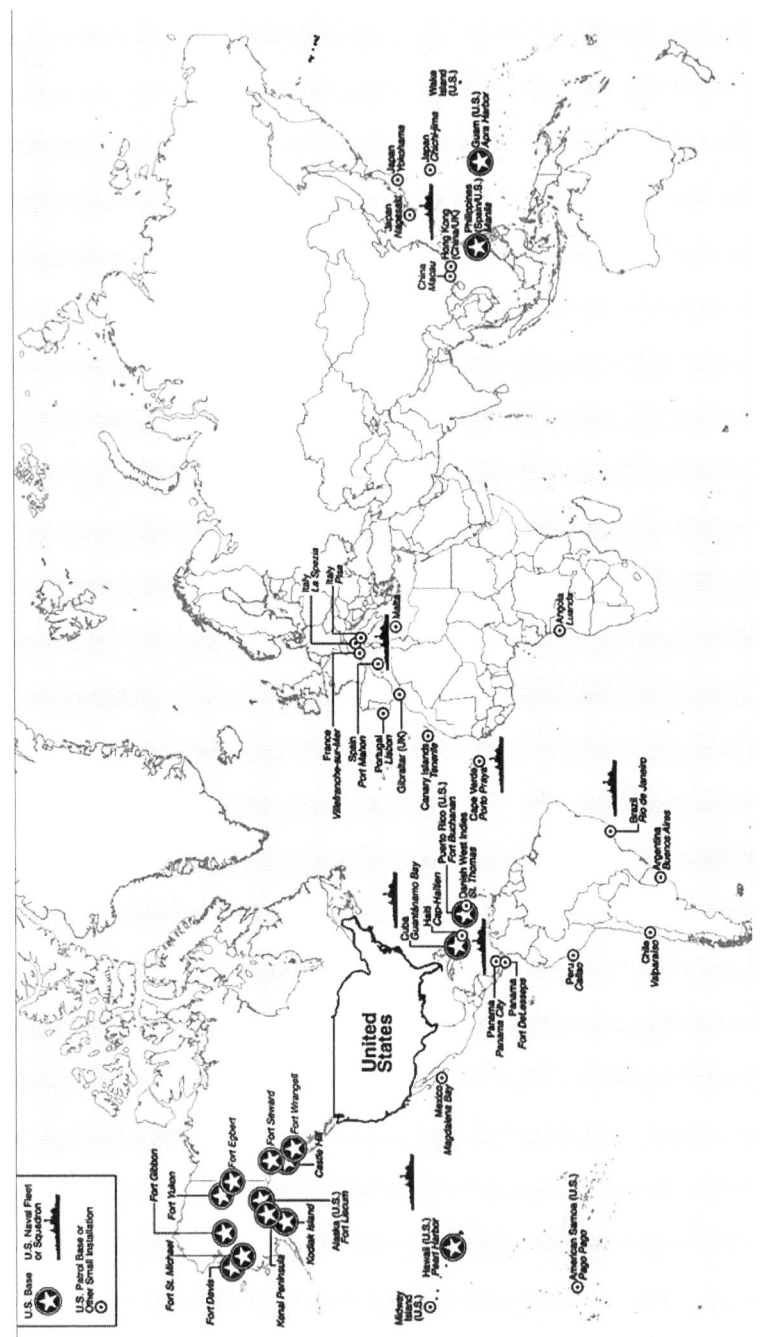

Figure 1.1

Map of U.S. military bases, 1776–1903.

Source: Map created by Kelly Martin/kmartindesign.com for David Vine, *Base Nation: How U.S. Military Bases Harm America and the World*

(New York: Metropolitan Books, 2015), https://www.basenation.us/maps.html.

In February 1907, the Commander of the Atlantic Fleet, Robley Evans, articulated the logic of ever-increasing coal consumption, even in peacetime. He said that "with the steady annual growth of the Navy, that must surely proceed if America is to retain her place among the World Powers, the coal consumption of the ships must necessarily increase. To be efficient, seamen must be trained, and that means that ships must be kept in commission and must burn much coal."[58] More installations that could be used for coaling were gradually consolidated. In 1903 the United States signed the Panama Canal Treaty, allowing the United States land rights around a proposed canal, and a treaty granting the United States the lease for a coaling and naval station in Guantanamo Bay Cuba for as long as it deemed the base necessary. The U.S. Congress also funded the fortified naval base in Pearl Harbor in 1907.

THE GREAT WHITE FLEET, COAL, AND THE TRANSITION TO OIL

A few months later, the global range of the U.S. Navy, and America's "place" among the great powers was demonstrated when President Theodore Roosevelt dispatched a group of sixteen battleships, painted their white peacetime colors, on a world tour under the command of Rear Admiral Robley Evans. Leaving Hampton Roads, Virginia, on December 16, 1907, the battleships, pictured in figure 1.2, were accompanied by ten other ships, including a flotilla of six torpedo boats, that were to cruise ahead of the battleships. Over a fourteen-month voyage, the "Great White Fleet" traveled over 42,000 miles with stops in the Southern Atlantic, including Rio de Janeiro, transit through the straits of Magellan, and more stops, including San Francisco, Hawaii, New Zealand, Sydney, Melbourne, Manilla, and Yokohama, and transited the Indian Ocean, the Suez Canal, and Spain, before returning to Virginia on February 22, 1909. The twenty visits to ports were not only to show the U.S. flag, but also for coaling: the ships were required to take on more coal every one to two weeks, a process requiring several days.[59] The United States estimated that the fleet would "consume upward of five hundred thousand tons of coal just on the voyage from the United States to the Far East."[60] Evans was an advocate of anthracite coal; perhaps to his disappointment, the Great

Figure 1.2

The Great White Fleet departs Hampton Roads, 1907. *Photo*: Naval Institute Archives, "December 16, 1907: The Great White Fleet Departs Hampton Roads for Circumnavigation," *Naval History Blog*, December 16, 2012, U.S. Naval Institute, https://www.navalhistory.org/2012/12/16/december-16-1907-the-great-white-fleet-departs-hampton-roads-for-circumnavigation.

White Fleet burned bituminous coal, creating plumes of black smoke that advertised the arrival of the fleet well before it reached any port. Because the United States had only six colliers, forty-one British-owned ships and eight other foreign colliers were hired to supply the coal.[61] Foreign colliers or fuel from ports were used for refueling 90 percent of the time. Britain supplied 85 percent of the world's coal at the turn of the century, and although U.S. dependence on British coal was somewhat embarrassing, its dominance was no surprise.[62]

The cover story of the February 20, 1909 edition of *Scientific American* praised the Great White Fleet's circumnavigation as a success in demonstrating American naval power. Indeed, the Great White Fleet's voyage comprised the entire issue of *Scientific American*, featuring drawings of the ships, maps of the voyage, and a comparison of the Great White Fleet with those that had participated in the Spanish–American War. Illustrations in the magazine showed how much larger the newer ships and their armament and engines were, and ranked the United States among the world's top naval powers. In sum, the coverage portrayed the voyage as a success.

The Great White Fleet's circumnavigation consumed 430,000 short tons of coal, nearly as much as the U.S. Navy had purchased from July 1897 through June 1898, a period that included the peak of U.S. naval operations in the Spanish–American War.[63] See figure 1.3. It was a striking demonstration of logistics. Yet, *Scientific American* noted "our great shortage of colliers." This meant that "had it not been for the foreign bottoms in which coal was shipped to the fleet at various points of rendezvous, it would have been impossible for this voyage to have been made."[64] The dependence on other nations' colliers had thus exposed what *Scientific American* called "the most important lesson" of the feat. "Had war flamed out at the shortest notice, when our fleet was, let us say on the coast of Australia, or at Suez, it would've been as helpless, and even more so, as a fleet of dismantled frigates in the days of sail power and smooth-bore. With coal declared the contraband of war; with no colliers of our own available to carry the necessary fuel; our 16 battleships, for all their tremendous fighting power, would have been as useless, as far as active operations on the high seas are concerned, as so many anchored floating batteries." Ultimately, the *Scientific American* article reached the same conclusion that members of Congress came to later that year: "Undoubtedly, the greatest need of the navy to-day is a fleet of large and fairly fast colliers, built expressly for naval purposes."[65] Of course, it was taken for granted that the colliers

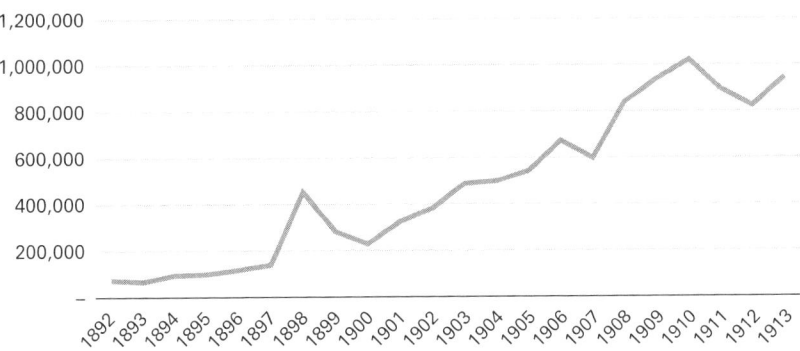

Figure 1.3

U.S. Navy coal purchases, in tons, 1892–1913.

Source: U.S. Navy Department, "Quantity of Coal Purchased at Home and Abroad," *Annual Reports of the Navy Department for the Fiscal Year 1913* (Washington, DC: 1913), 396, table 3. One short ton is 2,000 lbs.

were necessary because the United States intended to be a global power. And like Britain, which Mahan thought the United States should emulate, it was assumed that the United States would need a global network of naval bases from which it could project power and store the coal that enabled its presence. As figure 1.4. illustrates, British coaling stations dotted the empire.

As Steven Gray argues about the British Royal Navy at this time—"naval coaling infrastructure was, in fact, integral to Britain, empire and the world in the nineteenth century"—much the same could later be said of the U.S. Navy. "Fuel did not simply appear at stations across the globe, but was subject to multiple movements before arriving. It needed to be sourced, moved to ports, shipped and unloaded."[66] During the Mexican–American War, the voyage of Commodore Perry's fleet, the Civil War, and in expansion into the Pacific after the Civil War, the United States extended a network of coal refueling stations and commercial arrangements.

But the U.S. Navy's annual coal purchases peaked in 1910 at more than a million tons, and even as *Scientific American* was arguing that colliers were the U.S. Navy's "greatest" need, navy leadership was contemplating a transition to oil. Oil had already been taken up by the British Navy because it had many virtues compared to coal. Perhaps most important, because oil has twice the thermal content of coal, oil-powered ships could travel faster or nearly twice as far on the same weight of fuel. Further, oil was cheaper than coal in the Pacific. In addition, while the U.S. Navy did develop a means for refueling with coal at sea, ships powered by oil could be more easily refueled, and oil could be pumped into an engine, while tending a coal steam engine was hot, dirty, and heavy labor.[67] While Britain produced the world's most desirable coal, and had coaling stations all over the world, it lacked a large supply of oil in its home islands.[68] Transitioning to oil was thus a great risk for Britain and when it made the commitment to do so, at the urging of First Lord of the Admiralty Winston Churchill in 1911, in the short term, "Britain . . . created new problems for itself" since it lacked a secure supply of oil.[69] This started Britain in the business of owning oil fields in the Persian Gulf, with the government acquiring a 51 percent share of the Anglo-Persian Oil Company.

On the other hand, the United States had oil, which was discovered to be plentiful in the United States in the middle of the nineteenth century.

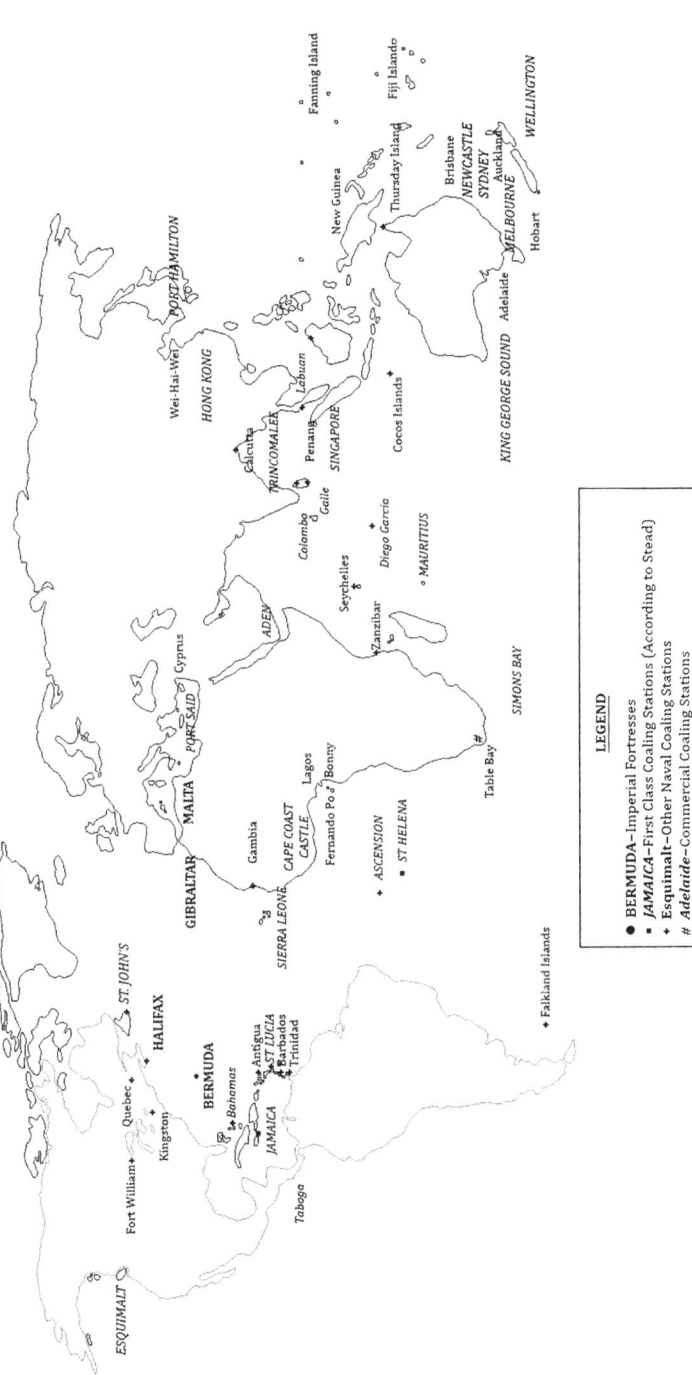

Figure 1.4

British Royal Navy coaling stations in 1914.

Source: Steven Gray, "Fueling Mobility: Coal and Britain's Naval Power, c. 1870–1914," *Journal of Historical Geography* 58 (October 2017): 92–103; 96.

Indeed, the United States dominated world oil production at the turn of the twentieth century. However, the U.S. Navy balked at making the transition to oil. Its first experiments with using oil on steamers had occurred in 1864, but the U.S. Navy, like the British, decided to primarily use coal as naval fuel over the next several decades. Discoveries of large oil fields in the 1890s in California and in Texas in 1901 spurred greater interest in oil and the navy received appropriations in 1898 for more tests. But while experiments through 1904 were encouraging, the navy was slow to change its fleet to liquid fuel. Its first gasoline-powered submarine, the *Holland*, was completed in 1900 but in 1910, out of the entire fleet of 196 active ships, the navy had only two battleships, four destroyers, and four submarines running on oil. And even then, oil was an auxiliary to coal in the battleships—with oil sprayed on the coal to reduce ash and increase efficiency.[70] In 1913, the U.S. Navy was still building ships using coal: eight battleships, one transport, and a supply ship that used both coal and oil. The navy had built or had under construction only four battleships running exclusively on oil.[71]

The decisive pivot to oil occurred in April 1913 when the U.S. Navy decided that its new Nevada class of ships would run solely on oil. Admiral Dewey, head of the General Board of the U.S. Navy, told Secretary of the Navy Josephus Daniels, "The military advantages of burning oil, the advantage to the United States in being the greatest oil producing country, and the added advantage that the Navy has its own oil-bearing lands, are all so great that the return to coal burning could only be viewed as a calamity."[72] In 1914 the *Annual Report of the Navy Department and the Secretary of the Navy* averred the "advent of oil"; "Each year marks more distinctly the passing of coal, with its attendant discomfort and engineering and military inferiority as fuel for ships of war."[73] From then on, only oil-powered vessels would be purchased.

WORLD WAR I AND THE WAVE OF OIL

While hot air balloons had been used by the United States in the Civil War and the Spanish–American War for reconnaissance, and by the Italians to drop bombs, balloons were not widely used in war. The internal combustion engine, invented in the middle of the nineteenth century, made possible first motorcycles, then automobiles, trucks, and eventually tanks. But where steam engines boiled water to make steam to turn a piston, internal combustion engines turned the energy of a liquid fuel directly into mechanical energy and tended to be much more efficient than steam engines. Aircraft, invented in the United States in 1903, took internal combustion to the sky. Thus, at the same time Theodore Roosevelt was exhorting Congress to fund more battleships, he was also urging the U.S. Army to purchase aircraft. Aircraft were considered immediately for bombing, reconnaissance, and long-range transport, and thus the massive mobilization that accompanied the United States entry into World War I led to the production of thousands of aircraft, although most of those were trainers. While horses were still part of the fight, trucks and tanks also made their appearance on a large scale. All of this required coal, oil, gasoline, and petroleum lubricants for machines and engines.

In 1914, Secretary of the Navy Daniels and Assistant Secretary Franklin D. Roosevelt took charge of navel fuel contracts and in May 1917, a month after the United States entered World War I, the preferred suppliers of navy "standard" coal were called to Washington to discuss how those companies would meet the enormous demand for coal and what the U.S. government was willing pay.[74] All told, the U.S. Navy was able to transport 130,000 tons of coal, 746,000 tons of fuel oil, and 12,000 tons of gasoline to Europe. At the same time, the non-naval demand for coal was unprecedented: coal-powered military industry and railroads transported troops and war material to ports, and bottlenecks and shortages prompted Congress in August 1917 to take control under the Food and Fuel Act.

Coal and oil played another role in the war as each side tried to limit their adversary's access to fuel. The Russians sent their army to protect access to their own oil from Austria-Hungary and the British sabotaged the oil fields and refineries of Rumania (today, Romania), significantly reducing their output before Germany invaded that country in 1916.[75] At sea, the Allies imposed a naval blockade on the Central Powers that included food, oil, and nitrates for gunpowder and fertilizer on the list of contraband—goods that a neutral power was forbidden trading with a belligerent. The aim of the Allies blockade was to squeeze the *entire* Germany economy by cutting off access to "industrial goods of all kinds, oils, metals and minerals, chemicals, precious metals."[76] In 1915, in response to the British blockade, Kaiser Wilhelm authorized German U-boat submarines to attack any merchant vessel—even neutral shipping—around Britain without warning. The Germans' submarine war on oil tankers supplying Britain became increasingly effective, and an average of one tanker per day was damaged or destroyed in the early months of 1917.[77] This reportedly "caused the usual six to eight months reserve of oil fuel to be reduced at one time to an eight weeks supply, a shortage which necessitated restricting the movements of the [British Royal Navy] Grand Fleet."[78] In October 1917, pleasure driving was halted in Britain. The Allies were also intent on cutting off German fuel supplies: to prevent the Germans from acquiring Romanian oil fields they sent a team to sabotage the oil fields and refineries in November 1916, which succeeded in destroying 70 refineries and about 800,000 tons of crude oil.[79] And although both Britain and Germany were short on food, Germany attributed its loss in World War I to shortages of food and fuel, while Lord Curzon, a member of the British War Cabinet said the Allies "floated to victory on a wave of oil."[80]

Much of the wave of Allied oil came from the United States, the world's largest producer of crude oil from 1914 through 1918. The United States provided 62 percent of the Allies' crude oil in 1914, and by 1917 the share provided by the U.S. increased to 86 percent.[81] The U.S. Navy also shipped 746,000 tons of fuel oil and 130,000 tons of coal to the front.[82] In 1917, with coal supplies too short and prices too high, the navy began to commandeer coal by requisitioning at the prices that it set.[83] Still, there was too little coal and in response to U.S. coal shortages, in January 1918 the administrator of the United States Fuel Administration, Harry Garfield, issued a "closing order" that "banned all but the most indispensable industries from consuming coal for the week beginning January 18 and for two months of Mondays thereafter."[84] The U.S. supplies of gasoline also included high-quality aviation gasoline that allowed Allied aircraft to operate at altitudes of 18,000 feet, safe from enemy anti-aircraft guns, and gasoline for use on the Western Front. As the Allies' gasoline requirements grew to between 520,000 and 720,000 barrels per month from January to September 1918, demand outstripped U.S. refinery production. Americans living east of the Mississippi were asked to ration gas for seven weeks, engaging in gasless Sundays, starting in September 1918. They apparently complied, and total consumption declined.[85]

The need to move material to ports to send to Europe underscored the limits of the railway system, and trucks were used to overcome bottlenecks in supply but, even then, the roads were insufficient, and the increased truck traffic led to their physical deterioration.[86] After the war, the U.S. Army's Motor Transport Corp Convoy traveled from Washington, DC, to San Francisco to test a road route across country. Starting on July 7, 1919, the trip took sixty-two days, at an average speed of about six miles per hour. One of convoy participants, Lt. Col. Dwight Eisenhower, the future general and later president, would play a pivotal role in setting up the U.S. interstate highway system in the 1950s. But action to improve roads was immediate: the Federal Highway Act of 1921, whose rationale was the need for a better road system in case of war, authorized the U.S. federal government support for state road construction. In 1922, the head of the Bureau of Public Roads commissioned General John J. Pershing to provide the "Pershing Map," as it was known, which identified 200,000 miles of public roads to meet the needs of the United States during wartime.[87] Railroad transportation peaked just after the war, as travel by automobile increased.

Domestic demand for oil grew during the war as the production of automobiles, trucks, and aircraft ramped up; U.S. refinery output of gasoline, kerosene, lubricants, and fuel oil reached a peak of almost 357 million barrels in 1919. On the one hand, production exceeded demand immediately after the war, but domestic demand, especially for gasoline and fuel oil, was growing. Surplus military equipment including trucks and airplanes was sold to Americans, who put them to use. On the other hand, the domestic oil industry was concerned that the United States was running out of crude oil reserves, since the output of oil fields already in production was declining.[88] The U.S. Navy was concerned that there would be an oil shortage within the next two decades and "the fear spread that other countries were about to acquire dominance or exclusive control over the most promising foreign sources of production."[89] In fact, before the war, the navy's leadership, ever anxious about the supply of fuel, had authorized three naval petroleum reserves, two in 1912, at Elk

Hills and Buena Vista Hills in California, and in 1915 at Teapot Dome in Wyoming.[90] Some in the United States were also concerned about the rapidly increasing price of oil. In late 1913, the General Board and the secretary of the navy were recommending that the navy not only own its own oil fields, to keep the price down, but also set up its own refineries and run its own pipeline. Part of Secretary Daniels's argument was that the British had gone into the oil business "to protect British naval supremacy."[91] Indeed, in 1914, the British bought a controlling interest in the Anglo-Persian Oil Company, which had the oil concession for Persia. Oil prices declined the next year, and Daniels's proposal to expand the navy's role in oil production was dropped.

But the question of access to oil arose again during and after World War I, eventually resulting in the Teapot Dome scandal—a complex affair involving private oil companies giving their support to 1920 presidential candidate Warren Harding in exchange for him appointing a secretary of the interior who would provide them favorable access to the U.S. Navy's oil reserves at Elk Hill and Teapot Dome.[92] The leases were granted to Mammoth and Pan American oil companies, but later invalidated by a federal district court in 1926. The Teapot Dome scandal and the persistent concern that the United States would not have enough oil led President Calvin Coolidge to form the Federal Oil Conservation Board in 1924, which consisted of the secretaries of the war, navy, interior, and commerce departments. Coolidge argued, "Developing aircrafts indicate that our national defense must be supplemented, if not dominated, by aviation. It is even probable that the supremacy of nations may be determined by the possession of available petroleum and its products."[93]

Rivalries over commercial access to Middle Eastern oil arose between the United States and the UK after World War I in what became known as the "oil war" of the 1920s. Competition for access to oil fields in the region intensified in part because after the Ottoman Empire entered World War I, it almost immediately lost control of southern Mesopotamia, now Basra, Iraq. Further, the British government confiscated the German's 25 percent share in the Turkish Petroleum Company during World War I and then rearranged ownership of the company, which now included France, Royal Dutch Shell (a combined British and Dutch company), and the Anglo-Persian Oil Company. U.S.

companies were given access to oil discovered in the Kirkuk region of Iraq. Members of Congress predicted that American production would decline, and that British and French companies would have a stranglehold on the resource in the Mesopotamian fields. The tension between the United States and the European oil companies only relaxed when new oil fields were discovered in the United States. Total U.S. crude oil production at major oil fields more than doubled between 1920 and 1929.[94]

THE CENTRALITY OF FOSSIL FUELS IN WORLD WAR II

As preparations for war in Europe and Asia grew in the late 1930s, the great powers were well aware that fossil fuel, especially oil, would be essential to power the fight if it came to war. The question in the late 1930s, was: Who had the oil? The U.K/Dutch firms (including Royal Dutch Shell), the Soviet government, and U.S. firms were dominant, with their crude oil production constituting, respectively, about 38, 28, and 24 percent of all production in 1938.[95] By contrast, Germany, which had secretly begun to rearm, was only capable of producing small amounts of oil in its own territory. However, the Germans had vast coal reserves and the German Defense Ministry invested in a process of producing liquid synthetic fuels (synfuel) from coal in the late 1920s as part of the government's drive for autarky, even though synthetic fuels were much more expensive than those produced from crude oil. By 1938, German synfuel companies were supplying 84,000 barrels of the country's daily consumption of 150,000 barrels, with the Germany military using about half that amount.[96] That year, the UK/Dutch, Soviet and, Americans were producing, respectively, about 810,000, 586,000, and 507,000 barrels of crude per day.[97]

Hitler's blitzkrieg strategy depended on mobility—their Panzer tank divisions, railroads, and aircraft—all of which were extremely thirsty for fuel. Starting in 1933, the Germans built an "autobahn" road network to facilitate troop movements. The German military moved to increase crude oil supplies by annexing Austria in 1938 and then, in September 1939, invading Poland, where the Poles attempted to sabotage their own oil wells and refineries as their army collapsed in the face of the Wehrmacht, the armed forces of Nazi

Germany. The Soviet Union had signed a nonaggression pact with Germany in August 1939 and as agreed between the two governments, the Soviets joined the German invasion of Poland in September 1939, where it captured and then controlled the majority of the Polish oil fields. "Germany had been counting on the entire production of Polish oil. It ended up controlling only 1,646 producing wells while Russia held 2,273."[98] In February 1940, the Soviets also promised to supply other raw materials to Germany, but they never supplied as much as the Nazis wanted.

The biggest fuel gain for the Nazis was an alliance in June 1940 with Romania which, at the time, was producing 60,000 barrels of fuel per day. The Germans also supplemented their supplies when they quickly overran Denmark and Norway in April 1940, and the Netherlands, Luxembourg, and most of France in the next months and stripped assets from the occupied territories—including food, vehicles, coal, steel, and oil. When the Germans occupied France, they captured the seven million barrels of fuel that the French had stockpiled for the war. Of this, two million barrels were aviation fuel, which the Luftwaffe, the German air force, immediately put to use in August 1940 when the Germans began the air assault that became known as the Battle of Britain. By the summer of 1940, half of the Nazi infantry was fueled by the oil that Germany had taken by conquest and the German war industry was in part powered by French, Dutch, and Belgian coal.

But given German ambitions and its military fuel consumption, the theft of those assets could never be enough. The Nazis' appetite for fuel only increased, so civilian oil and gasoline were rationed in Germany and the occupied territories. By 1941, the German military was using about 75 percent of all the liquid fuel either imported or produced in Germany.[99] As noted earlier, the Germans had been frustrated in their efforts to control Poland's oil fields when their ally the Soviet Union beat them to most of the oil. In July 1940, Hitler told Göring that it was necessary for Germany to get food from Ukraine, and he argued that "we must break through to the Caucasus in order to get possession of the Caucasian oil fields, since without them large-scale aerial warfare against England and America is impossible."[100] Thus, Hitler decided that he needed to betray his alliance with the Soviets and in June 1941, the Nazis invaded the Soviet Union in hopes of quickly capturing the oil. This invasion was a shocking surprise for the Soviets, who immediately stopped supplying the Germans with oil, and then later that month sabotaged Romanian refineries held by Germany.[101] The Caucasian oil was never secured; the German invasion stalled for lack of fuel.

Across the English Channel, although the British were self-sufficient in coal, they imported nearly all the liquid fuel they would need to fight the Nazis in tanker ships, mostly from the Western hemisphere. Even before the U.S. entry into the war, the United States supplied 53 percent of the aviation fuel consumed by the British in 1940.[102] In August 1940, the Nazis' air, surface, and submarine fleet began a blockade against Britain that sunk hundreds of tankers and merchant ships, dramatically reducing British oil and food imports.[103] It is hard to overstate the critical role of the United States at this point. But key to all of this was the transfer of fuel.

In May 1940 President Franklin Roosevelt launched a rearmament program and in September 1940, Roosevelt and Prime Minister Winston Churchill made the Destroyers for Bases deal that gave Britain fifty World War I-era destroyers in exchange for ninety-nine-year leases for U.S. military bases in eight British colonies. The bases in Newfoundland and Bermuda, in particular, became crucial elements of the North Atlantic and Mid-Atlantic Air Ferry Routes used by the United States to transit ships and supplies to the war.[104] The United States also expanded facilities at existing bases, including in Hawaii and the U.S. Virgin Islands, which it had purchased from Denmark in 1917. In March 1941, the Lend-Lease Act committed the United States to send food, fuel, warships, and aircraft to the nations allied against Germany, Italy, and Japan. Between 1939 and 1941, the United States increased domestic petroleum consumption by 20 percent, fueling the domestic arms industry. But because of the significant loss of tankers due to the blockade of Britain, too little oil was getting through to Britain, and in the spring of 1941, the British Navy estimated that, without imports, it had a two-month supply of fuel. Roosevelt then asked U.S. tanker owners to dedicate fifty tankers to supply Britain with fuel.[105]

On the other side of the world, Japan's domestic oil production was limited, and it was heavily dependent on oil imports from the United States and Royal Dutch Shell. In 1931, Japan had invaded Manchuria, in part to secure coal and in the hope that oil would be found there. Japanese oil consumption doubled between 1931 and 1939; 80 percent of that fuel was imported from California.[106] In early 1940, the United States was still supplying aviation and motor fuels to Japan, but in early 1941 Roosevelt was deciding whether or not to embargo oil sales to Japan. Roosevelt argued that if he kept supplying the Japanese oil, that would decrease their incentive to invade other countries in the Pacific for their oil. In July 1941, Japan invaded Indochina and the United States finally halted oil exports to Japan and froze Japanese funds. The Japanese attacked the U.S. naval base at Pearl Harbor on December 7, 1941, destroying or damaging hundreds of planes and sinking or damaging much of the fleet, including battleships, destroyers, and cruisers. However, the Japanese apparently left off their plans for destroying the U.S. Pacific Fleet's oil tanks, which contained about 4.5 million barrels of oil.[107] But certainly on the Japanese list was securing 70,000 barrels of oil per day by taking the oil wells in Borneo in January and the Royal Dutch Shell oil refinery in Sumatra, which they captured in February 1942.[108]

By the time the United States entered the war, the Roosevelt administration had prepared the United States to increase fuel production, and military procurement of refined crude oil products grew from 12 million barrels in 1939, to 14.5 million barrels in 1940 and 25.1 million barrels in 1941 (figure 1.5).[109] Axis powers attacked the United States' shipments of oil to the Britain and Russia. In January 1942 German submarine U-boats began patrolling the Atlantic close to the U.S. East Coast, destroying shipping, including oil tankers that were either moving from the Gulf Coast to power Eastern manufacturing industry oil or on the way to Europe. At the same time, the Imperial Japanese Navy began to attack U.S. shipping on the West Coast. The assaults were visible to civilians on both coasts. The desire to decrease vulnerability led the U.S. Petroleum Administration for War to construct the Big Inch and Little Inch pipelines. The public was also encouraged to conserve fuel: advertisements paid for by the Petroleum Industry War Council in 1942 warned "Oil Is Ammunition—Use It Wisely."[110] Nevertheless, the United States still dominated world crude oil production.

Figure 1.6, taken from the U.S Department of the Interior's *Minerals Yearbook* of 1946, illustrates an index of relative productivity of different economic sectors. In the first half of the twentieth century, in both world wars, growth

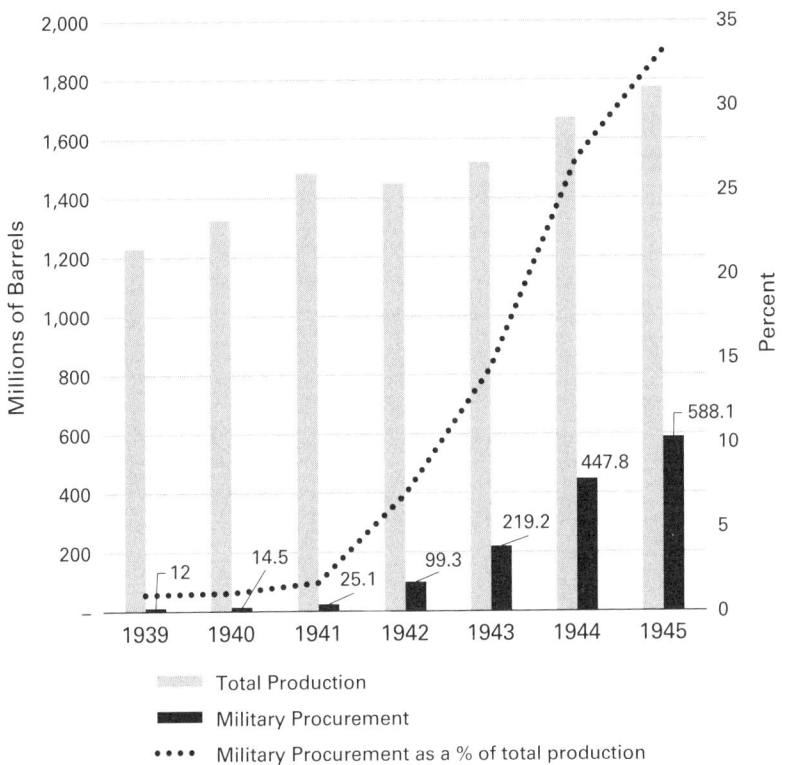

Figure 1.5

U.S. domestic production of refined petroleum products, military procurement, and military procurement as a percent of total production, 1939–1945. Military Procurement includes all military purchases of major products both in the United States and foreign countries, plus purchases made abroad for delivery to U.S. ports. Foreign purchases have been variously estimated that from one to four per cent of the total.

Source: Data in Harold F. Williamson, Ralph L. Andreano, Arnold R. Daum, and Gilbert C. Klose, *The American Petroleum Industry, Volume II: The Age of Energy 1899–1959* (Evanston, IL: Northwestern University Press, 1969), 772.

in agricultural, minerals, and industrial production corresponded to periods of war. The largest spike in production occurred in World War II.

The Allies armed forces "excluding Russia, used 22 billion gallons of petroleum products to conquer Germany, and this does not include the products used to transport men and materials to the scene of action."[111] One of the keys to Allied success was the U.S. provision of high 100-octane aviation fuel—enabling aircraft to fly faster, farther, and with greater acceleration

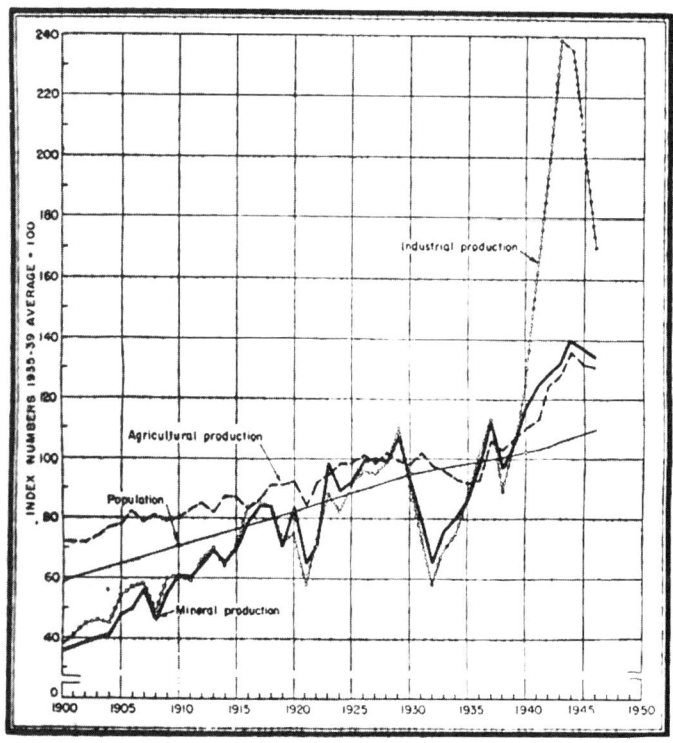

Figure 1.6

Comparison of physical volume of mineral production with agricultural, and industrial production and population, 1900–1946. Average productivity from 1935–1939 = 100.

Source: U.S. Department of the Interior, *Minerals Yearbook*, 1946 (Washington, DC: U.S. Government Printing Office, 1948), 3.

than the German and Japanese planes. The B-29 bombers required 10,000 gallons of fuel for a long-range round trip. When the Ninth Air Force was bombing Germany, "its planes were consuming an average of 634,000 gallons of 100-octane aviation gasoline each day" and during a month in 1944, the Far Eastern Air Forces burned 143,257,000 gallons of aviation fuel in its attacks on Japanese shipping.[112] By the end of the war, U.S. refineries provided 86 percent of the 100-octane aviation fuel the Allies used during the war.[113]

The war spurred a transition to petroleum-based products. Petroleum-based asphalt was used to pave runways at airbases and military roads. Petroleum-based plastics, for which there was little demand prior to the

Table 1.2

U.S. production of 100-octane aviation gasoline, 1941–1945, thousands of barrels

Year	Annual production
1941	21,110
1942	27,688
1943	62,044
1944	136,130
1945	124,215

Source: Harold F. Williamson, Ralph L. Andreano, Arnold R. Daum, and Gilbert C. Klose, *The American Petroleum Industry, Volume II: The Age of Energy 1899–1959* (Evanston, IL: Northwestern University Press, 1969), 789.

war, were engineered to replace copper, aluminum, steel, and zinc—those metals that the United States had "an urgent need to conserve" for use in ships, tanks, aircraft and other war material.[114] The rubber, essential for tank treads and tires, made from rubber trees was largely replaced with synthetic rubber, so that by the end of the war, "nine in ten pounds of U.S. rubber were factory made, mostly from oil."[115] When the United States began firebombing German and Japanese cities, the incendiary napalm was made by Standard Oil of New Jersey. Coal was used, along with petroleum, to produce the toluene that was a necessary ingredient of explosives. And while total production of coal and oil grew, civilian conservation enabled increased military procurement; 120,000 homeowners living in the Eastern and Midwestern United States participated in petroleum conservation efforts by switching their oil-burning furnaces to coal.[116]

The logistics of supplying fuel were so intricate and enormous that the military became much more systematic and centralized in procuring and distributing fuel during the war. Fuel requirements were set by the Army-Navy Petroleum Board created in July 1942 which itself was under the control of the Joint Chiefs of Staff. The Petroleum Administration for War, created in 1941, coordinated production and distribution within the United States. and the fuel made it to the front and the two services cooperated on its distribution through the efforts of the Army-Navy Petroleum Board working with the Army Services Forces.[117]

Production and distribution were only half the story. Denial was as important. Just as the United States was seeking to evade the German and Japanese attacks on its oil tankers, the United States set up a blockade of Japan with the U.S. Fifth Fleet. In December 1941, even while recovering from the attack at Pearl Harbor, the United States sank twelve Japanese merchant vessels; in 1942, the United States was able to destroy 229 ships, including two fuel tankers. In 1943, out of the 434 Japanese merchant vessels sunk, twenty-three were tankers. In 1944, out of a total of 969 Japanese merchant vessels destroyed, the United States sank 131 tankers.[118] The air forces dropped 10,600 tons of bombs on Japanese-controlled oil refineries out of its total of 656,400 tons.[119] The United States sank 103 oil tankers in 1945 out of a total of 701 Japanese merchant vessels that year. Of course, all that destruction required its own fuel: in just seven weeks during June and July 1944, the U.S. Fifth Fleet burned 630,000,000 gallons of fuel oil.[120]

Thus, by early 1945 Japan was completely cut off from crude and other oil imports, including diesel fuel.[121] Overall, the blockade was so successful that the Japanese imports of food, raw materials, and fuel declined dramatically over the course of the war as U.S. submarines and aircraft decimated the Japanese merchant fleet. As aviation gasoline became scarcer, the Japanese were forced to put a mixture of wood turpentine and alcohol in their aircraft, and by the end of the war they were attempting to manufacture aviation fuel from pine roots and oil from potatoes.[122] There was little fuel left for domestic needs, and the Japanese cut down forests to provide heating and cooking fuels for civilian needs.

In the European war zone, retarding Nazi transportation and the petroleum industry were top priorities. The United States and the Soviet Union stationed troops in Iran to facilitate the transfer of equipment to the Soviet Union as part of the Lend-Lease arrangement and deter the Germans from attempting to get the oil. In early 1945, the United States began construction of an airfield in Dhahran, in eastern Saudi Arabia, but the base was not completed before the end of the war.[123]

The American-British Combined Bomber Offensive put oil high on the list of targets. Indeed, German petroleum production facilities were at the top of the list of strategic bombing targets, along with aircraft factories and air bases.[124] See figure 1.7. The Allies attacked German-controlled oil fields in

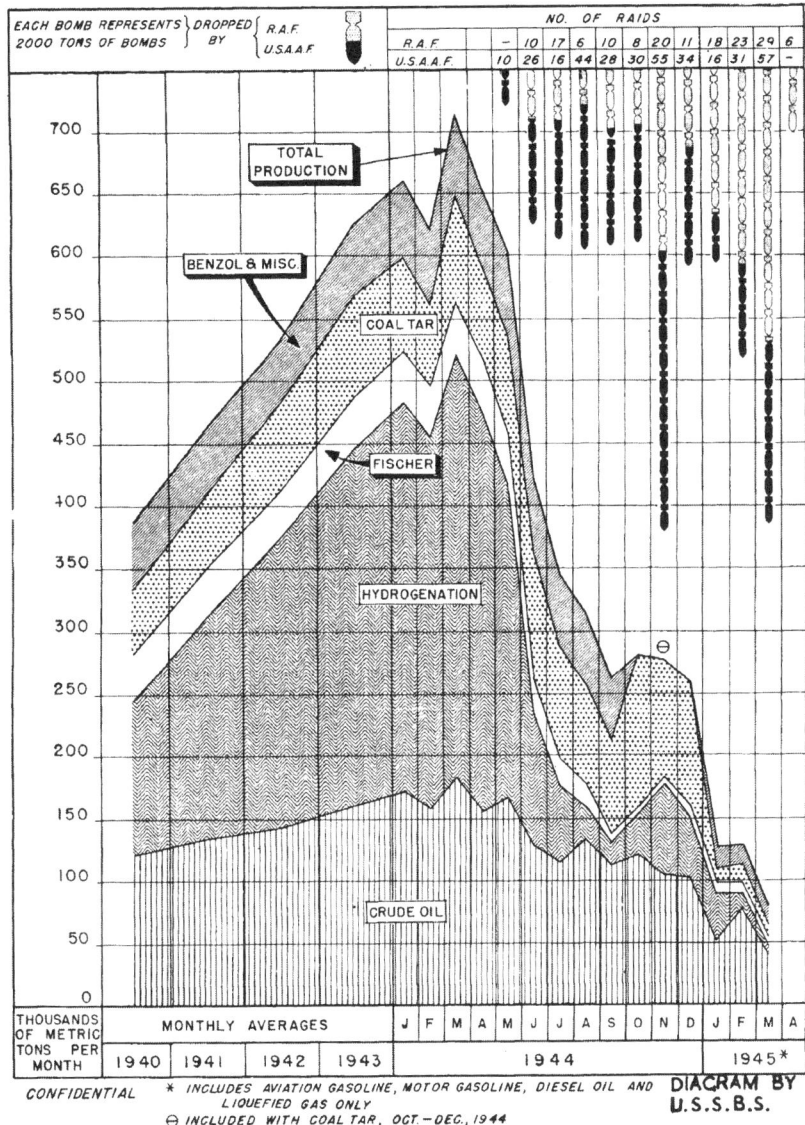

Figure 1.7
Estimated effects of combined bombing on German World War II petroleum production.
Source: Chiefs of Staff Committee, *Oil as a Factor in the German War Effort, 1933–1945* (UK Minister of Defense, March 1946), 16.

Romania in August 1943 and again in April 1944, but those attacks were no longer necessary after the Soviet Union occupied Romania in August 1944. The Combined Bomber Offensive also attacked the German synthetic oil and Fischer-Tropsch plants that produced liquid fuel from coal, and aviation gas production facilities, starting in May 1944, hitting every major plant by July. Each time the plants were hit, the Germans rebuilt them; nevertheless, synthetic fuel production in Germany declined dramatically.[125] In January 1945, Germany's Minister of Armaments and War Production Albert Speer complained: "We have sometimes managed to resume full production, even if it was only for three, four or five days, and that after a reconstruction period of six or eight weeks. Then the plant is smashed up again. We start again and reconstruct it, then again everything is smashed up and again everything has to be rebuilt."[126] Shortages of aviation gasoline fuel persisted, even as the Germans continued to switch to other fuels. The Germans were forced to curtail their operations, but even then, the military's consumption of oil "exceeded production from May 1944 on."[127] Further, because nitrogen was also produced at synthetic oil facilities, the attacks on oil production interfered with the German ability to produce explosives and ammunition.

CONCLUSION

After centuries of little change in the technology of war, the transitions wrought by coal and oil were comparatively rapid and dramatic. One hundred eighty years ago, navies and armies, and military industries began to make the transition from wood, wind, and water, to coal power. By the turn of the twentieth century, coal was king—powering railways, ships, and factories. By the middle of the nineteenth century, naval strategists argued that the United States needed to have regular access to coal, via coaling stations, or at least, to control coaling stations so possible U.S. adversaries would not have access to those stations. Strategists also solved the coal supply problem by building larger ships with bigger coal bunkers, increasing the efficiency of the steam engines, and ultimately acquiring reliable access to fuel in the Pacific through developing refueling stations in the ports of its new possessions: the Philippines and Hawaii.

The rise of coal is particularly instructive. As the United States and other great powers enlarged their steam-powered navies in the nineteenth century, they became increasingly concerned with the problem of coaling, specifically acquiring the bases for refueling, and providing for the security of supply. Coal was still the dominant fuel at the start of World War I for war and industry. By the end of World War I, while coal was still an important fuel for weapons factories and electricity generation, it was clear that liquid fuels burned in internal combustion engines—diesel and gasoline—were increasingly dominant, allowing for rapid mobility to and within war zones. The innovations begun and accelerated in World War I only further accelerated after the war.

The Allies not only believed that oil helped them win World War I, but they also pegged their postwar military power—their new warships and air forces—on access to it. Navies transitioned to oil power and air forces became more capable. The infant aviation industry of the United States blossomed. As with coal, strategists became increasingly concerned with the problem of petroleum supply, specifically acquiring the technologies and bases for refueling and denying the adversary their own oil supplies, refineries, and oil fields. The United States was less concerned about oil in the Middle East when it was guaranteed by British and U.S.-controlled oil companies, and when, starting in the 1920s, U.S. domestic oil production and reserves dominated the global market. Producing oil and denying it to the Axis powers was even more central to the Allied victory in World War II than in the first World War. The Germans shaped their strategy around oil and were limited in their war effort by inadequate supply of it. The United States and its allies were able not only to destroy German oil supplies and refineries, but also to deny the Japanese forces oil through an increasingly effective blockade.

War increased the military and civilian use of fossil fuels. Both World War I and World War II accelerated industrialization and the use of coal and oil in the industrial economies and oil-powered infrastructure that linked it all. War stimulated the growth of the infrastructure that allowed the United States to produce vast quantities of petroleum for industrial use and for fuel. The U.S. Navy was an early adopter of coal in the middle of the nineteenth century, and war proved a stimulus to the adoption of coal in U.S.

arms manufacturing. The navy also promoted oil in its transition to oil just before World War I. After both World War I and World War II, the desire to improve roads for military reasons also promoted the transition to liquid fuel for automobile transport in the early twentieth century. Also in both postwar eras, military surplus vehicles and subsidies that were part of the war effort helped the aircraft and automobile industries produce increasingly affordable vehicles, which in turn increased demand for fuel. As we shall see in the next chapter, all that demand created anxiety about supply.

2 THE LIFE BLOOD AND THE DEEP CYCLE: OIL AND U.S. MILITARY DOCTRINE SINCE WORLD WAR II

The economic lifeline of the industrial world runs from the gulf and we cannot permit a dictator such as this to sit astride that economic lifeline. To bring it down to the level of the average American citizen, let me say that means jobs. If you want to sum it up in one word, it's jobs. Because, an economic recession worldwide, caused by the control by one nation—one dictator if you will—of the West's economic lifeline, will result in the loss of jobs for American citizens.
—Secretary of State James Baker, November 1990[1]

Energy is the lifeblood of our warfighting capabilities.
—General David Petraeus, 2011[2]

Maintaining overseas military bases is a uniquely American preoccupation: the United States has approximately 800 military bases; France and the United Kingdom have roughly 12 each; and Russia, the adversary with the next most overseas bases, has about 9.
—John Glaser, 2017[3]

World War II heightened the conviction that oil was the key to victory and that airpower was essential. If there is one through line in American foreign policy for the past century, beginning in World War I, outlasting the Cold War and the long post-9/11 wars in Afghanistan and Iraq, it is the belief that the world needs access to petroleum from the Middle East. Thus, when World War II came to a close, the United States reached an agreement in September 1945 with Saudi Arabia to finish construction of the airfield at

Dhahran and continued to support the efforts of American oil companies to gain access to the rights to develop oil in the region.[4] Oil has usually been cheap to extract, plentiful, and convenient or even essential to make into a variety of products from gasoline to jet fuel to plastic. And the world's leaders assumed that they needed not only access to plenty of petroleum, but oil at the right price. U.S. domestic oil consumption expanded immediately after World War II and through the 1960s. "From the 1930s through the 1960s, the government accepted and used the U.S-based multinational oil companies as arms of U.S. foreign policy, to help ensure access to foreign oil."[5]

However, for at least the last fifty years, as the world became even more reliant on oil to fuel economic growth, the United States became concerned that oil would become scarce when domestic oil reserves and then global oil supplies reached a peak and then declined. This concern about the declining productive capacity of oil fields and smaller reserves coincided with Anglo-American owned corporations' gradual loss of control over oil production in the Middle East, Latin America, and then more recently discovered oil fields in Africa, and the growing political instability in those regions. There was no thought that at the root of some of the unrest and instability might be the U.S. presence and the fact that the United States was backing greedy and undemocratic leaders who often were disliked by the general population and clamping down on dissent. The big multinational oil companies still physically, for the most part, ran the oil wells and the refineries, but local governments were asserting more control and, in some instances, even nationalized the supply. The felt urgency to control access to oil has only been amplified since the British pulled back from their bases east of the Suez Canal in the late 1960s, through the oil embargos of the 1970s, the Iranian Revolution and the Soviet invasion of Afghanistan in 1979, and then the concern that Saddam Hussein's Iraq threatened to control the largest portion of known oil reserves. The United States, for many decades the world's largest petroleum producer, took on the burden of "protecting access" to Persian Gulf oil in the 1980s from the threats that it assumed were latent in the region: domestic instability and potential revolution that could put in place regimes hostile to the United States and the West; aggression by local powers in the region and external military intervention that could lead to a hostile power controlling a large share of the regions' oil reserves; and external political and economic influence by governments that were hostile to the United States.

In other words, the taken-for-granted belief that oil is essential for economic development was matched by the belief that oil supplies are vulnerable—either because the world will run out of oil, or at least oil that is inexpensive to extract, or because a state with bad intentions will attempt to monopolize and control supply, and hence price. These two assumptions—that oil is the *sine qua non* of modern life, and that its supply is dwindling or insecure—have driven U.S. policy, alliances, and force structure since World War II. Oil gradually came to be seen as a "vital" national security interest.

Further, U.S. decision makers since President Nixon have also talked about the importance of reducing U.S. oil imports, even as they have done little beyond exhortation to reduce oil dependency. Despite early gestures by President Nixon toward "Energy Independence," and President Jimmy Carter's urging in July 1979 to deal with the energy crisis—what Carter called a "clear and present danger to our nation"—through among other things, an investment in alternative energy sources and reducing oil imports, there was little interest in questioning any of the basic premises of the scarcity-insecurity logic.[6] If anything, until the policy began to shift in 2009, U.S. policies increased American's dependence on oil and other fossil fuels. And so, U.S. policy makers have responded to their anxiety about access to oil and natural gas in three ways: by attempting to conserve domestic capacity by using other's oil first; by diversifying domestic and regional energy production in part by introducing new technologies for drilling and refining natural gas and oil; and by an ever-increasing commitment and military capacity to protect the flow of foreign oil. The strategy for "protecting" access to foreign oil worked either directly, through a military presence in the Persian Gulf, or by arming and supporting U.S. allies in the region.[7]

In this chapter, I focus on the ways that concern about access to oil, and in particular, Persian Gulf oil, have structured U.S. doctrine and force posture for the last several decades. There has been a steady and escalating sense of the importance of the Persian Gulf to U.S. interest and a concomitant growth of U.S. forces there. There has also been a generally growing sense of dependency on Persian Gulf oil, which, even when it was not the case, kept the United States invested in defending oil access by military power. The other options—decreasing consumption through conservation and alternative fuels—were explored and promoted with comparatively anemic investments. In the two chapters that follow, I describe the growing alarm that grew from the U.S. military's understanding of the consequences of climate change for its installations and operations and for the potential for increased conflict.

POST–WORLD WAR II: CAPACITY

Americans were looking ahead to the postwar era even as they continued to fight. U.S. decision makers and military strategists took for granted that oil would be key to any future war and maintaining and increasing the infrastructure for fossil fuel production and bases for refueling was understood to be vital strategy and remained in place. In 1943, Secretary of the Interior Harold Ickes warned that the United States was "running out of oil."[8] In 1944, Herbert Feis summarized the common wisdom: "The Middle East is potentially important to us as a source of oil supply that supplements our own and prolongs our reserves. The war has established the facts that American military action may take place anywhere in the world, and that, particularly in any struggle involving the Pacific, control over these oil fields (and the political status of this area) might be of direct concern to us."[9]

By January 1944, the United States had decided a list of requirements for a global network of postwar air bases, and President Roosevelt had instructed his secretary of state to begin negotiations for bases and air transit rights. On December 11, 1944, Secretary of the Navy James Forrestal wrote to Secretary of State Edward Stettinius that it was "distinctly in the strategic interest of the United States to encourage industry to promote the orderly development of petroleum reserves in the . . . Persian Gulf, thereby supplementing . . . Western hemisphere sources and protecting against their early exhaustion."[10] Roosevelt's secretaries of war, the navy, and state wrote on December 22 to recommend the United States work out the details of an air base in Dhahran. There were two other arguments put forward by the undersecretary of the navy in favor of a base in Dhahran: it would deter some other power from trying "to occupy the oil fields" and "the mere existence of an American military airfield at Dhahran would contribute to the political integrity of Saudi Arabia and to the maintenance of our interest in the oil fields."[11] In May 1945, the military and State Department told President Truman that the United States should continue construction of the U.S. base at Dhahran even though the war in Europe was over, and the war with Japan soon would be. In August 1945, U.S. defense officials urged President Truman to encourage the use of Middle Eastern oil as a way to preserve U.S. oil reserves in case of another war.[12]

By the end of World War II, the United States had control or use of about "thirty thousand installations at two thousand base sites abroad."[13] U.S. leaders also believed that though some bases should be closed, others should remain active as a hedge against future conflict. Among those that the United States would keep, the U.S. Air Force continued constructing the air base in Dhahran near the oil fields originally leased by Standard Oil in 1931 and then controlled by companies that were partly owned by the United States and the Kingdom of Saudi Arabia.[14] The Dhahran base was completed in early 1946 and the United States remained there, under leasing agreements, until 1963. The United States also secured a ninety-nine-year lease on twenty-three bases and military installations in the Philippines. Further, Congress funded bases on locations that the United States had taken from the Japanese in the war including Okinawa (which had been the site of an intense ground battle), Christmas Island, Ogasawara (Bonins), Saipan, Tinian, Einiwetok, Majuro, Kwajalein, Truk, and Palau. Other locations that received funding already had a military presence, including Pearl Harbor, and included Johnston Atoll, which had been used as a refueling base for U.S. bombers and submarines during World War II.[15] The Wheelus Air Force Base in Libya, opened in 1943 and jointly operated with the British also remained. But, overall, U.S. military presence was consolidated. Within the next two years, about half of the 2,000 bases the United States had at the peak of World War II were closed or returned and in 1949, the United States had about six hundred bases.[16] One of those was the berthing space for two navy destroyers and a seaplane tender at HMS Jufair, the Royal Navy's base in Bahrain.[17]

Cold War military strategy dictated that the United States would require bases overseas to station and refuel its medium-range nuclear bombers in the event of a nuclear war, and that the United States had to have conventional forces stationed in Europe and Asia that would be capable of mounting a "forward defense" of the United States and its allies. Fuel was also necessary, and the Army-Navy Petroleum Board became the Army-Navy Petroleum Purchasing Agency after the war. Thus, U.S. military operations remained global. The islands in the Pacific would be used to test nuclear weapons. The long-range B-52 bomber the United States first deployed in June 1955 could leave bases in the United States and travel, with aerial refueling, to the Soviet Union and back. Increased range diminished the requirement for the medium range B-47s, which were retired in 1963, and some overseas bases used to refuel medium-range bombers were closed.[18] Okinawa and Guam remained essential bases during both the Korean and the Vietnam wars.

U.S. doctrine was ambivalent about how to protect oil. On the one hand, the military's working assumption was that the Soviet Union would gain control of oil in the Persian Gulf and the United States would have to rely on its own resources, which might be insufficient in a future war. In February 1947, the Joint Logistics Committee of the U.S. Joint Chiefs of Staff produced the report *The Problem of the Procurement of Oil in a Future War*, which encouraged the United States to import and use as much Middle Eastern oil as possible in order to conserve U.S. reserves and more easily defend U.S.-controlled reserves.[19] On the other hand, the U.S. Navy leadership was urging U.S. leaders to prepare to defend the oil and began to periodically send ships to the Persian Gulf to "show the flag." The Soviets had been slow to withdraw from Iran, only leaving in 1946. In 1949, the U.S. Navy established a regular presence in the region, the Middle Eastern Force, and U.S. forces in the region were already heavily relying on Persian Gulf oil.[20] A ring of bases around the Mediterranean would allow the United States and the UK to defend Persian Gulf oil, or alternatively destroy it, in order to deny the Soviet Union access.[21]

Meanwhile, oil demand dramatically increased throughout the entire U.S. economy. As noted above, industry not only invented new uses for oil during World War II, but also built the infrastructure and technologies to produce and deliver the oil, which remained after the war. These technologies, many of them based on oil, "flooded into society at large."[22] Oil was turned into plastics, fertilizer, and the very roads the new cars would travel. The World War II Big Inch and Little Inch pipelines were put to use transporting the fuel that took people to their new suburban homes. Surplus DC-3 and C-54 aircraft were sold to commercial airlines and wartime advances in aircraft technology boosted air travel. All of this was subsidized by the war just completed—or the next war the United States feared.

The emissions of nitrous oxide, another greenhouse gas, with a global warming potential about three hundred times that of carbon, also grew in the wake of World War II. Ammonium nitrate, an important ingredient in munitions, was in excess supply after the war. As Michael Pollen tells it, the use of ammonium nitrates in agriculture is a wartime legacy. "The great turning point in the modern history of corn, which in turn marks a key turning point in the industrialization of our food, can be dated with some precision to the day in 1947 when the huge munitions plant at Muscle Shoals, Alabama, switched over from making explosives to making chemical fertilizer. . . . Serious thought was given to spraying America's forests with the surplus chemical, to help the timber industry. But agronomists in the Department of Agriculture had a better idea: spread the ammonium nitrate on farmland as fertilizer."[23] When ammonium nitrate or other nitrogen-based fertilizers are spread on crops about half of the nitrogen is taken up by the plants. The rest of the nitrogen goes into streams and ground water, or is converted into gas, including the potent greenhouse gas nitrous oxide. In recent years, about 75 percent of all U.S. nitrous oxide emissions were from agricultural processes, specifically soil management.[24] Methane, which can be captured and sold as "natural gas," is also an important greenhouse gas. Industrial agriculture (and to a lesser degree organic agriculture) also tends to produce methane, and today, agricultural sources of methane account for about 36 percent of all methane emissions in the United States.

Nitrous oxide and methane are also products of fossil fuel combustion, which increased as automobile use skyrocketed in the 1940s and 1950s. President Dwight Eisenhower advocated a national highway system, in part to make evacuation from large cities easier in case of a nuclear war. Eisenhower said, "After seeing the autobahns of modern Germany and knowing the asset those highways were to the Germans, I decided, as President, to put an emphasis on this kind of road building. . . . The old [1919] convoy had started me thinking about good, two-lane highways, but Germany had made me see the wisdom of broader ribbons across the land."[25] In June 1956 the National System of Interstate and Defense Highways Act authorized a thirteen-year program of interstate highway construction, which promoted automobile travel and the auto industry, accelerated the pace of suburbanization, and led to an increase in trucking that displaced more efficient railway travel, all of which increased civilian demand for oil.[26] Past war thus required and then inspired greater fuel use and literally paved the way for future war. The U.S. oil industry was able to keep up with demand, and through 1970, the United States was the world's largest producer of petroleum, and the price of oil was extremely low.

GROWING CONCERN ABOUT OIL AND THE 1973 EMBARGO

In August 1956, the Egyptian leader Gamal Abdel Nasser, a nationalist with ties to the Soviet Union, nationalized the Suez Canal, which was then largely owned by British and French shareholders. In late October 1956, when negotiations to put the Suez Canal under a form of joint control were failing to come to resolution, the Israelis attacked Egypt and moved toward the canal. Britain and France then told both the Israelis and the Egyptians to retreat ten miles from it. Israel, which had secretly colluded with Britain and France in staging the invasion, withdrew. When the Egyptians failed to do so, the UK and France started bombing Egyptian targets. A UN ceasefire ended the crisis in early November, but Nasser's promotion of Pan-Arabism and U.S. concerns about growing Soviet influence in the region combined to increase the U.S. government's anxiety about stability in the Middle East. In early 1957, under the Eisenhower Doctrine, the United States articulated a policy of giving economic or military aid to countries in the Persian Gulf, and that it would even intervene militarily to support countries that faced a threat from international communism.[27] The United States supplied $200 million in economic and military aid to the region and the U.S. military drew up plans for intervention.

Three things increased U.S. anxiety about access to Persian Gulf oil and prompted an increase in the U.S military presence in the Middle East in the late 1960s and early 1970s. First, the British, who had exerted the predominant external military power in the Persian Gulf region since the withdrawal of the Ottomans in the early twentieth century, began to withdraw their forces from the Middle East and Southeast Asia and the configuration of international bases and forces changed. The UK announced in 1968 that it would withdraw all its forces from East of the Suez and declared its intention to leave the Gulf by the end of 1971, and the United States eventually took over many of those bases. This included expanding the U.S. naval base in Bahrain, which housed the approximately four hundred personnel in MIDEASTFOR (the Middle East Force), and in late 1971 the United States took over the British facility. The United States had already acquired basing rights from the UK on the island of Diego Garcia in the Indian Ocean in 1966, which allowed it to have long-range aircraft access to not only South Asia, but also Southeast Asia, the Persian Gulf, and Africa. However, the Qaddafi government ended U.S. access to Wheelus Air Base in Libya in June 1970. The Soviets sent three ships to the region and the United States decided it would need to remain in the area as a demonstration of its interest. The United States depended on its special relationships with the Shah of Iran, and Saudi Arabia, to keep the balance of power and to promote U.S. interests in the Persian Gulf.[28]

Second, output from oil wells in the United States peaked in 1970 and then began to decline in 1971. On June 4, 1971, President Richard Nixon warned that sharp increases in energy demand meant that the days of cheap energy were drawing to a close. He argued that the United States should use energy "more wisely," invest in "clean energy," and also speed up development of new fossil fuel technologies such as piloting programs to convert coal into gas and developing shale deposits and offshore drilling.[29] Also in 1971, the U.S. government established price controls on domestically produced crude oil that would remain in place until 1981, and in 1973 eliminated restrictions on imported oil. "By holding down domestic prices, they reduced incentives for domestic production. At the same time, they encouraged consumption, furthering U.S. demand for and dependence on imported oil."[30] As U.S. domestic petroleum production declined, less expensive Persian Gulf oil met increased U.S. demand.

Third, it was increasingly clear that the United States could not guarantee the price or the flow of oil imports as it had in the past. The U.S. oil companies that had the concessions in Iraq, Saudi Arabia, Kuwait, Qatar, Abu Dhabi, and Iran were not able to manage the increasingly independent Arab governments that were seeking a higher share of the profits from the oil being pumped and exported from their countries. Further, oil was a growing factor in Middle East politics. During the 1967 Six Day War in which Israel took huge swathes of territory, Iraq, Kuwait, and Saudi Arabia reduced the flow of oil to the countries that supported Israel—the UK, United States, and West Germany—and the Suez Canal was briefly closed. This embargo was ineffective since U.S. oil production could make up the difference, but it demonstrated the potential for such action in the future.

Thus, strategists feared that growing dependence on oil imports left the United States vulnerable to countries that might want to use the "oil weapon" against it. The U.S. Congress held hearings in 1972 and the House Subcommittee on the Near East, chaired by Lee Hamilton, produced a report that summarized the state of play and proposed a policy of a "low key" presence.[31] The House subcommittee assumed that the United States would need increasing quantities of petroleum, production would not increase without new discoveries, and the world oil market would move from a buyer's to a seller's market. Further, the subcommittee predicted that an increasing share, as much as two-thirds of the oil consumption of Western Europe and Japan, would come from the Middle East and North Africa. The United States was in a better position, but unless it decreased consumption or increased domestic oil production it would have to import about 50 percent of its oil needs.[32] The report also warned, "In an area of many and overlapping . . . disputes, we should not become identified too closely with any one state, when its causes might well become our causes. This is particularly true of Iran and Saudi Arabia . . . [and] the United States should avoid . . . becoming too closely identified with particular individual leaders."[33] The House committee report noted increased Soviet presence in the region but not an immediate Soviet threat, and recommended that a "low-profile policy should be designed to promote great power restraint and insulate the Persian Gulf from great power politics and competition. While the Soviet Union would like to, and probably will, increase its influence in the area, there is no evidence that the Soviet Union is embarked on a grand scheme to control the oil faucets in the Persian Gulf." Further, the report noted, the Soviets did not need the fuel, and in any case, if their allies in Eastern Europe did need to increase their imports of fuel, the Soviets lacked "the capability to refine, transport and market massive amounts of oil produced outside the Soviet Union."[34]

Despite the fears that Persian Gulf oil would be cut off, the actual flow of oil was guaranteed by commercial oil agreements. The flow was only jeopardized during the Arab oil embargo that began in October 1973. The reduction in oil imports and the tripling of the price from 1973 to 1974 caused a great deal of concern in Washington about what could be next.[35] By 1973, Saudi Arabia had a 21 percent share of world oil exports.[36] During the October 1973 Arab–Israeli War, Saudi Arabia and other Arab members of the Organization of Arab Petroleum Exporting Countries (OPAEC) agreed to reduce production and exports to Israel's allies. The war, which began in a surprise attack on October 6, took most observers by surprise. No matter, the United States was able to put in place a massive "air bridge" by October 13 that allowed the United States to deliver 1,000 tons of military equipment each day to Israel, "in particular the formidable Maverick antitank missile."[37] After the Nixon administration on October 19 requested $2.2 billion in emergency aid for Israel, Saudi Arabia completely cut off oil exports to the United States. While Israel, Egypt, and Syria reached a truce on October 25, the embargo remained in force until March 1974, dramatically increasing the price of oil from under $3 per barrel to nearly $12 per barrel.[38]

On November 7, 1973, with the embargo still in place, President Nixon called for dramatic action, including halting the conversion of coal to oil at power plants, and converting some power plants back to coal. Although oil was still flowing, there was a panic about future supplies and costs.[39] He proposed legislation that would relax environmental regulations "permitting an appropriate balancing of our environmental interests, which all of us share, with our energy requirements, which, of course, are indispensable."[40] Further, Nixon asked Americans to turn down home thermostats by at least 6 degrees and he urged the construction of new nuclear power plants. Nixon repeated his earlier calls for the United States to become energy self-sufficient, now called "Project Independence." In 1973, the Defense Fuel Supply Center, which was the descendent of the Army-Navy Petroleum Purchasing Agency, became charged with managing the storage, distribution, and sale of fuel to the services.

The United States was still fighting in Vietnam and thus, the U.S. military's need for oil remained high during the embargo and the accompanying production cuts. The Saudi Arabian government ordered Aramco and its affiliates (Exxon, Standard Oil of California, Mobil Oil, and Texaco) to withhold oil from U.S. armed forces in October 1973. The Saudis also threatened to withhold oil from refineries in the Philippines and Singapore, and this led to a cut-off of supplies to Pacific Command. According to Secretary of Defense James Schlesinger, the embargo caused the DOD to lose about 40 percent of its supplies.[41]

On October 27, 1973, White House Chief of Staff Alexander Haig told newly appointed Secretary of State Henry Kissinger that Secretary of Defense Schlesinger was considering "putting troops in crucial states to get oil." Kissinger replied, "He's insane." Haig continued, "He [Schlesinger] thinks forces should be put in."[42] Schlesinger then discussed the possibility of using military force to get Middle Eastern oil with members of NATOs Nuclear Planning Group from November 5–8.[43] The head of Britain's Joint Intelligence Committee sent a document to Prime Minister Edward Heath titled "Middle East—Possible Use of Force by the United States" that described possible British cooperation with U.S. military action.[44]

A November 30 memo from the Pentagon to Nixon's military assistant at the White House, Brigadier General Richard Lawson, said, "DoD petroleum shortage is very serious." Without finding more supplies, "we will soon be forced to begin standing down operational forces."[45] Deputy Secretary of Defense Bill Clements told the U.S. leaders of the Saudi oil company Aramco, "Find a way to get fuel to Vietnam. Our kids are dying out there fighting Communists."[46] The United States ordered twenty-two U.S. oil companies to transfer commercial stocks to the military during the crisis, under the Defense Production Act of 1950, thus diverting more than 825 million gallons of fuel.[47] The military also cut maneuvers, training, and ship steaming time for military reserves and fuel consumption was reduced by 16 percent in the second quarter of fiscal year 1974.[48] The pressure was reduced when the Saudis and the Americans came to a secret arrangement whereby oil was transported to countries that were not embargoed and then sent to U.S. naval ports.[49]

Thus, the 1973 oil embargo was more concerning than the 1967 embargo and moved the United States away from the idea of maintaining a "low-key" presence in the Persian Gulf. The United States increased arms transfers to Iran and in 1973 increased the U.S. naval presence, the Middle East Force (MIDEASTFOR) that had been originally established 1949. On January 14, 1974, Secretary of Defense Schlesinger said that although the United States did not foresee taking action to protect access to oil, it was "indeed feasible to conduct military operations if the necessity should arise." Such action, Schlesinger averred, would only occur in the case of the "gravest emergency and that we do not anticipate that the necessity will arise for us to conduct such operations."[50] On January 20, 1975, Secretary of State Kissinger responded to a question about the possibility of using force to secure access to oil by saying he did not think that would be a cause for military action. Asked a week later to elaborate, Kissinger said, "We should have learned from Vietnam that it is easier to get into a war than to get out of it. I am not saying that there's no circumstance where we would not use force. But it is one thing to use it in the case of a dispute over price; it's another where there is some actual strangulation of the industrialized world."[51] In late February 1975 President Gerald Ford said the United States would be willing to use force in case of "economic strangulation . . . we had to be prepared, without specifying what we might do, to take the necessary action for self-preservation."[52]

Concerned by the prospect, House of Representatives member Lee Hamilton requested that the Congressional Research Service (CRS) prepare a report, *Oil Fields as Military Objectives: A Feasibility Study*.[53] Hamilton's House Subcommittee on the Near East report in 1972 had urged a low-key presence in the Middle East, so it was no surprise that he asked for an evaluation of what would have been an about-face in the longstanding U.S. policy of maintaining a small naval presence. The CRS found that a sustained Organization of Petroleum Exporting Countries (OPEC) embargo of the United States did not constitute a threat to U.S. vital interests. However, while a tight and sustained embargo would be disruptive and would "degrade" U.S. security, CRS found that U.S. allies who were more dependent on oil imports than the United States would find such an embargo much more devastating. The CRS concluded that "military operations to rescue the United States (much less its key allies) from an air-tight OPEC embargo would combine high costs with high risks wherever we focused our efforts. This country would so deplete its strategic reserves that little would be left for contingencies elsewhere. Prospects would be poor with plights of far-reaching political, economic, social, psychological and perhaps military consequence the penalty for failure."[54] Worry that the United States might use force to obtain foreign oil faded in 1972, however, when a feared embargo never materialized.

Around the same time, the Nixon and Ford administrations began to consider other ways to respond to potential oil embargos. Notably, the important National Petroleum Council studies of the problem in 1973 did not suggest that United States should significantly reduce petroleum consumption even in the face of a shortfall in imports. Among the alternatives considered, including alternative fuels, the favored option was to create a petroleum reserve stockpile.[55] Other countries already had national petroleum stockpiles and in November 1974, the members of the newly created International Energy Agency formally established an International Energy Program (IEP) that obliged members to establish national petroleum stockpiles and in the event of an emergency shortfall, to participate in an Emergency Sharing System.[56] Under the IEP, signatories including the United States are required to keep petroleum stockpiles equivalent to ninety days' worth of their prior year's average daily net imports.[57] Reducing energy consumption was not on the table: as the deputy undersecretary of the interior said in October 1974, "It's not our job to force the country to change consumption patterns."[58]

Members of Congress did think it was their job, however, to promote greater energy efficiency. In its 1975 Energy Policy and Conservation Act, Congress established higher automobile fuel economy standards. It also established the Strategic Petroleum Reserve (SPR), a program to fund the DOD's acquisition of up to a billion barrels of crude oil to be stored in the Defense Fuels Supply Center's underground salt dome caverns. Interestingly, since its creation the SPR has never been filled to one billion barrels of crude—its maximum capacity in 2020 was 713.5 million barrels.[59] The Strategic Petroleum Reserve's first director, Robert L. Davies, came to the position from the Department of Defense, where "he had advocated oil stockpiling as an alternative to naval requests for more ships to protect the growing number of super tanker fleets in the world."[60] The United States purchased the first fuel for the SPR from Saudi Arabia. As Wayne Cuttrel, the director of oil acquisition for the SPR from 1979 to 1984, told Bruce Beaubouef, "The reasoning was that we really didn't want to store up our own oil; we should store up theirs, and use it as a bargaining chip."[61]

THE CARTER DOCTRINE AND THE RAPID DEPLOYMENT FORCE

The next period of "crisis" began in early 1979 when the Shah of Iran, a U.S. ally, was forced out of the country by anti-government demonstrations. In November 1979, Americans at the U.S. embassy in Tehran were taken hostage. On December 24, 1979 , the Soviet Union invaded Afghanistan to bolster their Afghan ally then under attack from U.S. backed "freedom fighters." The Iranian revolution, and the Soviet invasion of Afghanistan combined as a shock to the U.S. public.[62] Addressing the crises, President Jimmy Carter then announced that, if necessary, the United States would take action to protect Persian Gulf oil. Known as the Carter Doctrine, it is worth recalling Carter's exact words:

> The region which is now threatened by Soviet troops in Afghanistan is of great strategic importance: It contains more than two-thirds of the world's exportable oil. The Soviet effort to dominate Afghanistan has brought Soviet military forces to within 300 miles of the Indian Ocean and close to the Straits of Hormuz, a waterway through which most of the world's oil must flow. The Soviet Union is now attempting to consolidate a strategic position, therefore, that poses a grave threat to the free movement of Middle East oil. . . .
>
> Let our position be absolutely clear: An attempt by any outside force to gain control of the Persian Gulf region will be regarded as an assault on the vital interests of the United States of America, and such an assault will be repelled by any means necessary, including military force.[63]

President Carter, like Nixon before him, invoked the need for the United States to reduce its dependence on foreign oil. At the same time the Carter administration moved to make the potential for the United States to rapidly intervene in the Persian Gulf a reality. The administration already had plans underway for maritime prepositioning of ships and a Rapid Deployment Joint Task Force, later called the Rapid Deployment Force.

Yet, when Carter enunciated the Carter Doctrine in his State of the Union Address of 1980—"An attempt by any outside force to gain control of the Persian Gulf region will be regarded as an assault on the vital interests of the United States of America, and such an assault will be repelled by any means necessary, including military force"—the United States had already been working to enhance its military capabilities in the region by increasing its naval presence.[64] In October 1979, the president announced the formation of the Rapid Deployment Force that the administration had begun planning in 1977 as a global reaction force. Carter's secretary of defense ordered specially designed ships for the Marines in August 1979 for the purpose of maritime prepositioning and in December 1979, Carter told the Pentagon to look for bases in the Persian Gulf region.[65]

By March 1980, the administration had established the Rapid Deployment Joint Task Force. In the summer of 1980, a Near-Term Prepositioning Force of seven ships—"unmodified, off-the-shelf commercial ships dedicated to Navy and Marine Corps Operations and configured to support a MAB [Marine Amphibious Brigade] composed of 11,200 personnel"—was ready to go. A MAB was then comprised of "53 tanks, 36 artillery pieces, 96 heavy antitank weapons, 109 assault amphibian vehicles, [Raytheon MIM-23] Hawk and [Raytheon FIM-92] Stinger antiaircraft missiles, and more than 140 aircraft."[66] The navy assembled thirty days of supplies aboard Military Sealift Command ships stationed on the U.S. East Coast. To this, the United States added two other Maritime Positioning Ships (MPS) stationed in Diego Garcia, and Guam and Tinian in the Pacific. The region became more volatile in September 1980 when Iraq invaded an area of Southwest Iran called Arabistan or Khuzestan, and the Shatt al-Arab river, an important oil-producing region of Iran. During the Iran–Iraq War, both sides attacked each other's oil facilities and cities.

In 1981, in explaining a decision by the United States to sell AWACs aircraft to Saudi Arabia, President Ronald Reagan made it clear that the United States would "not permit" Saudi Arabia to become another Iran, stating, "There's no way that we could stand by and see that taken over by anyone that would shut off that oil."[67] The U.S. presence in the Middle East grew during the Reagan administration as the United States quietly supported Iraq in its war against Iran, making its support more overt in 1982. By early 1983, the Rapid Deployment Force renamed U.S. Central Command (CENTCOM) consisted of 220,000 U.S. Army, Navy, Air Force, and Marine Corps troops. Three aircraft carriers were part of the mix, each with six surface escort ships.[68] The commander of CENTCOM, Lt. General Robert Kingston was clear that the mission was to "ensure the unimpeded flow of oil from the Arabian Gulf."[69] The United States set about improving bases and capabilities in the region and even lengthened the runway at Diego Garcia so that it could accommodate U.S. long-range bombers. Figure 2.1 illustrates the CENTCOM area of responsibility.

The United States role expanded after Iran and Iraq attacked each other's merchant fleets in 1984, and in late 1986, Iran began to attack the ships of Iraq's allies, including Kuwait. In 1987–1988 the United States reflagged Kuwaiti oil tankers and escorted their passage in the Persian Gulf as a deterrent to Iranian attacks on them. In explaining the policy of escorting Kuwaiti tankers, and destroying Iranian assets that were impeding the flow of oil in the Gulf, Michael Armacost, the U.S. undersecretary of state for political affairs, said, "The unimpeded flow of oil through the Strait of Hormuz is a vital interest and critical to the economic health of the Western world; another very important interest is freedom of navigation for nonbelligerent shipping in and through the gulf, in line with our worldwide policy of keeping sea lanes open." The point, Armacost said, was also to limit Soviet "influence and presence in the gulf, an area of great strategic interest to the Soviets because of Western dependency on its oil supplies."[70] The Joint Chiefs of Staff posture statement in 1986 averred that its interest in the Middle East and Southwest Asia were to deter Soviet influence and expansion, and "ensure continued Western access to oil resources."[71] Although it bolstered U.S. military capabilities in region, the Iran–Iraq War ended in a ceasefire in late August 1988,

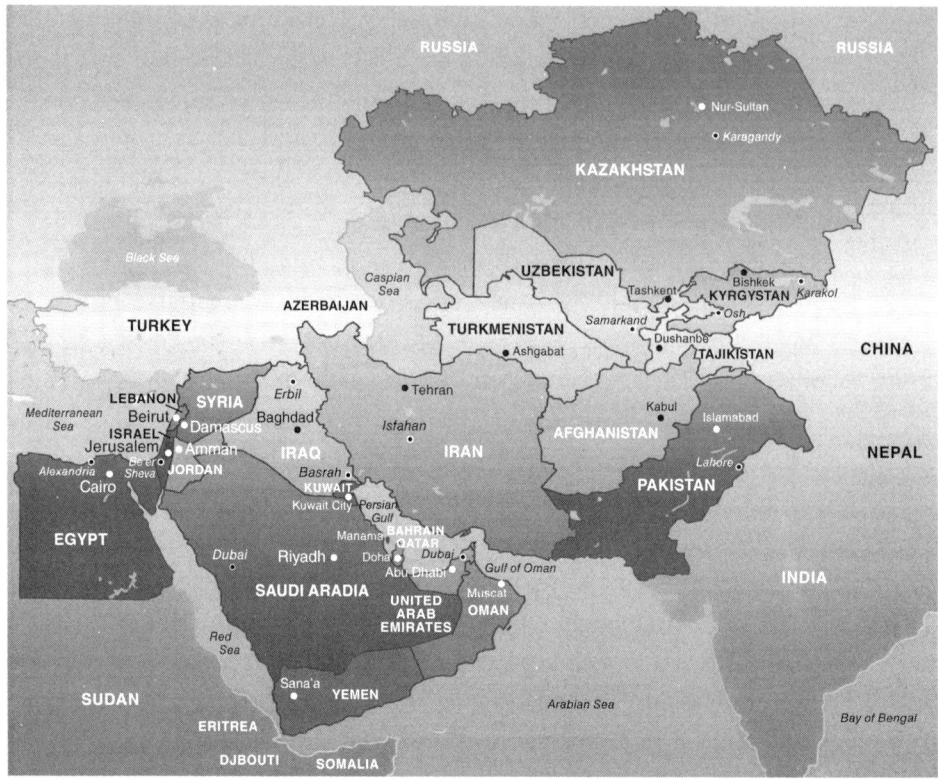

Figure 2.1

U.S. Central Command.

Source: U.S. Central Command, https://www.centcom.mil/AREA-OF-RESPONSIBILITY/.

at a cost of hundreds of thousands of lives in each country, with no change in the borders.

Meanwhile, in Europe the human rights and democracy movements working to end the Cold War by democratizing the Warsaw Pact states had some success. Inside the Soviet Union, President Mikhail Gorbachev's reform policies of *glasnost* (greater openness) and *perestroika* (economic and political reform and restructuring) accelerated the political change in the Warsaw Pact countries and decreased tensions with the United States. The Soviets also withdrew from Afghanistan in 1989. With tensions reduced, Gorbachev and Ronald Reagan, and then Reagan's successor George H. W. Bush came to significant agreements on nuclear and conventional arms control, which led to a reduction in conventional forces in Europe. Starting in 1988, the republics that made up the Soviet Union itself began to declare their intention to become independent states and proclaim their sovereignty. In November 1989, pro-democracy protesters breached the Berlin Wall, marking a symbolic end to the Cold War.

THE 1991 GULF WAR AND AN ENLARGED MIDDLE EAST PRESENCE

The *National Security Strategy of the United States, March 1990*, the first published after the fall of the Berlin Wall, reiterated both the Carter Doctrine and Reagan's view that protecting oil was a primary U.S. objective: "Secure supplies of energy are essential to our prosperity and security. The concentration of 65 percent of the world's known oil reserves in the Persian Gulf means we must continue to ensure reliable access to competitively priced oil and a prompt, adequate response to any major oil supply disruption. We must maintain our Strategic Petroleum Reserve at a level adequate to protect our economy against a serious supply disruption."[72] Interestingly, the 1990 *National Security Strategy* also mentioned "growing environmental concerns" in the same paragraph. "We will continue to promote energy conservation and diversification of oil and gas resources, while expanding our total supply of energy to meet the needs of a growing economy. We must intensify efforts to promote alternative sources of energy (nuclear, natural gas, coal, and renewables), and devote greater attention to reducing fossil fuel emissions in light of growing environmental concerns."[73] This was an early acknowledgment of the potential for climate change in a national security context, wrapped up in a concern for energy security.

On August 2, 1990, Iraq's military invaded Kuwait, using among other arguments the claim that the Kuwaiti government was stealing oil from an Iraqi oil field, and President Saddam Hussein announced that Iraq had annexed its southern neighbor. In just a few hours, Iraq nearly doubled its oil reserves, from 100 billion to 194.5 billion barrels.[74] The next day, President Bush announced the deployment of U.S. naval forces to the Persian Gulf, and within a week, 15,000 troops arrived in Saudi Arabia, begining Operation Desert Shield. Oil prices immediately increased as sanctions were imposed on the Iraqi regime, causing a reduction in oil exports from Iraq and Kuwait. The United Nations imposed a ban on importing oil from Iraq and Kuwait on August 6, which immediately removed four million barrels of oil per day from the world market.[75] This sent the price of oil from $21 to $28 per barrel.

On August 20, 1990, the Bush administration reiterated the importance of oil in the region in National Security Directive 45. "U.S. interests in the Persian Gulf are vital to the national security. These interests include access to oil and the security and stability of key friendly states in the region. The United States will defend its vital interests in the area, through the use of U.S. military force if necessary and appropriate, against any power with interests inimical to our own."[76] The United States and other governments were concerned not only that the Iraqi government had violated Kuwait's sovereignty and international law, but also that Iraq would not stop there, and would move into Saudi Arabia, which at the time held an estimated 255 billion barrels of oil. Indeed, in late September, Saddam Hussein threatened to destroy Saudi oil fields, sending the price of crude oil even higher.[77]

Over the next months, the United States deployed more than 300,000 troops to the region, a coalition of forces sanctioned by the United Nations mounted a naval blockade, and the United States led coalition threatened war unless the Iraqi military withdrew from Kuwait. When negotiations failed, the United States and its allies, now totaling more than 900,000 troops, launched an air assault, Operation Desert Storm, on Iraqi forces on January 17, 1991, followed by a ground invasion. A few days into the air strikes, the Iraqi military set 600 Kuwait oil wells on fire and then proceeded to dump oil into the Persian Gulf. The coalition began its land assault on Iraqi positions at the end of January and quickly overran the Iraqi military. By February 28, the United States had declared victory and Iraq announced that it would completely withdraw from Kuwait. The Kuwaiti oil fires burned for months, with the last extinguished in November 1991.[78]

Given the heavy reliance on airpower to move troops and equipment to the region, fuel needs for the thousands of aircraft used in the war were high. The United States dedicated more than 80 percent of its KC-10 and 44 percent of its KC-135 tanker force to supporting the war, a "total of 262 KC-135s and 46 KC-10s operating out of 21 locations in 10 countries."[79]

During Operation Desert Shield, many U.S. aircraft flew nonstop from the United States using an Atlantic "air bridge" on nonstop flights of between 15 and 16 hours, for example, with F-4G fighter aircraft flying from the East Coast requiring a maximum of 15 aerial refuelings. Flights from the George Air Force Base, in California required four additional refuelings.[80] U.S. aerial refueling tankers offloaded more than 441,017,600 pounds of fuel during Desert Shield, and during the 42 days of Desert Storm, aerial tankers offloaded 800,736,000 pounds of fuel to support a total of 69,399 combat sorties.[81] Total fuel use for U.S. and coalition aircraft amounted to 44,825,480 barrels.[82] But the United States did not supply most of that fuel; rather, Saudi Arabia, the United Arab Emirates, and Oman contributed all of the ground fuel and most (41.8 million barrels) of the jet fuel.[83] In addition, Saudi Arabia's trucks provided most of the inland distribution of fuel from refineries and depots to the bases.

While the Iraqi military was essentially defeated, the United States and United Nations maintained economic sanctions on Iraq after the war. Further, some U.S. and allied troops remained to enforce "no-fly zones" over first Northern and then Southern Iraq, beginning respectively in March and August 1991, as a way to keep the Iraqi military from attacking Kurds in the north and menacing civilians and Kuwait in the South. Although there was a significant drawdown of military forces, the U.S. Navy remained in the Persian Gulf, enforcing the sanctions against Iraq, naming its permanent naval presence, based in Bahrain, the Fifth Fleet. The Saudi government, which had agreed to host the U.S. troops for the duration of the 1990–1991 mobilization and war against Iraq, agreed to have U.S. troops and bases remain in Saudi Arabia. Further, the east coast squadron of the Maritime Prepositioning Force (MPSRON-1) was relocated to the Mediterranean.

In February 1991, the Warsaw Pact dissolved. The political unrest in the Soviet Union, culminated with the attempted coup against President Mikhail Gorbachev in August 1991. Although the coup was thwarted, the legitimation crisis precipitated the dissolution of the Soviet Union as its constituent republics, starting with Ukraine, seceded and were admitted into the United Nations. The Cold War was over. On December 25, 1991, the Soviet Union dissolved.

In August 1991, before the breakup of the Soviet Union, the Bush administration had argued in its *National Security Strategy* that the prospects of a post–Cold War era offered great hope. However, the report stated, "this hope must be tempered by the even greater uncertainty we face."[84] U.S. interests included promoting a "heathy and growing economy," and thus the nation would seek to "ensure access to foreign markets, energy, mineral resources, the oceans and space."[85] In other words, if anything, the end of the Cold War was taken to mean that threats had multiplied. The *National Security Strategy* reiterated the importance of the war against Iraq as reaffirming the U.S. role in the Middle East: "American strategic concerns still include promoting stability and security of our friends, maintaining a free flow of oil, curbing the proliferation of weapons of mass destruction and ballistic missiles."[86] The strategy also stated that "secure, ample, diversified and clean supplies of energy are essential to our national economic prosperity and security." But, in the next sentence it contended, "For the foreseeable future, oil will remain a vital element in our energy mix. For geological and economic reasons, U.S. oil imports are likely to increase in coming years."[87] The 1991 *National Security Strategy* also suggested that the United States might improve fuel efficiency and substitute "alternative fuels" as a way to slow the increase in oil imports. In sum, the George H. W. Bush administration strategy seemed to be not only to rely on market forces to keep oil prices down and supply assured, and to diversify sources of oil from other regions, but also to count on the capacity of military force to assure access. The watchwords were "security" and "stability" in a world where disruptions could occur.

> Security of oil supplies is enhanced by a supportive foreign policy and appropriate military capabilities. . . . We will also maintain our capability to respond to requests to protect vital oil facilities, on land or at sea, while working to resolve the underlying political, social and economic tensions that could threaten the free flow of oil. The stability of the Gulf region, which contains two-thirds of the world's known oil reserves, is a fundamental concern to us. Political and military turbulence in the region has a direct impact on our economy, largely through higher oil prices and potential supply disruptions.[88]

The Bush administration also argued that the United States should improve strategic oil stockpiles by filling the Strategic Petroleum Reserve to the level prescribed by law of one billion barrels. Further: "Our use of oil is the key source of our vulnerability to world oil supply disruption" and therefore the United States should "reduce this vulnerability" by decreasing oil consumption, using oil more efficiently. The administration amplified the previous year's *National Security Strategy*, saying, "We must intensify the development of alternative sources of energy (nuclear, natural gas, coal and renewables) and support aggressive research and development of advanced energy technologies to provide clean, affordable, reliable energy supplies we will need in the mid-21st century."[89] These were in fact nearly the same proposals that Nixon and Carter had made more than a decade earlier, which had scarcely been pursued with anything like the vigor of the United States developing a military capacity to intervene in the Gulf.

On the other hand, in the most explicit acknowledgment of the links between fuel use, climate change, and national security to date in any administration's national security strategy, the 1991 *National Security Strategy* mentioned greenhouse gas emissions: "To meet pressing environmental concerns, we must limit the harmful effects of energy production, transportation and use. The increased, safe use of nuclear power, for example, can significantly reduce greenhouse gas emissions."[90]

When President Bill Clinton took office in January 1993, his administration continued the emphasis on strategic uncertainty. Clinton's Secretary of Defense Les Aspin conducted a "Bottom Up Review" that was meant to take a fresh look at U.S. forces after the Cold War. In the end, however, Aspin reaffirmed the importance of a global U.S. presence. He saw no need to dramatically restructure U.S. forces after the end of the Cold War: "The presence of U.S. forces deters adventurism and coercion by potentially hostile states, reassures friends, enhances regional stability, and underwrites our larger strategy of international engagement, prevention, and partnership. It also gives us a stronger influence, both political and economic as well as military, in the affairs of key regions." The United States, Aspin's *Report on the Bottom Up Review* argued, must be capable of intervening anywhere at any time:

> By stationing forces abroad we also improve our ability to respond effectively to crises or aggression when they occur. Our overseas presence provides the leading edge of the rapid response capability that we would need in a crisis. Moreover,

our day-to-day operations with allies improve the ability of U.S. and allied forces to operate effectively together.

Finally, our routine presence helps to ensure our access to the facilities and bases we would need during a conflict or contingency, both to operate in a given region and to deploy forces from the United States to distant regions.[91]

In other words, although the Cold War had ended, the United States was committed to a global military presence of "permanent or long-term stationing of U.S. ground, air, and maritime forces." The "demise of the global Soviet threat," meant that this could be done with reduced levels of forces, but it must still be done.[92] The threat by Iran or Iraq to dominate Southwest Asia was considered a danger to global economic stability; "limiting oil supplies" was on the list of "new regional dangers."[93] The Clinton administration continued to enforce the no-fly zones against Iraq and maintained a large presence in the Persian Gulf.

The lesson the United States took from the Iraq War was that airpower could be used to great effect to soften up a major conventional enemy and bombing from high altitude posed little risk to U.S. combatants. Indeed, during the 1995 and 1999 interventions in Bosnia and Kosovo to resist Serbian aggression in the former Yugoslavia, the United States-led NATO Operations Deliberate Force and Allied Force restricted U.S. operations to air strikes, flying planes from aircraft carriers in the Adriatic Sea, and from Aviano Airbase, in Italy. In June 1996, militants bombed Khobar Towers in Saudi Arabia, a facility housing U.S. airmen and other coalition forces who were supporting Operation Southern Watch. Over four hundred people were injured, and nineteen airmen were killed. The Clinton administration made cruise missile attacks on targets in Sudan and Afghanistan following al Qaeda attacks on U.S. embassies in Dar es Salam, Tanzania, and Nairobi Kenya in 1998 and the assault on the USS *Cole* in October 2000. Airpower was seen as the leading edge of American military power.

THE POST-9/11 WARS

Al Qaeda's attack on September 11, 2001, was of a whole other scale than the strike on the *Cole*. The war in Afghanistan the United States began after the attack and the larger Global War on Terrorism (GWOT) were meant to eliminate the threat from al Qaeda, which was then largely based in Afghanistan.

Although al Qaeda leader Osama Bin Laden argued that the attacks on New York and Washington, DC, were a response to U.S. presence in the Middle East, the U.S. war in Afghanistan was not about oil. Indeed, the George W. Bush administration, which had a number of advisers from the earlier Republican administration of his father George H. W. Bush, was very clear that there was no oil there. As one senior Bush administration official told the *New York Times* in February 1989 after the Soviets departed, "We have to be realistic. Afghanistan is not Iran. It has no oil reserves and isn't located on the Persian Gulf. It's not a particular strategic prize that has to be guarded at all costs."[94] Yet because the United States had long prepared to defend oil in the Persian Gulf, the military was well positioned to move military forces into the region within weeks after the 9/11 attacks. U.S. air strikes began on October 7, 2001, and in November the U.S moved a small force of 1,300 troops in to support the Northern Alliance, then fighting the Taliban. By December 2002, the United States had about 9,700 troops in Afghanistan and was preparing for a war against Saddam Hussein's regime in Iraq.

The Iraq War began in March 2003 under the argument that the United States had to rid the Iraqi regime of weapons of mass destruction that Iraq was said to be hiding and eliminate the regime of Saddam Hussein in favor of a democratic government. The Bush administration explicitly denied that the war was about oil. In November 2002 President Bush's Secretary of Defense Donald Rumsfeld answered questions about the potential war with Iraq on a radio talk show by saying, "Five days or five weeks or five months, but it certainly isn't going to last any longer than that." Rumsfeld insisted, "It has nothing to do with oil, literally nothing to do with oil." Rather, he said, the war would be about weapons of mass destruction (WMD). And, in response to the possibility that the United Nations weapons inspectors that were scheduled to continue looking for weapons of mass destruction in December 2002 would find no WMD, Rumsfeld said, "What it would prove is that the inspections process had been successfully defeated by the Iraqis if they find nothing."[95] Of course, as was soon discovered, there were no such weapons.

On the other hand, as Alan Greenspan, chair of the U.S. Federal Reserve from 1987 to 2006, said in 2007, "I am saddened that it is politically inconvenient to acknowledge what everyone knows: the Iraq War is largely about oil. Thus, projections of world oil supply and demand that do not note the highly precarious environment of the Middle East are avoiding the eight-hundred-pound gorilla that could bring world economic growth to a halt."[96] Even if the Iraq War was not primarily about oil, the idea that the United States has to protect the global flow of oil, and more specifically oil from the Persian Gulf, was largely taken for granted by U.S. military and national security experts. The architects of the 2003 war were many of the same people involved in the 1991 war, which was driven in large part by concerns about Iraq controlling too much oil. Oil was still the main prize in the region. A report by the Council on Foreign Relations noted in 2006, "Until very low levels of dependence are reached, the United States and all other consumers of oil will depend on the Persian Gulf."[97] In late 2008, to the concern that a local state might try to control the flow of oil from the Persian Gulf, President George W. Bush added that extremists might control oil and try to blackmail the United States: "You can imagine them saying, 'We're going to pull a bunch of oil off the market to run your price of oil up unless you do the following. And the following would be along the lines of, well, 'Retreat and let us continue to expand our dark vision.'"[98] By the time the United States had withdrawn from Iraq in 2011 and had begun concentrating on winning the war in Afghanistan, the price of oil had increased.

Whether or not the war in Iraq was about oil, the post-9/11 wars *required* a great deal of it. The energy intensity of U.S. warfighting was astounding. I noted earlier that during World War II, an army soldier's average operational fuel use was about one gallon per day.[99] The intensity of operational fuel use increased during the Vietnam War to more than seven gallons per day. There were never as many soldiers in Iraq and Afghanistan at the peak of troop deployments as there were in Vietnam, which had more than 500,000 troops at the height of the war—U.S. forces in Iraq peaked about 165,000 in 2007 and 2008 during the "surge" and U.S. forces in Afghanistan peaked at about 100,000 in 2011. In fact, the combined total "boots on the ground" deployments in the post-9/11 wars peaked in fiscal year 2008 at about 187,900.[100] However, as figure 2.2 illustrates, there were more forces outside the main war zones in the

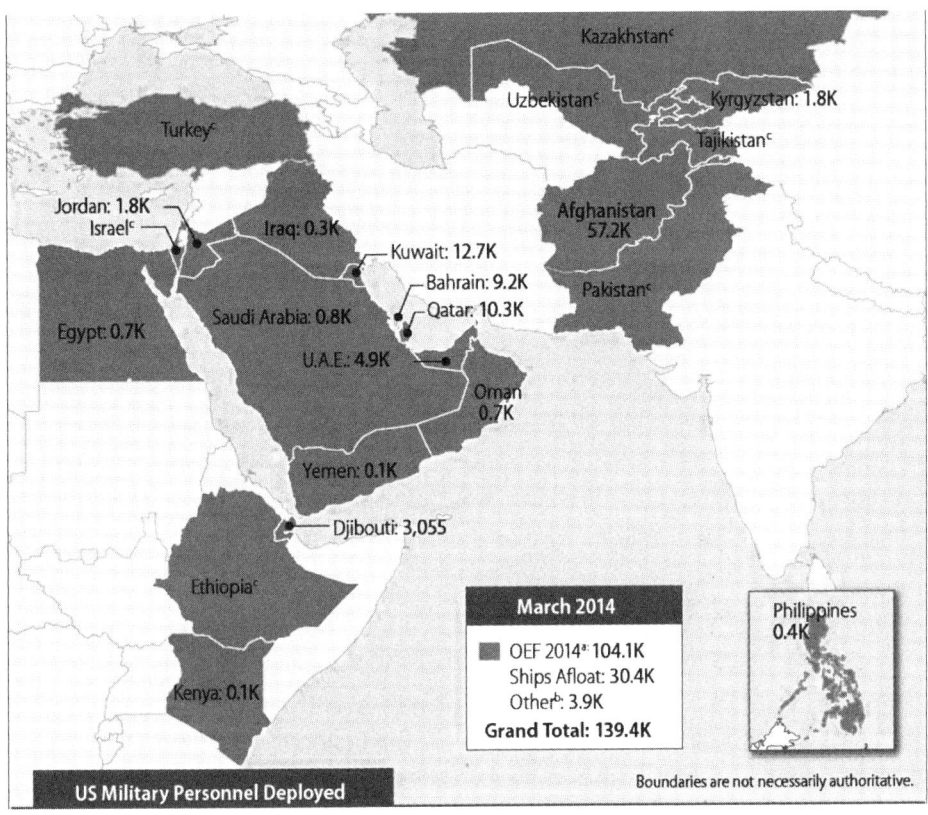

Figure 2.2

U.S. troops deployed for Afghanistan, Operation Enduring Freedom, March 2014.

Source: Amy Belasco, "The Cost of Iraq, Afghanistan, and Other Global War on Terror Operations Since 9/11," Congressional Research Service, RL33110, December 8, 2014, 13, figure 2, https://fas.org/sgp/crs/natsec/RL33110.pdf.

region and each soldier at the front used comparatively more energy in Iraq and Afghanistan than in previous wars.[101] In 2007 a soldier's average operational fuel use in Operation Enduring Freedom in Afghanistan and Operation Iraqi Freedom was fifteen to twenty gallons per day. A 2009 study found that every "forward operating base (FOB) in Afghanistan requires a minimum of 300 gallons of diesel fuel per day to satisfy its requirements. A typical USMC combat brigade alone requires over 500,000 gallons of fuel per day."[102]

After the 9/11 attacks, the U.S. military relied heavily on airpower for each of its major wars, never moving as many troops to either Afghanistan or Iraq as it had moved to the Gulf in the 1991 Gulf War. In neither case, however, did the United States deploy as much airpower as its military had in the 1991 Gulf War. The United States began the war in Afghanistan against the Taliban and al Qaeda on the night of October 7, 2001, with air strikes against the Taliban government's early warning radar, command and control facilities, and airfields in Afghanistan, a total of thirty-one targets. In 2001, the United States made more than 6,500 air strikes with at least one weapon being released in Afghanistan. The air war against Iraq in 2003 involved more than 20,700 air strikes with at least one weapon release.[103] And again each of these wars entailed the rapid mobilization of troops to the respective war zones, including airlifting troops and equipment. While the United States deployed some ground forces, the primary use of force in the war against ISIS in Iraq and Syria was air strikes: the U.S. military made about 36,600 air strike sorties with at least one weapon release (see figure 2.3) from August 2014 through 2021 in both Iraq and Syria.[104]

Further, oil became a target in the Iraq War and later the anti-ISIS war in Iraq and Syria that began in 2014 as each side destroyed oil in order to deprive the other side. The centrality of oil in U.S. calculations about its role in the war against ISIS was underscored when, after the United States withdrew some of its troops from Syria, U.S. Secretary of Defense Mark

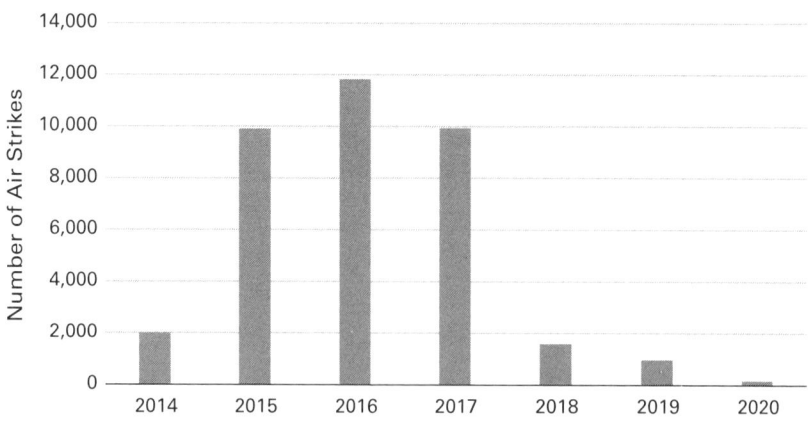

Figure 2.3

Operation Inherent Resolve air strikes in Iraq and Syria, 2014–2021.

Source: Air Force Central Command, https://www.afcent.af.mil/About/Airpower-Summaries/.

Esper said in October 2019 that the United States would defend oil fields in Syria to "ensure that we can deny ISIS access to the oil fields."[105] President Donald Trump expanded on the reasoning for keeping some troops there the next day, in a statement putting forth three arguments about the importance of the oil: First, that the U.S. presence kept ISIS from accessing the oil by denying territory and destroying machinery; second, the Kurds would be able to get access to the oil, which they had been able to benefit from before ISIS invaded; and third, that U.S. companies deserved to have access to the oil in the region because they could "properly" manage oil extraction. Trump explained these arguments in his address:

> Look, we don't want to keep soldiers between Syria and Turkey for the next 200 years. They've been fighting for hundreds of years. We're out. But we are leaving soldiers to secure the oil. And we may have to fight for the oil. It's okay. Maybe somebody else wants the oil, in which case they have a hell of a fight. But there's massive amounts of oil. And we're securing it for a couple of reasons. Number one, it stops ISIS, because ISIS got tremendous wealth from that oil. We have taken it. It's secured. . . . I don't want to leave 1,000 or 2,000 or 3,000 soldiers on the border. But where Lindsey [Graham] and I totally agree is the oil. The oil is, you know, so valuable. For many reasons. It fueled ISIS, number one. Number two, it helps the Kurds, because it's basically been taken away from the Kurds. They were able to live with that oil. And number three, it can help us, because we should be able to take some also. And what I intend to do, perhaps, is make a deal with an ExxonMobil or one of our great companies to go in there and do it properly. Right now it's not big. It's big oil underground but it's not big oil up top. Much of the machinery has been shot and dead. It's been through wars. But—and—and spread out the wealth. But no, we're protecting the oil, we're securing the oil. Now that doesn't mean we don't make a deal at some point.[106]

Oil was also core to the Biden administration policy on the Persian Gulf. The U.S. CENTCOM posture statement by General Kenneth F. McKenzie Jr. in 2021 emphasized that its first line of effort is to deter Iran from regional aggression, noting, "The Iranian regime demonstrated both the capability and willingness to employ all of these offensive weapons in complex attacks against Saudi Arabia's oil facilities in 2019, and again against U.S. forces in Iraq in 2020." McKenzie emphasized that China's interests in the region were substantial and included fossil fuels. "China's long-term goals are not just to cultivate trade relationships, economic investment, and comprehensive partnerships among regional states, but to exert coercive influence and eventually establish a permanent military presence in an area from which it imports nearly 50 percent of its crude oil and roughly 40 percent of its natural gas."[107]

Thus, in anticipation of increased Chinese military presence in the region, the U.S. military argued that it should continue a presence in the Persian Gulf and arm its allies there. Even as the United States withdrew from Iraq and Syria and reduced its presence in the region, starting in 2019, and completed its withdrawal from Afghanistan in August 2021, U.S. arms sales to allies in the Middle East remained high. From 2016 to 2020, 47 percent of U.S. arms sales went to the Middle East, with Saudi Arabia receiving the most weapons during this period (see figure 2.4).[108] There was, apparently, no notion that the U.S. bases, troops, and arming of U.S. allies could be seen as a threat to China.

THE DEEP CYCLE: FOSSIL FUEL USE, ECONOMIC GROWTH, AND STRATEGY

Recall the idea that security strategy is supposed to make governments and even populations more secure. But strategy can be counterproductive. Thus, Duane Chapman argues that a "cycle of violence and political radicalization . . . seems to be arising within the current Gulf security framework."[109] Chapman says the cycle looks like this: "An autocratic government supported by the US reduces the strength of secular opposition. Civil dissent and criticism are marginalized and rendered ineffective. Then Jihadist organizations fill the space of opposition to the autocracy, further radicalizing dissent. In Saudi Arabia, this process gave rise to the Qaeda attacks in the US in September 2001."[110] At the root of this process is the combination of dependency and anxiety about the supply and price of oil from the Persian Gulf region.[111]

However, this is a longer cycle involving more features and positive feedback. As I suggested in the introduction and in chapter 1, for more than 170 years the U.S. economy and military have been on a path that has become a deep and self-reinforcing long-term cycle of economic growth, fossil fuel

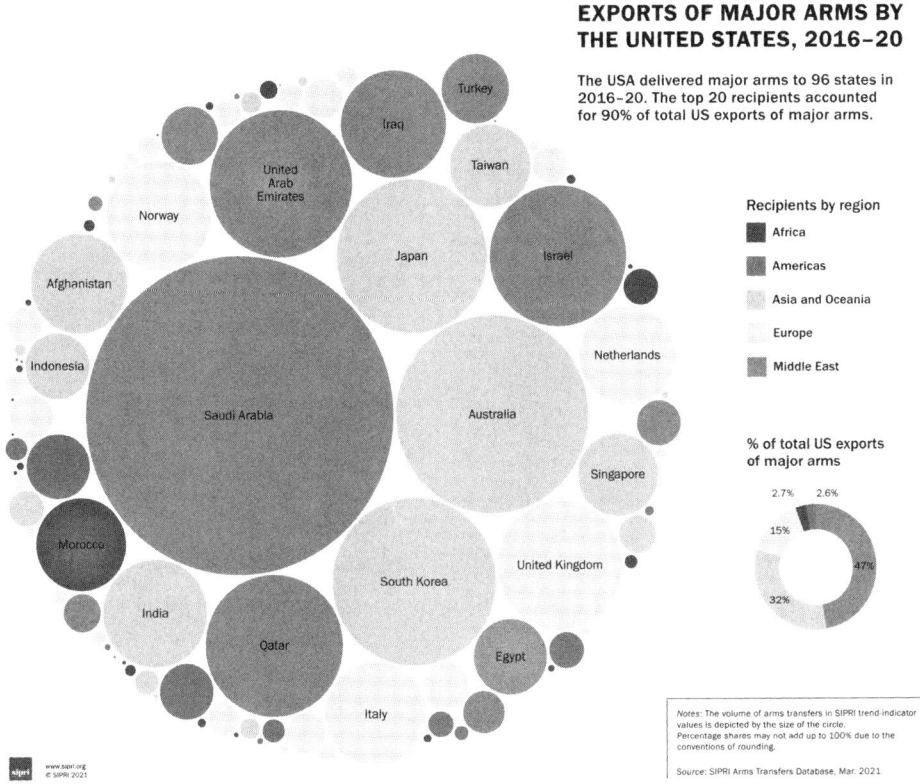

EXPORTS OF MAJOR ARMS BY THE UNITED STATES, 2016–20

The USA delivered major arms to 96 states in 2016–20. The top 20 recipients accounted for 90% of total US exports of major arms.

Recipients by region
- Africa
- Americas
- Asia and Oceania
- Europe
- Middle East

% of total US exports of major arms

2.7% 2.6%
15%
47%
32%

Notes: The volume of arms transfers in SIPRI trend-indicator values is depicted by the size of the circle.
Percentage shares may not add up to 100% due to the conventions of rounding.

Source: SIPRI Arms Transfers Database, Mar. 2021.

www.sipri.org
© SIPRI 2021

Figure 2.4

U.S. arms exports, 2016–2020.

Source: Stockholm International Peace Research Institute, https://twitter.com/SIPRIorg /status/1371463863355473929/photo/1.

use and dependency, expanding military power, and increasing military spending. The cycle began when the U.S. Navy became reliant on coal for powering steamships in the 1840s. Coal-powered steamships were useful in the 1848 war with Mexico and essential in the Civil War. When the search for markets led the United States to extend its power to Asia in the 1850s and again at the turn of the century, coaling stations were necessary, as were ports, most notably in Hawaii and Japan. The system of alliances, arms sales, military bases, and on occasion war, was meant to secure supply if and when the market system was weak or entirely failed.

Coal and increasingly oil powered the U.S. Navy in World War I and U.S. fuel supplied British and French airpower. By World War II, preventing the adversary's access to oil, or making sure the U.S. supply was secure, helped determine U.S. strategy. As in World War I, war-driven increases in the productivity of U.S. oil reserves allowed Britain and other allies to keep fighting in World War II. Although petroleum products were an essential element in United States and allied mobility in World War II, as well as powering the factories that produced war material and explosives, coal also powered U.S. industrialization through World War II. Indeed, coal was still the dominant source of U.S. energy until 1949, when it was surpassed by petroleum consumption.

The necessity for refueling determined the location of U.S. bases during World War II and the desire to keep oil out of German and Japanese military hands in part determined U.S. bombing and blockade strategy. German and Japanese strategy were also driven by the need to acquire oil, since neither country had oil reserves sufficient to fuel their ambitions. Oil also determined to some extent, the U.S. relationships with Persian Gulf countries in the 1950s. It was also in part anxiety about securing access to Persian Gulf oil that led the United States to take over Britain's role in the Middle East as they withdrew in the late 1960s.

Further, as the United States became more dependent on Persian Gulf oil, it became increasingly concerned in the 1970s and 1980s about both access to oil and the potential use of the "oil weapon." This sense of increasingly insecurity about access to oil prompted the United States to form political alliances with the oil-producing countries, including Saudi Arabia and Iran, that they thought could keep order and especially ensure the orderly flow of oil. These relationships also included military assistance and arms, as well determining where the United States would enlarge existing bases or acquire new facilities. All the while the U.S. military industry and the military itself were powered by a steady supply of fossil fuels.

In sum, the economy, foreign policy beliefs, and military doctrine institutionalized greater demand for fossil fuels. The deep cycle of oil demand, consumption, militarization, and conflict begins with demand for oil and increasing consumption. Then, when U.S. policy makers feel anxious about guaranteeing oil supplies in the face of dependency, or are concerned about

Table 2.1

U.S. bases in Middle East and Persian Gulf region, September 2021

Country	Number of U.S. military bases and ships	Estimated number of personnel
Bahrain	12	4,603
Djibouti (Africa Command)	2	126
Diego Garcia	2	3,000
Egypt	1	259
Iraq	6	2,500
Israel	6	127
Kuwait	10	2,054
Jordan	2	211
Oman	6	25
Saudi Arabia	11	693
Syria	4	900
Turkey (European Command)	13	1,758
Qatar	3	501
United Arab Emirates	3	215
Total	81 bases	16,972
United States Fifth Fleet	c. 10 ships including 1 aircraft carrier; 1 cruiser; 1 destroyer; 1 landing helicopter dock; 1 amphibious assault; 2 dock landing ships; 1 mine warfare; 3 logistics support ships; 1 guided missile submarine; and 1 Los Angeles class submarine	c. 10,000

Source: David Vine, Patterson Deppen, and Leah Bolger, "Drawdown: Improving U.S. and Global Security Through Military Base Closures," *Quincy Brief no. 16*, Quincy Institute for Responsible Statecraft, September 2021, https://quincyinst.org/report/drawdown-improving-u-s-and-global-security-through-military-base-closures-abroad/. See also Fifth Fleet, United States Navy, https://www.cusnc.navy.mil/Task-Forces/.

the price of oil, they advocate to increase the U.S. military presence and U.S. political and economic support for regional allies in the Persian Gulf and greater Middle East, as occurred in 1946 and 1949, in 1957, in 1973, in 1980, and in 1990. The United States played favorites within the states that had large oil reserves, even if some of those government leaders were autocratic, such as the Shah of Iran, Saddam Hussein, and the leaders of Saudi Arabia. Much could be said of U.S. allies in other oil-producing countries.

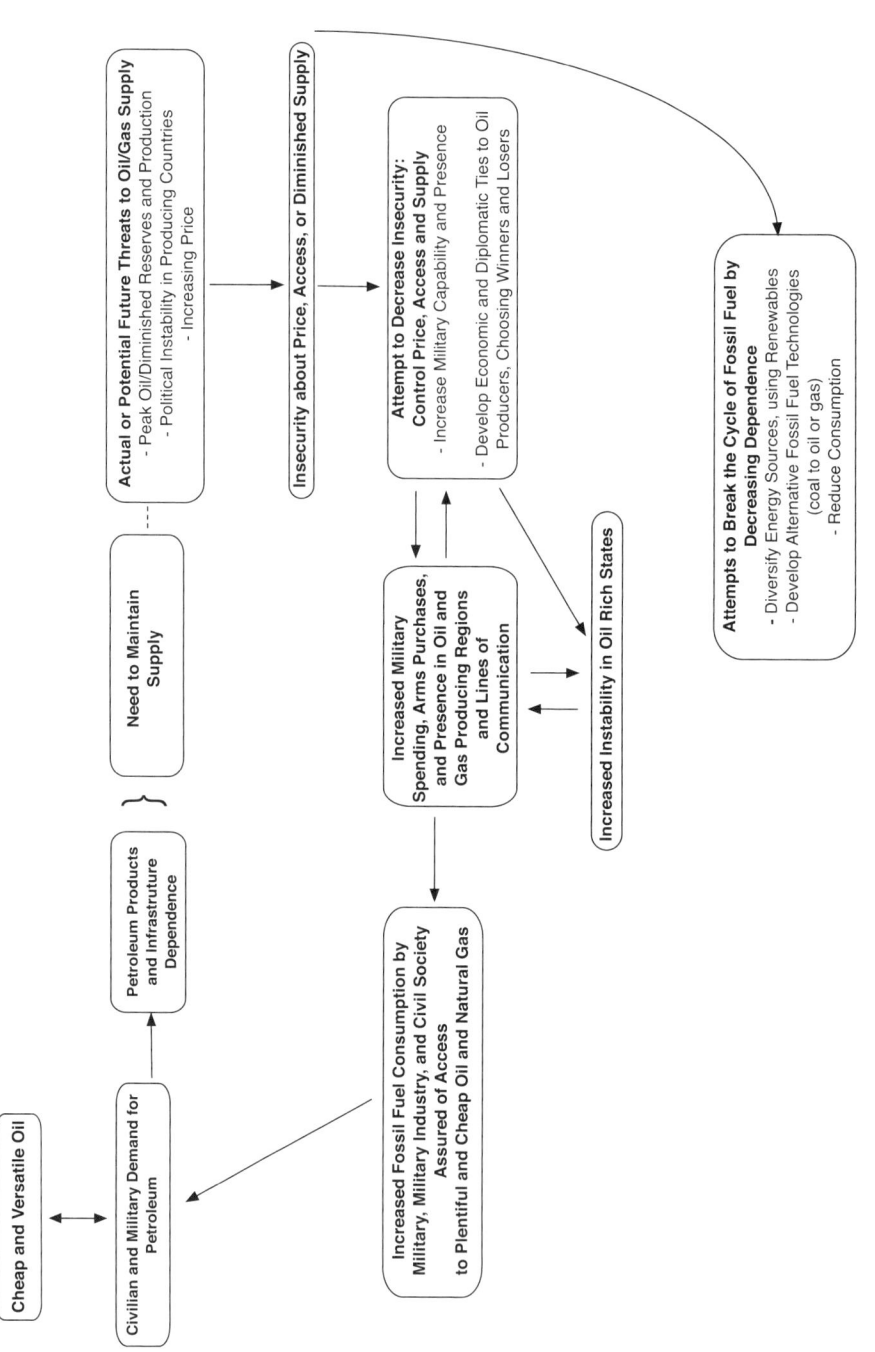

Figure 2.5

Institutional deep cycle: oil demand, consumption, militarization, and conflict.

Yet, the risk of supporting authoritarian regimes is that those regimes are increasingly unstable as the citizens who demand more say in their governments push back against authoritarian kings, emirs, and shahs. When challenges to the undemocratic, autocratic or kleptocratic rulers within states with large oil supplies occurred, or there were external challenges, the United States sometimes backed the leaders or system that it thought could bring stability. Thus, with the Eisenhower Doctrine, the United States backed Saudi Arabia's King Saud and Crown Prince Faisal as a way to balance against Egyptian leader Gamal Abdel Nasser. At times, as in the case of backing the Shah of Iran in the 1970s and Saddam Hussein in Iraq in the 1980s, these alliances backfired. As domestic unrest continued militants sometimes challenged a regime's relationship with the United States, protesting bases in the region and U.S. policies. This in turn increases the sense among U.S. elites that the Middle East is a volatile region that needs U.S. intervention to remain stable. While the deep cycle first occurred in the nineteenth century, figure 2.5 summarizes this deep cycle as it has occurred since the 1950s.

The deep cycle does not have to proceed in this manner. Indeed, there is an opportunity to go from insecurity about the access, price, or supply of energy to less insecurity by decreasing dependence on fossil fuel. In fact, the United States has at times attempted to decrease dependency on Persian Gulf oil—such as in the 1970s, when President Nixon and then President Carter included conservation in the mix of U.S. energy policy. However, the United States has mainly attempted to decrease dependency on individual foreign suppliers by diversifying global suppliers and increasing domestic production of oil and natural gas. It has not, until market forces and worries about climate change began to make wind and solar energy much less expensive, moved toward significant reductions of Persian Gulf oil supplies or fossil fuel consumption more generally. Because the beliefs about how the flow of oil can be best guaranteed are deeply embedded in U.S foreign and military leadership and are institutionalized in military doctrine that assumes the United States must be prepared to use force to preserve access to oil, it is difficult to break the cycle of dependency, arms, alliances, and the stationing of military force in the region.

II THE U.S. MILITARY AND CLIMATE CHANGE

3   CLIMATE CHANGE SCIENCE
AND THE POLITICS OF COUNTING
MILITARY EMISSIONS

> Let's face it, vast swathes of our military are big carbon emitters—tanks, Jeeps, humvees, jet planes—and of course much of our navy is not nuclear-powered, so [the Paris agreement] could be used as a trojan horse. This might be a good opportunity for people concerned with national security to go to congress and get some type of legislative exemption in the same way as was done during the Kyoto time period.
> —Steven Groves, Heritage Foundation, 2015[1]

The military and military-industrial utilization of and dependency on fossil fuels tracked and, in some cases led, civilian fossil fuel use. Furthermore, the growth in fossil fuel use paralleled scientific understanding of the atmospheric warming effects of carbon dioxide. The science developed in the nineteenth century, but the decisive link between accumulating emissions of carbon dioxide and the "greenhouse effect" occurred in the mid-twentieth century.

While basic research continued, the growing understanding of the risks of climate change motivated international negotiations to curb emissions growth. To be clear, the goal that negotiators set at the conference that established the 1997 Kyoto Protocol and before that, at the Rio Earth Summit in 1992, was for countries to keep total global emissions below a threshold that if crossed would lead to catastrophic consequences for the environment and hence human survival. The method of achieving those reductions relied on having a clear understanding of a number of factors, including the science of global warming (e.g., the types and quantity of greenhouse gas emissions and the effect of each gas and the quantity of sinks that absorb emissions), and the total national emissions of a country party to the treaty. The countries also needed a reference point—emissions levels in a particular year—since their goal was to keep under a certain threshold. The entire scheme rests on accurate and transparent reporting of emissions.

But the main mechanism for documenting emissions, a process whereby the parties produce national inventories of "all" greenhouse gas emissions, as articulated as part of the Kyoto Protocol, deliberately omitted large portions of military emissions from reporting requirements. The effort to keep military emissions out of the accounting process was led by the United States military and supported by Congress, including, at the time, Senators Joe Biden and John Kerry. As a result, there was no complete accounting of military emissions—either in the U.S. inventory or in the inventories of other countries—required after Kyoto.

In this chapter, I describe the development of the scientific understanding that certain gases act like a blanket on the earth's atmosphere. Important contributions to the science were funded by Congress directly or funneled through the Department of Defense and the Office of Naval Research. I also show how and why the United States kept total military emissions out of the national inventories, and I describe how U.S. emissions are reported as part of U.S. submissions as a consequence of decisions taken at Kyoto. Although the U.S. reports follow the rules, they are incomplete and not transparent, omitting a major category of emissions—from international bunker fuels and war. The Paris Agreement in 2015, which committed parties to the treaty to emissions reductions, included military emissions for the first time. However, full disclosure and military emissions cuts are optional, left to the discretion of the parties.[2] Futher, even the December 2021 Biden Administration order that federal agencies reduce their emissions allows that the "The head of an agency may exempt from the provisions of this order any vehicle, vessel, aircraft, or non-road equipment that is used in combat support, combat service support, military tactical or relief operations, or training for such operations."[3] The U.S. Department of Energy and the DOD have only recently started to report U.S. military emissions. The lack of a full accounting makes it necessary to do an independent estimate of U.S.

military emissions. In the next chapter, I summarize the estimates of recent U.S. military and military-industrial emissions that I made using Department of Energy data and the emissions data provided by some of the largest U.S. companies in the defense industry.

ENERGY TRANSITIONS AND THE NAVY'S INTEREST IN CLIMATE SCIENCE

Although 7.3 million tons of bituminous coal were burned in steam engines in the United States in 1860, wind power and waterwheels were still producing more than three times the amount of horsepower than generated by coal. Military power and military-industrial output grew as hydropower, wind, wood, whale oil, and draft animals were gradually replaced starting in the mid-nineteenth century with coal and then petroleum crude oil. By 1870, coal-powered steam engines, which consumed 16.2 million tons of bituminous coal, were producing slightly more horsepower than wind and water.[4] Of course the impetus for transitioning to coal, moving away from wood, water, and wind, was that if it could be transported to where it was needed, the energy content of coal was higher than that of wood and it did not depend on weather; oil was even more energy intensive than coal. (See table 3.1.)

These rapid energy transitions paralleled leaps in the scientific understanding of the effects of greenhouse gas emissions. The evidence that increasing atmospheric concentrations of carbon dioxide (CO_2) could warm the earth has been accumulating since 1856, when Eunice Foote reported on

Table 3.1
Fuel energy content

Fuel	Energy content in Megajoules/Kilogram
Wood (softwoods, 15–50% water content)	21–23
Anthracite coal	31–33
Bituminous coal	20–29
Crude oil	42–44
Natural gas	20–39 Mj/m^3

Source: Vaclav Smil, *Energy and Civilization: A History* (Cambridge, MA: MIT Press, 2018), 165 and 227.

her experiment showing that "moist air" and carbonic acid (carbon dioxide) had a significantly higher warming potential than dry air or other gases. Foote concluded, "An atmosphere of that gas [carbonic acid] would give to our earth a high temperature."[5] The understanding of this process, and the other gases that also caused atmospheric warming, slowly accumulated over the next century as more scientists, most famously John Tyndall in 1861, Svante Arrhenius in 1896, and G. S. Callendar in 1938 made similar discoveries and more precise predictions. The global warming effects of other gases, specifically methane, nitrous oxide, and chlorofluorocarbons, and their major sources, were also identified.[6]

The science of climate change as it evolved in the 1950s and 1960s—specifically understanding the link between emissions from fossil fuel production and combustion, agriculture, and industrial processes and global warming—was in large part funded in the United States by Congress for military purposes. The link was in some senses natural. The science received a great boost during the International Geophysical Year (IGY, 1957–1958) that involved scientists from dozens of nations working together to conduct basic research on the planet. The DOD was very supportive of the plans for the IGY and continued research on the atmosphere and oceans because it would yield information useful information for the military.[7] Roger Revelle, of the Scripps Institution of Oceanography, one of the leading climate scientists to document global warming in the 1950s, had served in the U.S. Navy for seventeen years before and after World War II. As a U.S. Navy Commander, Revelle was involved in documenting the effects on fish and waves of the 1946 nuclear weapons tests at Bikini Atoll during Operation Crossroads.[8]

In Revelle's 1956 testimony before the House Committee on Appropriations he described the necessity of understanding whether carbon dioxide was causing warming that would "cause a remarkable change in the climate." Revelle linked warming to the Cold War competition between the United States and the Soviet Union. As warming continued, Revelle said, "the Arctic Ocean will become navigable and . . . then the Russian Arctic coastline will be really quite free for shipping. . . . This would have the effect, if it does happen, of changing the character of the Russians opposed to ourselves. We are now the greatest maritime nation on the earth. . . . If the Russian coastline increases by something like 2,000 miles or so, the Russians will become a great maritime nation."[9] In 1957, Revelle's work, conducted with funding from the U.S. Office of Naval Research, documented rising levels of carbon dioxide in the atmosphere. Revelle and coauthor Hans Seuss wrote that "human beings are now carrying out a large scale geophysical experiment of a kind that could not have happened in the past nor be reproduced in the future. Within a few centuries we are returning to the atmosphere and Oceans the concentrated organic carbon stored in sedimentary rocks over hundreds of millions of years."[10]

The interest the U.S. military had in understanding atmospheric science led to the work of several other scientists receiving funding that advanced the understanding of climate change. For example, Gilbert Plass, whose work was funded by the Office of Naval Research, contributed several important papers in the 1950s that showed the relationship between CO_2 concentration and temperature. He warned, "The latest calculations of the influence of CO_2 on the infrared flux show that if the CO_2 concentration in the atmosphere doubles the average temperature rises 3.6° C and if it falls to half of its present value the average temperature falls 3.8° C."[11] Charles Keeling, with funding secured as part of IGY from the National Science Foundation, the United States Weather Bureau, and the Office of Naval Research, was able to acquire data from both Mauna Loa in Hawaii and the South Pole, which demonstrated that there were both seasonal fluctuations in atmospheric CO_2 and a steady rise in atmospheric CO_2.[12] The science became clear. Thus, in February 1965, Von Arx of Woods Hole Oceanographic Institution concluded, "Man has become a *force of nature*. He . . . is altering the radiation balance . . . and by vigorous consumption of fossil fuels the concentration of carbon dioxide."[13]

Thus, from the moment when the highest levels of the United States government were informed about global warming in the 1950s and 1960s, through research funded in part by the Office of Naval Research, scientists and strategists turned to the international security consequences of a warming planet. At the same time that each U.S. presidential administration over the last several decades has grappled with whether and how to use military force to ensure access to oil, the administrations have known about the role of greenhouse gas emissions as a cause of potentially catastrophic climate change.[14] The President's Science Advisory Committee summary gave the Johnson administration notice in 1965 that putting carbon into the atmosphere by burning fossil fuels was going to warm the climate with potentially disastrous consequences. In November 1965, the committee summarized the research on atmospheric carbon dioxide in a report on pollution. An appendix to the report that explained the science was coauthored by several renowned climate scientists, including Revelle and Keeling, who wrote, "Through his worldwide industrial civilization, Man is unwittingly conducting a vast geophysical experiment. Within a few generations, he is burning the fossil fuels that slowly accumulated in the earth over the past five hundred million years. The CO_2 produced by this combustion is being injected into the atmosphere." The scientists warned, "The climatic changes that could be produced by the increased CO_2 could be deleterious from the point of view of human beings."[15] They were confident in their analysis of the cause of rising carbon dioxide, but they urged more study of the consequences and what could be done to slow or halt the process. President Nixon was warned about the connection between carbon dioxide and global warming in 1969 by his urban affairs adviser Daniel Patrick Moynihan, who told Nixon's Chief of Staff "this very clearly is a problem. . . . [Rising CO_2 levels] could increase the average temperature near the earth's surface by 7 degrees Fahrenheit. This in turn could raise the level of the sea by 10 feet. Goodbye New York. Goodbye Washington, for that matter. We have no data on Seattle."[16] President Carter received briefings on global warming as well and his science advisers had come to a consensus that although "there was not a debate over whether a clear climate threat existed, but there were uncertainties about how quickly different climate change harms would occur."[17]

As atmospheric scientists continued to accumulate evidence and present data at international conferences, they converged on the view that it was vital to reduce greenhouse gas emissions. The World Meteorological Organization convened the first wide-ranging World Climate Conference in 1979, drawing scientists from all over the globe who concluded:

> We can say with some confidence that the burning of fossil fuels, deforestation, and changes of land use have increased the amount of carbon dioxide in

the atmosphere by about 15 per cent during the last century and it is at present increasing by about 0.4 per cent per year. It is likely that an increase will continue in the future. Carbon dioxide plays a fundamental role in determining the temperature of the earth's atmosphere, and it appears plausible that an increased amount of carbon dioxide in the atmosphere can contribute to a gradual warming of the lower atmosphere, especially at high latitudes. Patterns of change would be likely to affect the distribution of temperature, rainfall and other meteorological parameters, but the details of the changes are still poorly understood.[18]

The emerging consensus was articulated in June 1988 by James Hansen of NASA in his testimony before Congress that global warming "is already happening now."[19] Hansen's conclusions received widespread attention, including on the front page of the *New York Times*.

Climate change and national security were more firmly linked together in 1990. As noted earlier, President George H. W. Bush's March 1990 *National Security Strategy of the United States* mentioned the need to "devote greater attention to reducing fossil fuel emissions in light of growing environmental concerns" in the same paragraph as the need to ensure access to the oil in the Persian Gulf.[20] But this *National Security Strategy* was not focused on climate change. The Cold War was just ending and there were other national security concerns that were more urgent. In May 1990, the Intergovernmental Panel on Climate Change (IPCC) released its first assessment report with predictions about the patterns and rate of global warming over the next several decades.[21] The document, representing the consensus analysis of hundreds of scientists, was praised by Ronald Reagan's good friend, Britain's Prime Minister Margaret Thatcher, who called their findings "a report of historic significance . . . an authoritative warning system."[22]

And also in May 1990, the same month that the IPCC's first assessment report was released, a U.S. Naval War College study described "significant effects" of global warming "on the facilities, infrastructures, and operations of the Navy." The author, Terry P. Kelley, said that changes in the salinity, chemistry and temperature of water would affect acoustic characteristics of oceans, impacting the performance of sonar—critical for anti-submarine warfare. Indeed, the navy had already been conducting research on how the speed of sound in water increases with increases in temperature. Kelly also described how changes in ocean currents, high winds, greater turbulence, and the alteration of the shape and underwater character of coastlines would affect operations. Sea level rise would expand the reach of shallow water, and thus potentially "extend some target beaches 100 km inland," and possibly require changing amphibious assault equipment and tactics. Arguing that of all the armed services, the effects of climate change would be greatest for the navy, Kelley recommended that the U.S. Navy "place itself in the lead for DOD in supporting research and analysis" of global warming and develop a long-term plan to reduce its infrastructure and facilities and redesign its bases and ships to deal with likely challenges.[23] While, like Revelle, Kelley foresaw potential opportunities for navigation with Arctic ice melting, he also noted the potential for boundary tensions to become "severe" in areas where climate change caused drought and flooding and hindered the ability of people to have fresh water or grow food. Kelley warned that climate change would stress the nation's budget, and with increased political pressure for military spending to be reduced—a "peace dividend." Kelley argued that the Defense Department as a whole, and "the Navy, in particular, must insure that policy formulation, planning, and analysis processes adequately address the impact of global climate change before other sectors take precedence in resource allocation."[24]

Given the already wide scientific consensus that global warming was occurring, there are a number of ways the United States and the U.S. military could have responded to climate change in the 1990s. The military might have decided to radically rethink its bases, operations, and modernization plans. Indeed, the early 1990s was an unusually fluid time because the military and U.S. foreign policy were undergoing a reassessment following the end of the Cold War. But there was no major rethinking. Rather, the U.S. military did what organizations tend to do: it kept doing what it was doing before, and as Kelley urged, it protected itself from any challenges to its budget share.

THE KYOTO PROTOCOL: KEEPING TOTAL MILITARY EMISSIONS OFF THE BOOKS

The first major climate change negotiations occurred in Rio de Janeiro, Brazil, in 1992, at what became known as the "Earth Summit." One hundred seventy-two governments sent representatives, including 108 heads of state or government. Ten thousand journalists attended, as did thousands of environmental activists. Principles 6 and 7 of the Rio Declaration on Environment and Development that emerged from the summit laid out the principle of common but differentiated responsibilities. "The special situation and needs of developing countries, particularly the least developed and those most environmentally vulnerable, shall be given special priority. International actions in the field of environment and development should also address the interests and needs of all countries. . . . In view of the different contributions to global environmental degradation, states have common but differentiated responsibilities." In other words, developing countries were essentially exempted from making greenhouse gas emissions reductions at the 1992 Rio summit. Notably, war was mentioned in the Rio Declaration: "Warfare is inherently destructive of sustainable development. States shall therefore respect international law providing protection for the environment in times of armed conflict and cooperate in its further development, as necessary."[25]

While the Rio Declaration was important as a summary of principles and intentions about the environment and development, the adoption of the United Nations Framework Convention on Climate Change (UNFCCC) set specific goals. These included the commitments of the parties to: "Develop, periodically update, publish and make available to the Conference of the Parties, in accordance with Article 12, national inventories of anthropogenic emissions by sources and removals by sinks of all greenhouse gases not controlled by the Montreal Protocol, using comparable methodologies to be agreed upon by the Conference of the Parties." Article 12 specified that each party should provide "A national inventory of anthropogenic emissions by sources and removals by sinks of all greenhouse gases not controlled by the Montreal Protocol, to the extent its capacities permit, using comparable methodologies to be promoted and agreed upon by the Conference of the Parties."[26] While the UNFCCC stated that *all* anthropogenic greenhouse gas emissions ("carbon dioxide and other greenhouse gases not controlled by the Montreal Protocol") by source should be included in the national inventories, the specific gases to be included are not named and the methods for accounting were to be agreed upon later. Further, the UNFCC did not resolve how emissions would be limited. The systems for limiting emissions and meeting the targets was to be decided in the negotiations that led to the Kyoto Protocol, named for the Japanese city in which it was adopted in December1997. The specifics of the Kyoto Protocol negotiations and the results—including that the United States prevailed in setting up an emissions trading system that was modeled on the Montreal Protocol and that many details about implementation were left to be decided in subsequent negotiations—have been well discussed elsewhere. The focus here, however, is on the question debated at the conference of whether all greenhouse gas emissions, including those from the military and its operations, would be reported in national inventories.

In September 1997, a U.S. Department of Defense background paper given to the White House warned of dire consequences that could follow from the Kyoto Protocol. DOD leadership suggested that reporting military emissions would lead to emissions cuts and that "imposing greenhouse gas emissions limitations on tactical and strategic military systems would . . . adversely impact operations and readiness." The DOD used a 10 percent cut in fuel use as an example of the consequences for the U.S. military. It said such a cut would affect operations and readiness and reduce the U.S. advantage in tactics and training, which would then reduce U.S. combat effectiveness. A 10 percent reduction would be dramatic, they said: for the army, it would cut 328,000 miles from tank training; for the navy, it would cut 2,000 steaming days per year from training and operations; for the air force, it would cut 210,000 flying hours per year such that fighter and bomber crews would "be unable to maintain full combat readiness." Such a cut would also reduce air lift and aerial tanker capacity.[27] The background paper argued that "advanced capabilities" such as armaments for the F-15E and B-52 "would be lost."[28]

The background paper argued that military force would be required not only to protect U.S. interests, but also to save people from the worst consequences of climate change. Its authors said that "while global climate change may be a serious threat to the nation's long-term interests, there are other threats we must not forget. We must not sacrifice our national security or our ability to offer humanitarian assistance to those in need to achieve reductions in greenhouse gases. We must not see this as an issue of being able to achieve either national security or protection of the global climate. The United States must pursue both objectives."[29] The last sentence of the background paper states, "To accomplish this, the DOD strongly recommends that the United States insist on a national security provision in the climate change protocol now being negotiated."[30]

In sum, even though the 1997 DOD background paper framed the values of climate and security as *both-and*, it proposed to privilege the protection of the military's capabilities and therefore military emissions. Specifically, the DOD proposed to insert language in the treaty that would "exclude from its measurements of anthropogenic emissions attributable to military tactical or strategic systems used for military operations in support of national or collective security, humanitarian activities, peacekeeping, peace enforcement, or United Nations, NATO, and other international actions and any such emissions attributable to military tactical or strategic systems used in training to maintain readiness for conducting military operations." It defined military tactical or strategic systems as "equipment, vehicles, aircraft, and vessels designed or produced for use in military operations."[31] So when the Clinton administration went to the Kyoto Protocol negotiations in December 1997, there were two options on the table. The states that would be party to the treaty could have been required to give a full accounting of military emissions for national submissions, or there could have been a blanket exemption for military emissions, something like the language that the Pentagon proposed. A State Department cable prepared in advance of the negotiations said that "In reviewing the U.S. emissions levels, we note that the federal government is the biggest single user of energy, and much of that energy use comes from defense installations and training operations."[32]

When U.S. negotiators arrived in Japan in December 1997 to participate in the talks that would yield the Kyoto Protocol, the top U.S. negotiator, Ambassador Stuart Eizenstat, had specific instructions that were well known at the time. First, the Clinton administration wanted the "meaningful participation" of developing countries in dealing with climate change. Developing countries had been essentially shielded from greenhouse gas emission reduction targets on the argument that their inclusion would stifle their economic development. The United States wanted to include goals and timetables for developing countries to reduce their emissions and the Senate had made it clear, through the Byrd-Hagel Resolution, sponsored by Democrat Robert Byrd and Republican Chuck Hagel, that this was a condition for Senate ratification. The treaty would need sixty-seven votes for ratification in the Senate; the Byrd-Hagel Resolution had passed 95–0. While developed countries had put more greenhouse gases in the air than most developing countries combined, the argument was that it was unfair for developed countries to curb their emissions under binding targets if developing countries like China were allowed to continue emitting.

Second, the United States wanted to make sure that "flexible market mechanisms," not mandatory measures, would be used to achieve emissions targets and reductions. The target was a 7 percent reduction below 1990 levels by 2010. On this the United States succeeded, or as Eizenstat said, prevailed. "We firmly opposed and succeeded in opposing mandatory harmonized policies that were desired by others and that would have imposed on us uniform ways to reach our targets. We prevailed. The protocol leaves to the parties themselves to decide how best to meet their targets based on national circumstances. If somebody else wants to do it by heavy carbon taxes or heavy central regulation, that is their business. We are going to do it by market driven mechanisms alone."[33]

But there was another, less well-known, goal. The DOD wanted to make sure that military emissions would not be constrained by the treaty. Ambassador Eizenstat and the White House, however, were concerned that a complete exemption of military emissions from the treaty would not be acceptable to other countries. The State Department proposed that the United States try to get the support of other nations' militaries for a national security provision in the treaty. The U.S. delegation's approach to the Russian delegation, for example, was framed as a concern that the treaty not prevent rapid decisions about training and employing multilateral forces: "This Protocol must not create a conflict between our collective ability to preserve peace and our desire to reduce greenhouse gas emissions."[34] The United States' negotiating position appears to have been the result of an internal compromise between DOD and State, which resisted a blanket exemption. Ultimately, the U.S. position was that military greenhouse gas emissions, including fuel sold to ships and aircraft for international transport and for multilateral military operations—"bunker fuels"—should not be counted against a country's total emissions. Nor would emissions from overseas bases necessarily be counted against the U.S. total—they could be counted in the inventories of the host nation. The DOD apparently took the view that it would be to the United States' advantage to count overseas bases as part of its domestic emissions, however, since when the United States closed those bases, as was anticipated, it could show a significant reduction.[35] The State Department described the outcome in a confidential memo as a major victory: "In the form of a decision of the parties, DOD's three objectives were achieved: exempting emissions from bunker fuels; exempting emissions resulting from multilateral operations pursuant to the un charter; and explicitly permiting countries to decide how to account, among themselves, for emissions relating to multilateral operations (which, for example, would allow the U.S. and Germany to agree that the U.S. would count emissions related to U.S. training in Germany). This was a major victory."[36]

Ambassador Eizenstat testified before the Senate's Committee on Foreign Relations in February 1998 to explain and to a certain extent, defend the Kyoto agreement.[37] With respect to military emissions, his prepared statement was quite clear: the United States achieved what it wanted regarding military emissions.

> At the Kyoto Conference, the Parties took a decision to exempt key overseas military activities from emissions targets, including exemptions for "bunker fuels" (those used in international aviation and maritime transport) and for emissions resulting from a wide range of multilateral operations, such as peacekeeping and humanitarian relief. This exempts from our national targets not only multilateral operations expressly authorized by the U.N. Security Council (such as Desert Storm or Bosnia), but also multilateral operations that the United States initiates pursuant to the U.N. Charter without express authorization (such as Grenada). Countries may also decide among themselves how to account for emissions relating to multilateral operations (e.g., U.S. training in another NATO country) without going through emissions trading.[38]

Eizenstat's statement to the Senate Foreign Relations committee also made it clear that the U.S. position on military emissions had been shaped by the Pentagon: "I would like, Mr. Chairman and members of the committee, to deal very briefly with some misperceptions. The first is that the Kyoto Protocol will imperil the ability of our military to meet its worldwide responsibilities. This is absolutely untrue. We took special pains, working with the Defense Department and with our uniformed military, both before and in Kyoto, to fully protect the unique position of the United States as the world's only superpower with global military responsibilities. We achieved everything they outlined as necessary to protect military operations and our national security."[39]

While the chairman of the hearing, Senator Chuck Hagel, was very critical of the results of the negotiations at Kyoto, Senators Joe Biden and John Kerry were much more pleased. Senator Biden, then the ranking member of the Foreign Relations Committee congratulated Eizenstat for his work at Kyoto: "Stu, I think you did an incredible job." Biden was satisfied that Eizenstat was able to negotiate a treaty where the Pentagon's emissions were not counted. Biden said, "I just want to say to you that I thought it was remarkable that you went to Japan, you prevailed on the timetable, you prevailed on the average emissions, on the budget period, you prevailed on inclusion of the six gases that contribute to the greenhouse effect, and *you prevailed on dealing with the military for multilateral operations as well as ones we initiate*, and you also essentially laid in and accepted a provision for market based practices for trade in emissions. These are significant."[40] Eizenstat replied to Biden, "Our military—and I am proud that they are sitting behind me—our military has been the leader in energy efficiency. They are already very close to meeting the 1990 targets. Now, admittedly some of it is because of base closings. But they have put in real energy efficiency. So I don't want you to misunderstand. Obviously, those would be going on because they are good practices, not because we are subverting Senate will."[41] Eizenstat was correct, military emissions were down, as I describe in the next chapter. The base closings were a consequence of a Base Realignment and Closure process after the end of the Cold War, and they were significant.

For his part, Senator John Kerry used his time to ask Eizenstat for reassurance that the U.S. military would not be limited by the Kyoto Protocol. Kerry said, "Recently, there was a letter, the so-called Compass letter to the

President from a number of former national security personnel, and this letter, signed by Jeanne Kirkpatrick, Richard Cheney, Caspar Weinberger and others, suggested that the Kyoto Treaty threatens to limit the exercise of American military power: 'by exempting only U.S. military exercises that are multinational or humanitarian, unilateral military actions, as in Grenada, Panama, and Libya, will become politically and diplomatically more difficult.'"[42] Kerry continued, "I know that this is not, in fact, the case. But I would like to ask you, for the record, if you would address that question so we can try to answer it."[43] The letter Kerry referred to, produced by the Committee to Preserve Security and Sovereignty (COMPASS), had been sent to President Clinton and printed in Washington newspapers on the morning of the hearing. COMPASS, which also included former secretaries of state Lawrence Eagleberger and Al Haig, asserted that the Kyoto Protocol "will hamstring American military operations around the world."[44]

For a diplomat, Ambassador Eizenstat's reply to Kerry's question was caustic. "First of all, there is not one person on that list that I do not respect. Each has served his or her country in very positive ways and what they say has, therefore, to be taken seriously. It is, unfortunately, completely incorrect. Everybody is entitled to at least one mistake in life."[45] He continued, "If they would have checked with our uniformed military and with our Defense Department, they would know that every requirement the Defense Department and the uniformed military who were at Kyoto by my side said they wanted they got." Eizenstat underscored that the Kyoto Protocol left the United States plenty of room to act unilaterally. "It covers not only multilateral operations expressly authorized by the Security Council, like Desert Storm in Bosnia, it also covers multilateral operations we initiate pursuant to the right of self-defense. Under the U.N. Charter, it does not have to be authorized by the U.N. . . . We got bunker fuels covered."[46] He also said, "There was also a concern by our military . . . that somehow, countries [where the United States had bases] would want to throw out U.S. troops, because they would be concerned that somehow the emissions coming from bases and so forth would count against them. We got that taken care of by allowing a negotiation. We are willing to assume that ourselves." Eizenstat concluded by asserting, "I think it is a fair statement to say that our military really believes that we did what they wanted, that we produced what they wanted."[47]

In his prepared statement for the hearing Senator Hagel said he was convinced that "the treaty will have a severe impact on U.S. national security." Hagel said, "We all know that our armed forces are the largest users of fossil fuels in the U.S. government." He was aware that the Pentagon wanted an "exemption" for military emissions. "Before Kyoto, the Department of Defense asked for our negotiators to request a blanket exemption for our armed forces. The White House reportedly refused to seek such an exemption, afraid that other U.N. negotiators would not agree to such a demand." In his statement Hagel conflates the national accounting of military emissions with the *possibility* that military operations would be restricted because military emissions would be restricted. "The Administration has claimed that the Kyoto [P]rotocol does, in fact, exempt our armed forces. It does not. All it does is exempt multilateral operations approved by the United Nations. If America should have to take military action alone, or without the approval of the U.N., does that mean the use of the United States military will be limited by the amount of greenhouse gases they would emit? Since when do U.N. bureaucrats set our national security and national defense policy? Clearly, the entire concept is ludicrous. But so is the Kyoto Protocol.[48] Senator Hagel was prompted by Eizenstat's response in the hearing to Senator Kerry to again raise the issue of whether the Pentagon got a blanket exemption on their emissions. Eizenstat repeated his assertion that the U.S military was pleased with the outcome of the negotiations. "I know that there was in Kyoto a representative of the Joint Chiefs of Staff, who represented himself as representing the views of the Joint Chiefs of Staff, and I wouldn't say he kissed me, but he came darn close to it in terms of telling me how much he appreciated what we had done and that they were satisfied with what was done."[49]

Eizenstat's assertions were a slight exaggeration. A memorandum to the president in March 1998 by White House advisers Jim Sternberg, Todd Stern, Kathleen McGinty, and Gene Sperling acknowledged that while the Pentagon was "reasonably satisfied" with the military exemptions that were achieved at Kyoto, the DOD "remained concerned" about how the United States would implement emissions reductions. Specifically, the DOD accepted a program of domestic emissions trading that included facilities, but wanted operations and training exempted from cuts. "Noting the unique and often unpredictable nature of its mission, the Pentagon argues that subjecting military operations and training to greenhouse gas emissions limits could compromise military readiness." The advisers urged President Clinton to *preempt the critics by announcing that we would oppose emissions limits on military operations and training.* Noting that operations and training emissions were only .8 percent of total U.S. emissions, they argued that "downsizing together with fuel efficiency improvements may lead to reduced emissions from these sources in any event."[50]

Criticism of the Kyoto Protocol by Republicans on national security grounds continued. A few months later, former Secretary of Defense Frank Carlucci, also a member of the COMPASS group, charged in the *Washington Times*, "By signing the treaty, the administration has agreed to scale back fuel use by the U.S. military—a dangerous commitment that could have a disastrous impact on force readiness."[51] But these fears were not realized.

Although the United States signed the Kyoto Protocol, the Senate never ratified the treaty. Further, as passed in October 1998, the U.S. fiscal year 1999 National Defense Authorization Act included a prohibition on restriction of U.S. armed forces under the Kyoto Protocol: "Notwithstanding any other provision of law, no provision of the Kyoto Protocol to the United Nations Framework Convention on Climate Change, or any regulation issued pursuant to such protocol, shall restrict the training or operations of the United States Armed Forces or limit the military equipment procured by the United States Armed Forces."[52] However, the law also stated, "Nothing in this section shall be construed to preclude the Department of Defense from implementing any measure to achieve efficiencies or for any other reason independent of the Kyoto Protocol."[53] In sum, even though the Senate did not ratify the 1997 Kyoto Protocol, the United States military got the accounting system it preferred and the legal commitment that its operations, training, and procurement would never be restricted under the Kyoto Protocol. In March 2001, President George W. Bush announced that he would withdraw the United States' signature from the treaty.

IPCC GUIDELINES AND U.S. MILITARY
EMISSIONS REPORTING

Since 1994, when the United Nations Framework Convention on Climate Change went into effect, the countries that ratified UNFCCC, including the United States, have been obliged to submit their greenhouse gas emissions in what are called national inventories. The accounting scheme currently in effect requires each country to submit a National Inventory Report and a submission in a Common Reporting Format (CRF) so that the categories of emissions reporting for each country are comparable. The Intergovernmental Panel on Climate Change (IPCC) guidelines for governments to report their greenhouse gas emissions in their National Inventory Reports follow the conventions set at the Kyoto Protocol, whose provisions the United States helped write, and which have been further specified in subsequent IPCC guidelines.[54] The United States uses the Environmental Protection Agency (EPA) *Inventory of Greenhouse Gas Emissions and Sinks* as its National Inventory Report and the EPA inventory is thus the basis for the U.S. submission in the Common Reporting Format.

The EPA follows the IPCC conventions in its inventory and thus, the decision taken at Kyoto regarding military emissions has had lasting consequences for understanding the scale of military emissions. According to the IPCC guidelines, bunker fuels and military emissions from multilateral operations "shall not be included in national totals." Specifically, the relevant two paragraphs of the "Decisions" on methodological issues taken at the Kyoto Protocol in December 1997 are these:

4. *Recalls* that, under the Revised 1996 Guidelines for National Greenhouse Gas Inventories of the Intergovernmental Panel on Climate Change, emissions based upon fuel sold to ships or aircraft engaged in international transport should not be included in national totals, but reported separately. . . .

5. *Decides* that emissions resulting from multilateral operations pursuant to the Charter of the United Nations shall not be included in national totals, but reported separately; other emissions related to operations shall be included in the national emissions totals of one or more Parties involved.[55]

In other words, the IPCC treats military emissions, whether they occur on a country's own territory or abroad, differently than nonmilitary emissions. Under the guidelines, some emissions are to be counted as part of the national total; other emissions are not included in the national emissions totals but should be counted separately. Depending on where the military-related emissions occur, and their purpose, they are counted as part of the national inventory or separately enumerated.[56] Specifically, the military emissions of a country that occur within that country are considered domestic emissions and therefore counted. When states make their national submissions, the accounting scheme includes only military emissions for domestic operations and installations.

But it can be hard to tell from the national inventories which emissions are military or civilian. This is deliberate. "Due to confidentiality issues," the IPCC guidelines say, "many inventory compilers may have difficulty obtaining data for the quantity of military fuel use."[57] Under IPCC rules, governments may aggregate their emissions from military fuel use "with another source category" and indeed it is considered "*good practice* to aggregate military fuel use with another source category" to satisfy the need for "confidentiality" regarding military fuel use.[58] These guidelines, further specified in 2006, are the current guidelines for reporting annual inventories as affirmed at the Conference of the Parties in 2013.[59]

In sum, emissions from activity at overseas bases and multilateral operations *are excluded from national totals*.[60] As Eizenstat's testimony implied, emissions from overseas bases are supposed to be counted as part of the base host nation's greenhouse gas emissions inventory unless the two nations agree otherwise.[61] Any emissions resulting from military operations that are undertaken under the Charter of the United Nations are not included in national totals and should also be counted separately. Fuel given to a country for its operations (such as when the United States received fuel from Saudi Arabia during the Gulf War) should be counted: "The national calculations should take into account fuel delivered to the country's military, as well as fuel delivered within that country but used by the military of other countries."[62] In addition, "other emissions related to operations (e.g., off-road ground support equipment) should be included in the national emissions totals in the appropriate source category."[63] Further, as Alex Michaelowa and Tobias Koch note, decisions taken at Kyoto left out or left ambiguous how to count a number of other sources of military emissions, including those that resulted from "damages inflicted by enemy parties (burning of cities, oilfields, etc.)."[64] The DOD interpreted the IPCC rules in a way that meant many of its operational emissions would not be subject to reporting:

> The DoD interprets the term "multilateral operations" to mean operations involving more than one country, and may include providing combat forces, logistics and other support, or any combination of these. The DoD interprets the phrase "pursuant to the UN Charter" to mean multilateral operations that are consistent with the UN Charter, including not only multilateral operations expressly authorized by the UN Security Council, but also multilateral operations not expressly authorized but consistent with the UN Charter.[65]

The interpretation of "international bunker fuels"—consumption for international military transport, whether in the air or on the sea—was similarly broad and included civilian ships and aircraft. "Bunker fuels are defined as fuels sold to ships or aircraft engaged in international transport. The DoD interprets this definition to include fuel sold to military aircraft and ships for use in all military operations and training activities that involve flying or cruising in international airspace or waters, i.e., outside the territorial sea of any country or the airspace over the territorial sea. including those that begin and end within the same country in support of operations in international waters or airspace, and to activities that involve direct flying or cruising between two countries."[66] The DOD calculated that in 2000 more than 40.4 percent of its total navy aviation fuel use counted as bunker fuels. The U.S. Air Force calculated in 2000 that 13.2 percent of its aviation counted as bunker fuel. And the U.S. Navy estimated that between 1990 and 1999, 87 percent of its marine fuel use counted as time away, and therefor as bunker fuel use. In 2000, which the navy considered to be an anomalous year, it determined its maritime bunker fuel percentage to be 79 percent.[67]

The Environmental Protection Agency, which compiles the *Inventory of U.S. Greenhouse Gases and Sinks* receives its information for this report from the Department of Defense and the EPA inventory follows IPCC conventions when it publishes military emissions data. Overseas emissions and bunker fuels are not counted in its totals of the U.S annual submissions.[68] In addition, jet fuel consumption figures do not include military emissions that occur on U.S. territories.[69] Many of the U.S. territories—including

American Samoa, Guam, Commonwealth of the Northern Mariana Islands, Puerto Rico, U.S. Virgin Islands, and other U.S. Pacific Islands that are not permanently inhabited, such as Wake Island—have a long history of military basing. As the EPA notes, "The United States does not collect energy statistics for its territories at the same level of detail as for the fifty states and the District of Columbia. Therefore, estimating both emissions and bunker fuel consumption by these territories is difficult."[70] Further, the EPA does not consistently differentiate domestic military emissions from other domestic emissions and as a consequence, the EPA inventory does not distinguish emissions from power generation at military bases and installations from the emissions of civilian power plants. Finally, the EPA acknowledges sources of uncertainty with respect to military emissions that it does report: "Additionally, there are uncertainties in historical aircraft operations and training activity data. Estimates for the quantity of fuel actually used in Navy and Air Force flying activities reported as bunker fuel emissions had to be estimated based on a combination of available data and expert judgment. Estimates of marine bunker fuel emissions were based on Navy vessel steaming hour data, which reports fuel used while underway and fuel used while not underway."[71]

Like other parties to the United Nations Framework Convention on Climate Change, the United States makes two submissions of its greenhouse gas emissions, the National Inventory Report, which is the source of the data in the second submission, made in the form known as the Common Reporting Format. The official U.S. national inventory submission under the UN framework is the EPA report. Yet, as I have shown, because reports of military emissions are vague, and some categories of military emissions are not reported, it is not possible to use the U.S. national inventory compiled by the EPA as a source of data for comprehensive military emissions. The U.S. submissions in the Common Reporting Format for the UNFCCC includes military emissions in the category "other (not specified elsewhere)" for stationary and mobile emissions categories, but the submission is not detailed and it is not clear that it is complete.[72] For example, the data in the U.S. report of military emissions submitted to the IPCC Common Reporting Format in 2021 for mobile liquid fuel emission in the "other" and bunker fuel categories are reproduced in table 3.2. The 2021 submission reports emissions to be more

Table 3.2

Data from "Table 1.A.5 Other" and "Table 1.D International Aviation and International Navigation (International Bunkers)" and multilateral operations of the 2021 U.S. common reporting format (CRF), emissions data for mobile military emissions

Greenhouse gas source and sink categories	Aggregate activity data	Implied emission factors			Emissions		
	Consumption	$CO_2^{(1)}$	CH_4	N_2O	$CO_2^{(2)}$	CH_4	N_2O
	(Terajoules, TJ)	(t / TJ)	(kg / TJ)	(kg / TJ)	(kt)		
b. Mobile (*please specify*)	235,601.26				17,211.27	0.3	0.48
Military	235,601.26				17,211.27	0.3	0.48
Liquid fuels	235,601.26	73.05	1.29	2.04	17,211.27	0.3	0.48
International aviation (aviation bunkers)	569105.50				38033.60	NO, NA	1.22
Jet kerosene	569105.50	66.83	NO	2.15	38033.60	NO	1.22
Aviation gasoline	NA	NA	NA	NA	NA	NA	NA
Biomass	NO	NO	NO	NO	NO	NO	NO
International navigation (marine bunkers)	921795.46				65428.97	6.53	1.66
Residual fuel oil	755104.06	71.17	7.13	1.81	53744.34	5.38	1.37
Gas/diesel oil	166691.40	70.10	6.86	1.74	11684.63	1.14	0.29
Gasoline	NO	NO	NO	NO	NO	NO	NO
Other liquid fuels (*please specify*)	NO				NO	NO	NO
Gaseous fuels	NO	NO	NO	NO	NO	NO	NO
Biomass	NO	NO	NO	NO	NO	NO	NO
Other fossil fuels (*please specify*)	NO				NO	NO	NO
Multilateral operations(3)	IE	IE	IE	IE	IE	IE	IE

Source: United States, Common Reporting Format Submission to the UNFCCC 2021, "National Inventory Submissions 2021," https://unfccc.int/ghg-inventories-annex-i-parties/2021.

(1) The implied emission factors (IEFs) for carbon dioxide (CO_2) are estimated on the basis of gross emissions, i.e. CO_2 emissions + amount captured.

(2) Final CO_2 emissions after subtracting the amounts of CO_2 captured.

(3) "Parties may choose to report or not report the activity data and implied emission factors for multilateral operations, consistent with the principle of confidentiality stated in the UNFCCC reporting guidelines. In any case, Parties should report the emissions from multilateral operations, where available, under memo iItems in the summary tables and in the sectoral report table for energy. . . .

"Data on jet fuel expenditures by the U.S. military and estimates of the percentage of each services' total operations that are international operations were obtained from the Department of Defense (DoD). Military aviation bunkers include international operations, operations conducted from naval vessels at sea, and operations conducted from U.S. installations principally over international water in direct support of military operations at sea. Data on fuel delivered to the military within the U.S. was also provided by the DoD. Together, the data allow the quantity of fuel used in military international operations to be estimated."

than 17.2 MMTCO$_2$e, based on the EPA report for 2019. Other military emissions data is included within other categories in the Common Reporting Format, but this information is not disaggregated from those larger categories and is therefore not useful for analysis of U.S. military emissions. Similarly, there is no data given in the Common Reporting Format submission for "multilateral operations."

As discussed in the next chapter, to comply with the National Defense Authorization Act for Fiscal Year 2021, the DOD released data in their "Report on Greenhouse Gas Emission Levels" on their total scope 1 and scope 2 emissions from fiscal year 2010 through fiscal year 2019 in late August 2021.[73] Although short, and not very detailed, it was the DODs most comprehensive accounting of its recent emissions and the data corresponds to the information available from the Department of Energy.

CONCLUSION: TOWARD A FULL ACCOUNTING

As part of the United States Cold War competition with the Soviet Union, the U.S. military funded important scientific work in the 1950s and 1960s that demonstrated the accumulation of carbon dioxide in the atmosphere and the oceans, and the link to rising temperatures. The U.S. military, in particular the navy, and civilians within the Department of Defense, were also attentive to the implications of climate change well before many members of the general public. We have seen that the U.S. Navy was early not only to fund climate change research, but also to understand the implications of global warming. Recall that Terry P. Kelley's May 1990 paper for the Naval War College, "Global Climate Change Implications for the United States Navy," emphasized threats that climate change posed to naval operations, facilities, and systems and his analysis focused its recommendations on monitoring and adapting to climate change. The navy did just that. In July 2000, the U.S. Navy held a meeting at its Naval Ice Center that recommended the establishment of a commission to analyze climate change in the Arctic. The resulting Arctic Ocean Climate Change Commission used four models to understand the effects of global warming. The commission found that while all the models predicted a reduction in ice extent and thickness, some models "predict that the Arctic ice will significantly reduce in area and volume or possibly disappear during summer months as a result of increased greenhouse gases. The sea-ice albedo feedback is used to explain such a scenario. It implies that at warmer temperatures there will be less sea ice in the Arctic, which will allow an increased absorption of solar radiation due to decreased albedo, which will result in even warmer temperatures, and so on."[74] As I show in chapters 5 and 6, this awareness led the U.S. military to begin taking steps to adapt and decrease its vulnerability at bases and in operations and to consider the potential implications of climate change for conflict.

However, when it came time to take the first steps in reducing emissions—namely, documenting the source and scale of emissions—the DOD saw an accounting of the full greenhouse gas emissions associated with the military and war as undesirable and inconvenient. Therefore, the DOD resisted including military emissions in the 1997 Kyoto Protocol. Thus, it is fair to say that while some within the Department of Defense were quick to understand the implications of climate change, the institution as a whole has been slow, sometimes resistant, to acknowledging its part in the problem and the imperative to change equipment, operations, and practices. However, the connection was clear, and some made it. In 1998, President Clinton's Secretary of Defense William Cohen said, "Effective and efficient utilization of energy is essential because conservation not only saves money but also reduces greenhouse gases and counters global warming."[75]

The fact that the U.S. military's position, which left some military emissions out of the national inventory of greenhouse gases, prevailed at Kyoto has had a lasting impact on the way military emissions are counted. The U.S. military's resistance at the time of the negotiation of the Kyoto Protocol to the full accounting of greenhouse gas emissions in the UN Framework Convention on Climate Change set the standards for counting military emissions that are in effect at the time of this writing. The decisions to obscure some military emissions for reasons of "confidentiality," and to not include international bunker fuels and other military and war-related emissions in national emissions inventories also have important consequences for understanding and evaluating military greenhouse gas emissions of other countries.

4 A GUIDE TO U.S. MILITARY AND MILITARY-INDUSTRIAL EMISSIONS SINCE 1975

> DoD is the largest single energy-consuming entity in the United States, both within the Federal Government and as compared to any single private-sector entity. DoD operational and installation energy consumption represents approximately 77 percent of total Federal energy consumption, more than fifteen times the total energy consumption of the next closest Federal agency (the United States Postal Service).
> —Office of the Assistant Secretary of Defense for Sustainment, September 2021[1]

In August 2021, in a rare if not unprecedented accounting, the U.S. Department of Defense reported its greenhouse gas emissions in a six-page report in response to a Congressional requirement in the fiscal year 2021 National Defense Authorization Act.[2] Table 4.1 reproduces the data for military emissions that the DOD provided Congress in that report. The U.S. Department of Energy (DOE) has also provided greenhouse gas emissions data for all government departments, including the DOD, for fiscal year 2008 and continuously after 2010. Both the DOD and DOE report that U.S. total DOD greenhouse gas emissions in 2019 and 2020 were respectively about 55 million metric tons of CO_2e ($MMTCO_2e$) in fiscal year 2019 and 52 $MMTCO_2e$ in 2020. As noted above, DOD emissions are less than 1 percent of total U.S. national emissions and about 80 percent of U.S. federal government emissions.[3]

But, as I show in this chapter, these aggregate numbers are just the beginning of a larger story that includes historical emissions and details about

Table 4.1

DOD scope 1 and 2 emissions, CO_2e, for Fiscal Years 2010–2019

DoD Agency Wide Emission (MTCO2e)	2010	2011	2012	2013	2014	2015	2016	2017	2018	2019
Total Emissions	76,523,243	74,433,358	69,332,305	63,597,848	61,876,821	62,599,140	59,308,145	58,393,599	55,406,644	54,772,262

Source: Office of the Assistant Secretary of Defense for Sustainment, "Report on Greenhouse Gas Emission Levels" (Department of Defense, August 2021).

how those emissions are produced by the military and military industry. The DOE and DOD reported only the scope 1 and 2 emissions of the military. However, as I discussed in the introduction, there are other, scope 3 emissions that include U.S. military-industrial emissions (since 2017, federal agencies were not required to report scope 3 emissions). The DOE also provides information on fossil fuel consumption after 1975, and in combination with data from the DOD, it is possible to estimate U.S. military greenhouse gas emissions and to describe the general trends in U.S. military fuel use and emissions over the past several decades.[4] Further, there are additional war-related emissions that are difficult to quantify. And finally, there are emissions associated with the deep cycle of military force, war, and industrialization.

With a global network of about 750 bases and high-performance weapons, the U.S. armed forces are the most capable in the world. But unless you are on the wrong side of the military, the scale and power of U.S. armed forces, especially the U.S. Air Force, may be hard to picture, which is in part why there is something of a tradition in the United States of military aircraft flyovers at major sporting events, air shows, and culturally important ceremonies such as on the Fourth of July. In an unusual display, the U.S. Air Force flew all three versions of its long-range strategic bombers—the venerable B-52, the B-1B, and the B-2—over the Super Bowl in Tampa Bay in February 2021. In fact, the air force does about a thousand flyovers each year "as a way to showcase the capabilities of its aircraft while also inspiring patriotism and future generations of aviation enthusiasts."[5] Since 1946, the U.S. Navy's "Blue Angels" have streaked across the sky in formation or performed aerobatic maneuvers in thousands of airshows designed to showcase the "professionalism, excellence and teamwork" of the U.S. Navy and Marine Corps and the "thrill and magic of flight."[6] The current Blue Angels squadron consists of F/A-18 "Super Hornets" capable of flying nearly twice the speed of sound.

Fuel provides the U.S. military with unprecedented speed and range, capacities that are fossil-fuel intensive and therefore greenhouse-gas intensive. During each mission, aircraft put hundreds of tons of CO_2 in the air, not to mention emissions from the support activities of naval and ground-based assets for the air missions. A B-2 bomber on a mission from Whiteman Air Force Base in Missouri would be refueled many times to reach a target in Libya. For example, on January 18, 2017, two B-2 bombers, accompanied by fifteen KC-135 and KC-10 aerial refueling tankers, made a thirty-hour round-trip mission from Whiteman Air Force Base to drop bombs on ISIS targets in Libya.[7] The United States' oldest long-range bombers, B-52s, in the U.S. arsenal since the 1960s, were deployed to the Middle East during the 1991 Gulf War, flying from bases in the United States and Saudi Arabia to drop 1,000-pound bombs and air-launched cruise missiles. The B-52s returned in 2016 to make air strikes against ISIS in Iraq.[8] Later, B-52s deployed several times to Afghanistan during the U.S. war there, dropping large payloads of precision weapons. In April and May 2021, six B-52 bombers were deployed to Qatar from Minot Air Force Base in North Dakota to protect U.S. forces as they withdrew from Afghanistan.[9]

As I have shown in detail in chapter 3, under the Kyoto Protocol and IPCC emissions accounting rules, a significant portion of greenhouse gas emissions from military operations and installations have been discounted and obscured, by omission or consolidation with civilian emissions, from national inventories of greenhouse gases. Specifically, in the case of the United States, the Environmental Protection Agency showed in 2002 that exclusion of marine bunker fuels from national inventories would lead to the omission of between 79 and 87 percent of U.S. emissions in a year that the United States was not at war. The exclusion of aviation bunker fuels could lead to omitting more than 40.4 percent of naval aviation emissions and 13.2 percent of air force aviation emissions.[10] But bunker fuel use itself varies with whether the United States is at war and the type of operations that are dominant during a particular war. Most wartime emissions are also excluded from national inventories. Put another way, while the U.S. Navy Blue Angels certainly count as military aviation when their demonstrations occur in U.S. airspace, the emissions of more than 40 percent of all naval aviation may not be reported as part of the U.S. national inventory. The Paris Agreement in 2015 included provisions for states to voluntarily report their military emissions and include them in emissions cuts, as noted earlier, but there appears to have been no practical change in the way the United States reports its military emissions.[11]

If a country's report of its national inventory of military emissions is both incomplete and opaque, how is it possible to know the scope of U.S. military emissions and how they have changed? This chapter focuses on quantifying and describing military and military-related greenhouse gas emissions—from fuel use for installations and operations, war, and U.S. military industry. Using Department of Energy data for military fuel use and energy consumption to estimate the military's greenhouse gas emissions over the last forty-five years and more recently released data, I show that U.S. military greenhouse gas emissions track U.S. war. I also estimate the emissions of the largest U.S. defense industries and their supply chain and show that military-industrial emissions are a major source of greenhouse gases.[12] Indeed, I estimate that the scale of military-industrial emissions is on par with military emissions.

U.S. military emissions have declined in recent years due to both greater fuel efficiency and base closures following the end of the Cold War. The military, under President Biden, has also made emissions reductions a part of their climate strategy. However, U.S. military emissions are still larger than the emissions of most countries in the world and in fact directly drive the emissions of U.S. military industry and indirectly drive the military and military-industrial emissions of other countries that are both allied and in competition with the United States. The deep cycle, fossil fuel use, a foreign policy designed to protect access to fuel, and military industrialization that I described previously, continues. This chapter concludes with reflections on the larger relationship between military and civilian emissions, including the evidence that war and military spending are associated with higher civilian emissions. It is also the case that in the same way the battle for regional influence and arms races tend to drive increases in military spending, so too does international military competition tend to increase the greenhouse gas emissions of other states. This is also part of the deep cycle of military mobilization, war, and greenhouse gas emissions.

MOVING FUEL: DLA ENERGY

In many respects the U.S. Department of Defense gets its energy for mobility much differently than it did in the nineteenth and early twentieth centuries.

In the days of coal-powered steamships, ships would carry some of their own coal, but had to make frequent stops to refuel, either from coal tenders that traveled along with them or at ports where they could buy local coal, or coal that had been delivered for them in advance. For example, for the voyage of the Great White Fleet, coal was purchased from private U.S. companies or from foreign suppliers and distributed as needed. "It would take several days to coal a ship. Afterward, the crew would spend several more days cleaning the ship, inside and out, fore and aft, since coal dust settled everywhere."[13] The coal soot that came from the ship's smokestacks, and the coal dust that settled after each refueling, was regularly washed from the white hull and when necessary, the ship was repainted. During World War I and World War II, the system became much more centralized, with the federal government working more closely with private companies in the coal and petroleum industries to supply fuel needs.

While the U.S. military still uses fuel that it acquires abroad from other governments or commercial suppliers, in most years, the vast majority of U.S. fuel for vehicles is centrally supplied by the Defense Logistics Agency (DLA) Energy command.[14] As the DLA says, "through evolution in mission and name changes, DLA Energy is committed to providing world class support to the Warfighter."[15] DLA Energy has the capacity to deliver fuel for operations in any region of the world and provides energy for DOD facilities by either producing it on site from fuel provided by the DLA or purchasing steam or electricity from a local energy company. Figure 4.1 illustrates the global DLA footprint.

The process of moving fuel to where it is needed entails a vast logistical effort. It begins when the DLA purchases fuel for vehicles in bulk from refineries and oil companies at market rates. In 2008, and 2009, for example, BP, Shell, Valero, Abu Dhabi National Oil Company, the Kuwait Petroleum Corporation, and Exxon Mobil were among the Pentagon's top suppliers of petroleum.[16] The DLA, which manages all logistics support, stores fuel at its nearly 600 bulk fuel storage facilities strategically located all over the world, then charges each of the services for fuel at a price that is supposed to account for the cost of delivery, at the "standard price of fuel." Table 4.2 shows the standard price that the DLA set for the DOD's military customers to pay for fuel in fiscal year 2021. The DLA then delivers fuel to the armed services

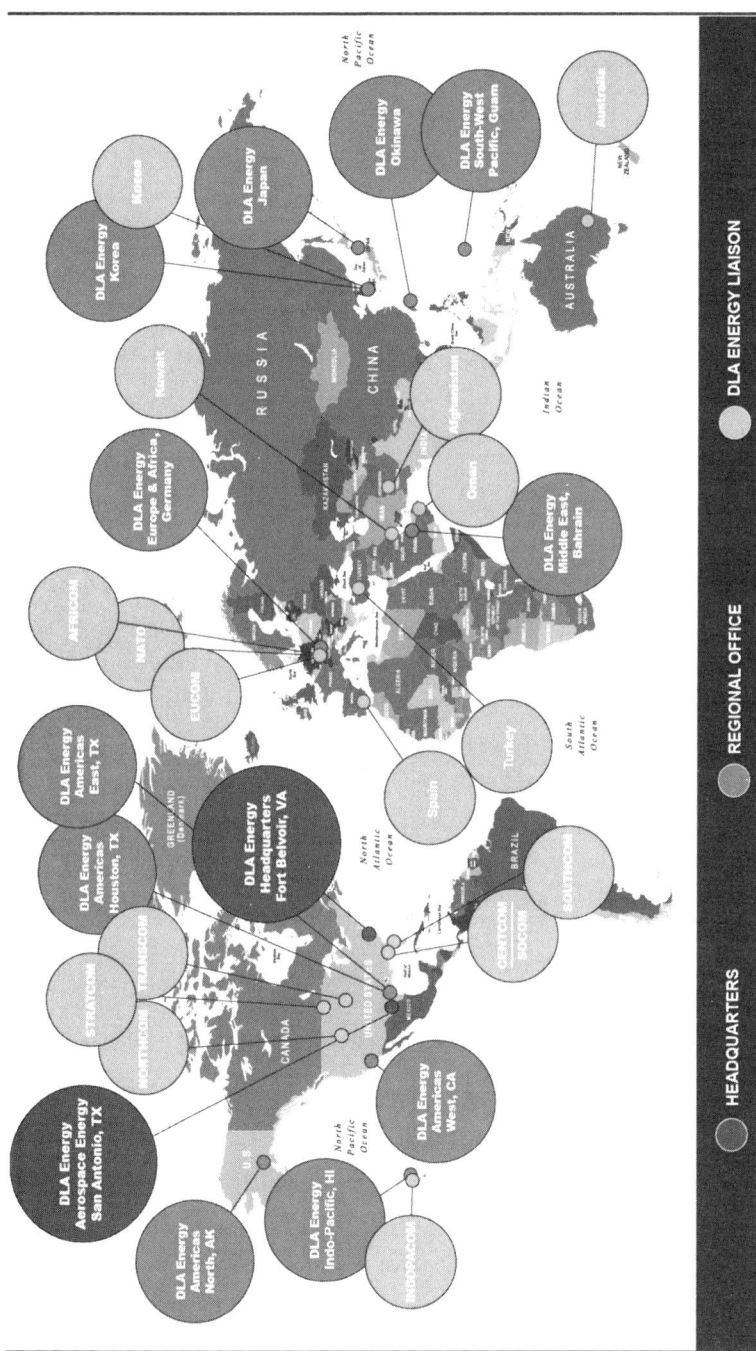

Figure 4.1

Defense Logistics Agency "Worldwide Locations" in 2020.

Source: Defense Logistics Agency Energy, *Fiscal Year 2020 Fact Book*, 7, https://www.dla.mil/Portals/104/Documents/Energy/Publications /DLAEnergyFactBookFY20_lowres2.pdf?ver=VE-mCUImzFiKKnG1uajkxg%3d%3d.

Table 4.2
Defense Logistics Agency standard prices: jet fuel, diesel distillates, and gasoline, FY2021

Fuel	Purchase cost in dollars per gallon	Purchase cost in dollars per barrel
Jet fuel type JP8	2.37	99.54
Diesel distillates	2.39	100.38
Regular unleaded gasoline	2.32	97.44

Source: Undersecretary of Defense, Defense Logistics Agency, "Fiscal Year (FY) 2021 Standard Fuel Price Change," Effective October 11, 2021, September 17, 2020, https://www.dla.mil/Portals/104/Documents /Energy/Standard%20Prices/Petroleum%20Prices/E_2020Oct1PetroleumStandardPrices_200930.pdf?ver=1ob Qukd5sd1MspwQlW0ghg%3d%3d.

wherever it is needed in the world—via truck, pipeline, rail, barge, tanker, and shipping containers called "sea vans." In fiscal years 2019 and 2020, the DLA spent, respectively, $8.2 million and $9.8 million transporting gasoline and $298.3 million and $366.2 million transporting jet fuel worldwide.

If the cost of acquiring fuel on the market is lower than the cost to the armed services, the DLA keeps the excess in a working capital fund. If the market price is higher than what it charges each service, the DLA draws on the working capital fund so that it can remain solvent.[17] The DLA tracks the quantity of fuel that it purchases and sells to its military customers and in any one year, the total amount the DLA purchases in each fuel category is about the same as the amount that it sells. Thus, some fuel is always available: at any one time the DLA has an inventory of about fifty to sixty million barrels of bulk fuel of various types on hand and in transit.[18]

TRACKING MILITARY AND WAR-RELATED EMISSIONS

The 1907–1908 voyage of the Great White Fleet consumed 430,000 short tons of coal.[19] Depending on the type of coal the United States was able to purchase at home and at foreign ports, the Great White Fleet emitted about 803,000 metric tons of CO_2e.[20] Because CO_2, methane, nitrous oxide, and fluorinated gases remain in the atmosphere for decades until they are reabsorbed into the carbon cycle in the case of CO_2 or break down, past emissions are just as important as current emissions. Specifically, carbon dioxide—the majority of emissions from coal—remains in the atmosphere for 300 to 1,000 years. Thus most of the emissions and the heat trapping effects caused by the Great White Fleet are still with us.

Ideally, the Department of Defense would provide fuel consumption data and a full accounting of all of its emissions in a form that was usefully detailed and disaggregated. While some of this data is available in separate DOD reports, the most systematic information is collected and presented by the Department of Energy, which tracks and reports the energy consumption after 1975 for all federal agencies and departments. Energy consumption and military emissions have changed quite a bit over the last several decades. It is important to understand the entire profile of U.S. military energy use in order to contemplate the source of greenhouse gas emissions and the progress that has or has not been made at reducing those emissions.

But there is no publicly available time series of DOD fuel consumption for the post–World War II era through 1974. More than sixty years ago, energy analysts Sam Schurr and Bruce Netschert essentially threw up their hands and said they could not estimate the U.S. military's fuel consumption as a portion of total government consumption: "The military segment, for obvious reasons, is difficult to handle."[21] Further, they said that "in the main, these amounts are probably reported under government, so what is involved is the distribution of the government total between military and nonmilitary."[22] While all consumption in the economy is accounted for, Schurr and Netschert said, "The one possible exception to this statement is the production and consumption of power generated and consumed by military establishments which, according to available evidence, is not reported in any available statistical series."[23] The DOD emphasized its own lack of knowledge about fuel consumption in 2000.

In 2001, a Defense Science Board report noted that the DOD "does not maintain comprehensive fuel consumption data necessary to determine emissions directly," and "no single source collects or maintains a complete and accurate data set for all fuel types and end uses." The report also stated, "Because of the difficulty of obtaining accurate data, DoD should consider establishing a centralized energy and fuel data collection and analysis function."[24] The DOD acknowledged that their own estimates of their emissions were incomplete when it calculated that the department emitted forty-one million metric tons of carbon dioxide in 1996, down about 20 percent from their estimate for DOD emissions in 1990. That estimate for 1996 counted most greenhouse gases (carbon dioxide, methane, and nitrous oxides) and included some bunker fuels.[25] Even though some information was available, the Defense Science Board argued that the DOD should "track and document its emissions and remain aware of the potential for accruing carbon credits by reducing those emissions."[26] Yet, despite the Defense Science Board recommendation, there was no publicly available, comprehensive, and official Department of Defense accounting for *all* military fuel use and for the DOD's direct and indirect greenhouse gas emissions until late 2021.

To fill this gap, several scholars have estimated DOD greenhouse gas emissions based on fuel consumption data and military spending data for individual years.[27] For instance, in 2010, Adam Liska and Richard Perrin estimated U.S. military emissions of 85.4 MMT CO_2e based on data for fuel use and military spending in 2009. Liska and Perrin estimated that "upstream emissions" for procurement that year were an additional 86.8 MMT CO_2e.[28] As we shall see below, Liska and Perrin appear to have overestimated total DOD emissions. Oliver Belcher, Patrick Bigger, Ben Neimark, and Cara Kennelly, a group of geographers based in the United Kingdom, made an estimate of DOD emissions of 23.4 MMT CO_2e for 2017 based on U.S. military fuel use data they acquired via a Freedom of Information Act request, a significant underestimate compared to the DODs 2021 report of total agency-wide emissions.[29]

There are several major sources of greenhouse gas emissions related to war and preparation for it. I focus here on: (1) *total U.S. military emissions* for installations and war and nonwar operations, such as peacetime training, from the combustion of fossil fuels; and (2) *military-industrial emissions*, including the emissions associated with the production of weapons, ammunition, and other war material. But even here, the data on military industries are only partial. I focus on the top twelve U.S. defense industries.

There are a number of war and war-related emissions that should also be accounted for, but that are beyond the scope of this book. These include *military contractors' emissions in war zones*. Military contractors are increasingly providing many functions that militaries used to perform such as security for convoys, fueling services, meal preparation, equipment maintenance and repair, telecommunications, and the construction of infrastructure used at overseas bases.[30] For example, in late 2020, there were 43,809 people working as contractors for U.S. Central Command, including 27,388 working in Afghanistan, Iraq, and Syria. The total U.S. troop presence at the time in both countries was about five thousand total service members.[31] Further, while I will describe emissions caused by the *direct targeting of petroleum*, namely the deliberate burning of oil wells and refineries by all parties, I do not estimate the size of those emissions. Nor do I attempt to estimate the emissions from *explosions and fires* due to the destruction of nonpetroleum targets in war zones and from *fire suppression and extinguishing chemicals*, including halon, a greenhouse gas. And if the aim is to enumerate all the emissions in a particular war, the military *emissions of other belligerents* including U.S. allies and adversaries should be counted, as well as energy consumed and emissions due to *reconstruction* of damaged and destroyed infrastructure. More than sixty countries joined the United States in their war in Afghanistan, thirty-seven countries fought with the United States in the Iraq War, and more than sixty countries were allied with the United States in the war against ISIS in Iraq and Syria. If we were to add the emissions that are the result of these activities together, they would be substantial.

The Department of Defense classifies energy use by purpose—"installation" and "operational" energy. "Installation" energy powers buildings and operations at installations including nontactical vehicles at "enduring locations." Depending on whether the United States is at war, installation use is about 25–30 percent of all Pentagon energy consumption in any year. Installations currently account for about 30 percent of all Pentagon energy consumption.[32] "Operational" energy is used "in direct support of military operations, and in training that supports unit readiness for military operations, to include the energy used at non-enduring locations (contingency bases)."[33] This is the fuel required for the Pentagon's fighting "tooth" in war zones and its logistical "tail."

By contrast, the Department of Energy employs three broad categories for reporting U.S. government departments' greenhouse gas emissions—standard operations, nonstandard operations, and biogenic emissions.[34] Biogenic emissions, from natural sources such as the decomposition of vegetation or combustion, are much smaller than standard and nonstandard emissions and I am not tracking those here. By Department of Energy definition, nonstandard operations are "vehicles, vessels, aircraft and other equipment used by Federal Government agencies in combat support, combat service support, tactical or relief operations, training for such operations, law enforcement, emergency response, or spaceflight (including associated ground-support equipment). Non-Standard operations also include the generation of electric power produced and sold commercially to other parties."[35] Standard operations are everything else that a department does to accomplish its functions, roles, and missions, and most government departments have only emissions from standard operations.

DOD'S ENERGY CONSUMPTION

Military fuel use likely peaked in the period between 1942 and 1968, the height of the Vietnam War, when the DOD was still using significant quantities of coal at installations and the United States was engaged in three successive major wars—World War II, the Korean War, and Vietnam—as well as mobilizing at home and abroad for the Cold War. Following the energy crises of the 1970s, and the recurring emphasis on the United States achieving energy independence, the U.S. government gradually decreased its overall energy use. As figure 4.2 illustrates, Department of Defense emissions are just one part, albeit the largest, of U.S. government energy consumption. While the DOD remained the U.S. government's largest energy consumer, the Pentagon's share of U.S. government energy consumption decreased from 87 percent of all U.S. federal energy consumption in fiscal year 1975, to 76 percent in 2020. Since most government energy consumption is fossil fuel based, greenhouse gas emissions track energy consumption. For example, total U.S. government emissions were about 68.6 million metric tons of CO_2e in fiscal year 2020 (excluding biogenic emissions), or about 1.2 percent of total U.S. national emissions that year; DOD emissions were about 51.5 million metric tons of CO_2e, or 75 percent of total government emissions.[36]

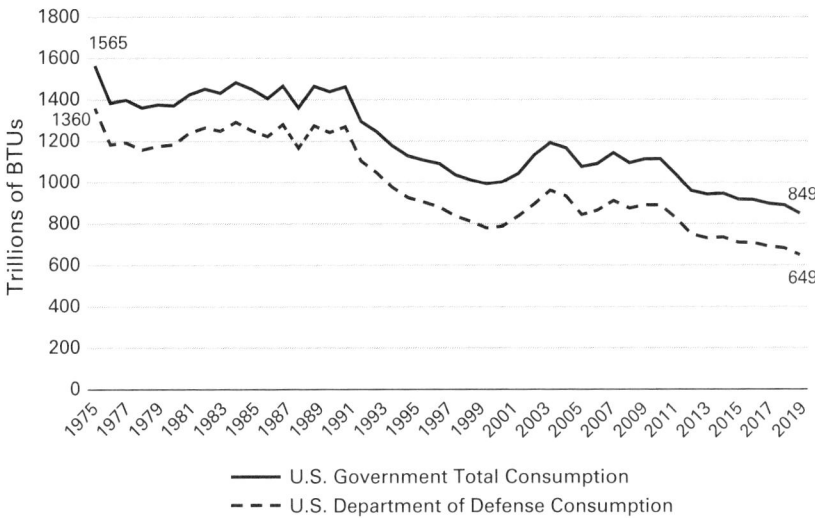

Figure 4.2

DOD and Total U.S. federal government energy consumption, FY1975–2020, in trillions of British thermal units (BTUs).

Source: Data from U.S. Energy Information Administration, https://www.eia.gov/totalenergy /data/monthly/pdf/sec2_14.pdf, accessed September 16, 2021.

As figure 4.3 illustrates, the Department of Defense leads the consumption of U.S. government in jet and diesel fuel use, the two most consumed fuels. For example, in Fiscal Year 2020, the DOD used 98 percent of all government jet fuel and 91 percent of diesel fuel. The DOD is also the leading consumer of electricity, fuel oil and natural gas for facilities. And, at the time of this writing, the DOD is the only remaining U.S. federal government user of coal.

U.S. military greenhouse gas emissions are a function of fuel use. Even though there is no single public database that provides consistent numbers for total U.S. and military energy consumption prior to 1975, it is safe to assume that military energy use was higher in the decades prior to 1975.[37] By David Vine's reckoning, at the height of World War II, the United States controlled more than 2,000 bases and 30,000 installations overseas. Many of those were closed or abandoned at the end of the war, but the U.S. military footprint grew again during the Cold War and the major U.S. wars in Korea and Vietnam.[38] The U.S. troop presence in Vietnam peaked in 1968

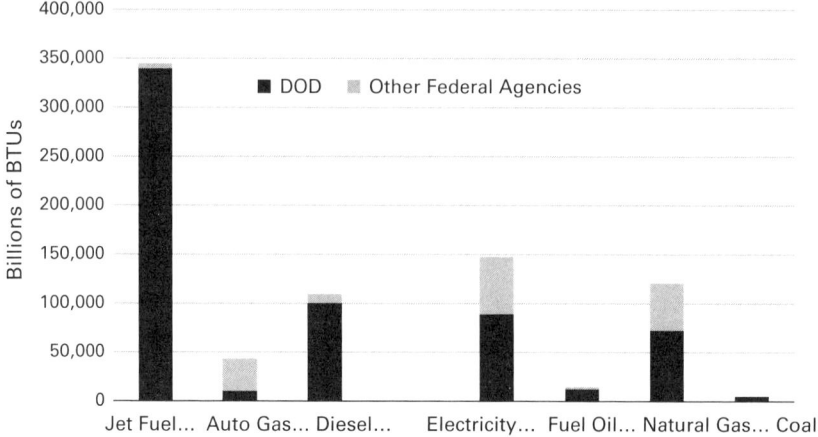

Figure 4.3

Major categories of energy consumed by the U.S. government and DOD, FY2020 in billions of British thermal units (BTUs).

Source: Data from U.S. Department of Energy, "All Agency Energy Consumption Data by End Use Sector in FY2020 (Billion BTU)," Comprehensive Annual Energy Data and Sustainability Performance, https://ctsedwweb.ee.doe.gov/Annual/Report/AgencyMasterDataByYear.aspx.

at 536,000 troops and air operations only declined dramatically after 1972.[39] In January 1973 U.S. combat in Vietnam halted with the signing of the Paris Peace Accords, and after most U.S. troops left Vietnam. The United States finally completely exited Vietnam in April 1975, when the last Americans and a number of Vietnamese were airlifted by helicopter from the roof of the United States Embassy in Saigon and flown to U.S. Navy ships in the South China Sea.

Installation Energy Use

It is well known that the Pentagon building itself, at 6.5 million square feet, is the world's largest office building. But that is the admittedly large tip of the iceberg. In fiscal year 2020, the DOD reported that it had had more than 281,000 buildings.[40] In fiscal year 2020, the DOD spent about $12.7 billion to heat, cool, and provide electricity to its facilities, amounting to 1.9 trillion square feet of space—more than half of all government facilities' square footage.[41] The DOD currently supports U.S. military operations and power projection capability on more than seven hundred bases and smaller "lily pads."[42] Although military installations in the United States and abroad necessarily contribute to operations, the DOD tracks installation energy use separately.[43] As the Pentagon notes, "In many ways, installation energy supports warfighter requirements through secure and resilient sources of commercial electrical energy, and where applicable, energy generation and storage, to support mission loads, power projection platforms, remotely piloted aircraft operations, intelligence support, and cyber operations."[44]

Installation energy use also includes the fuel for about 160,000 nontactical fleet vehicles (cars and trucks). And although the army is the most energy intensive at its installations, energy consumption at installations is relatively equally shared by the service branches. While DOD installations consume more energy than any other part of the U.S. government, as figure 4.4 illustrates, the total energy consumed by DOD facilities has declined more than 50 percent from fiscal year 1975 to fiscal year 2020, from 439,228 BTUs to 197,983 BTUs. The most significant cause of the decline in installation energy

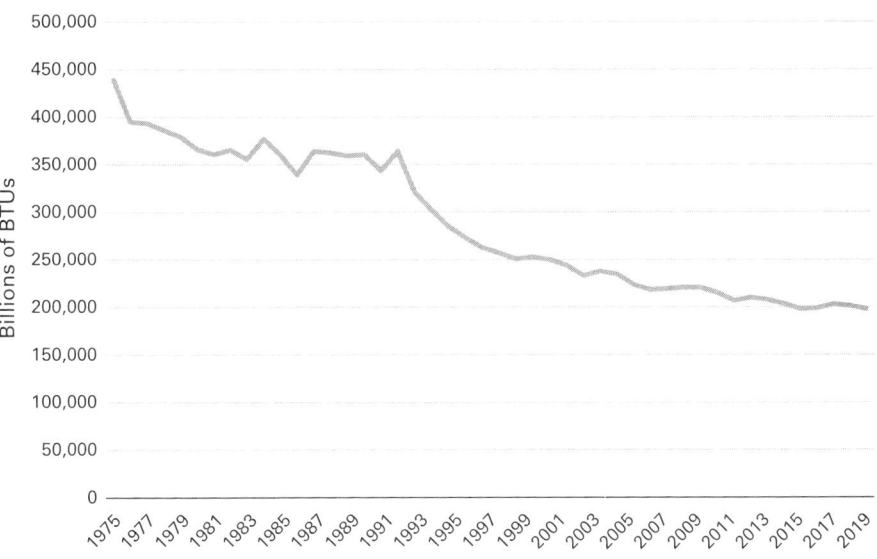

Figure 4.4
Total DOD facility energy use, billion BTUs, FY1975–2020, all sources.
Source: Department of Energy, all DOD facilities combined, https://ctsedwweb.ee.doe.gov
/Annual/Report/HistoricalFederalEnergyConsumptionDataByAgencyAndEnergyTypeFY1975
ToPresent.aspx.

consumption is that the overall size of the DOD decreased at the end of the Cold War from about two thousand bases to the current seven hundred-plus.

Further, the DOD changed its mix of power sources for installations, as illustrated in figure 4.5. Coal and fuel oil consumption—the most GHG-intensive fuels—dramatically decreased, while natural gas, with its lower greenhouse gas emissions intensity, was a greater proportion of energy use. Specifically, the two most greenhouse gas-intensive fuels—coal and fuel oil—together

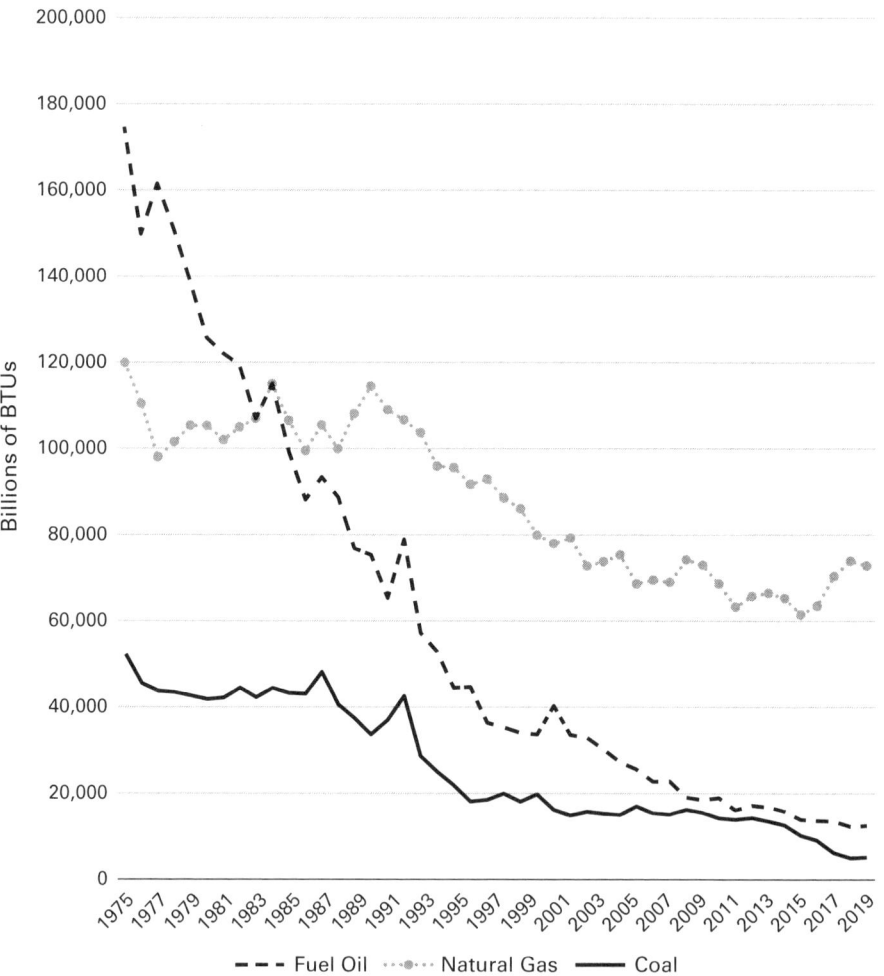

Figure 4.5

DOD facility energy use FY1975–2020: coal, fuel oil, and natural gas in billions of BTUs.
Source: Department of Energy data.

constituted more than 52 percent of the energy produced at DOD facilities in 1975: fuel oil (40 percent) and coal (12 percent). By 2020, fuel oil and coal were producing less than 10 percent of DOD's facilities energy, respectively 6 and 3 percent, while the proportion produced by natural gas increased. In fiscal year 1975, coal consumption yielded 52,305 BTUs of energy. In fiscal year 2019, coal produced 4,989 BTUs, and in fiscal year 2020, 5,185 BTUs. Although coal use grew slightly in 2020 over the previous year, this was, overall, a more than 90 percent reduction since 1975. The drop in fuel oil consumption was similarly dramatic, declining more than 90 percent from nearly 175,000 BTUs in1975 to about 12,500 BTUs in 2020.

Thus, the mix of energy used at installations and facilities has diversified in the last several decades and become less greenhouse gas intensive. There is one exception. The military had nuclear power generation capacity as part of its Army Nuclear Power Program, which constructed and operated eight nuclear power reactors from 1953 to 1976. All those reactors were deactivated and the U.S. military does not currently run nuclear power reactors in its facilities.[45] However, the military has also, over the last several years, become interested in reintroducing nuclear power into the mix of energy sources. The Pentagon awarded three small contracts in 2020 for the design of mobile, small nuclear reactors to be used in the United States and abroad.[46] I return to the issue of the DOD acquiring small nuclear reactors later. Figure 4.6 illustrates a snapshot of the types and amount of fuel and energy produced in fiscal year 2020.

While the facilities' total energy production declined from fiscal year 1975, the proportion of electricity and steam generation, source unspecified, increased. In fiscal year 1975, electricity accounted for 20 percent of facilities energy use; by fiscal year 2020, electricity accounted for nearly 50 percent of facilities energy use. While the DOD had renewable energy systems before 2008, renewable energy was added to the DOE's reporting of the mix of energy sources at facilities with on-site production in 2008, and purchases of renewable energy for DOD facilities beginning in fiscal year 2012. At this point, neither the military nor any other federal agency is a significant producer or purchaser of renewable green energy. I return to the scale of U.S. military renewable energy consumption in chapter 7.

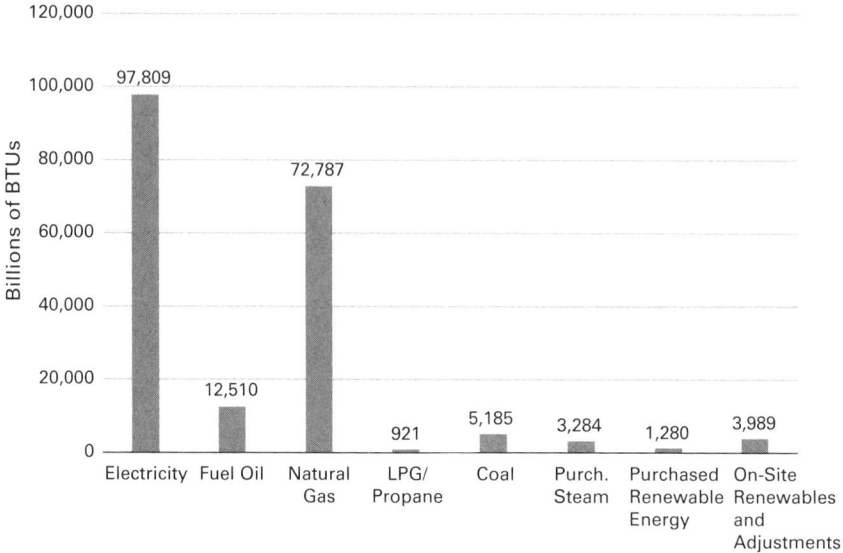

Figure 4.6
DOD facilities energy consumption by major type FY2020 in billions of BTUs.
Source: Department of Energy data.

Operational Energy

The profile of fossil fuel energy consumption looks different when we consider "operational" energy use, which the Pentagon defines as the energy "required for training, moving, and sustaining military forces and weapons platforms." Most of the Pentagon's energy for vehicles is provided by fossil fuels. Although the military does not count nuclear energy when it totals operational energy use, nuclear power is a significant source of operational energy. The U.S. Navy began a transition from diesel and battery power to nuclear-powered submarines in 1955 when it launched the *Nautilus*. Thereafter, the United States gradually transitioned to powering its entire submarine fleet—both ballistic missile-carrying submarines and attack submarines (which search for an adversary's submarine)—by nuclear energy. The navy's aircraft carriers fleet began a transition to nuclear power in 1961 when the USS *Enterprise* was commissioned. The last conventionally powered aircraft carrier left service in 2009, and the entire fleet is currently nuclear powered.[47] The U.S. Navy uses more than 180 nuclear reactors to power more than 140 submarines and surface ships, including all 11 U.S. aircraft carriers, and 70 submarines.[48]

Operational energy accounts for about 60 to 70 percent of DOD energy consumption, not including nuclear energy.[49] Operational energy use varies, depending on what the U.S. military is doing in any particular year—its ongoing and occasional missions. When the United States is engaged in war, consumption of jet and diesel fuels increase, as one would expect. Their ratio will depend on the types of operations the military is performing—whether the war or particular phase of the war is land or air intensive. Figure 4.7 tracks the amount and spending on operational fuel consumption in a period when the United States was at war. Average fuel use was nearly

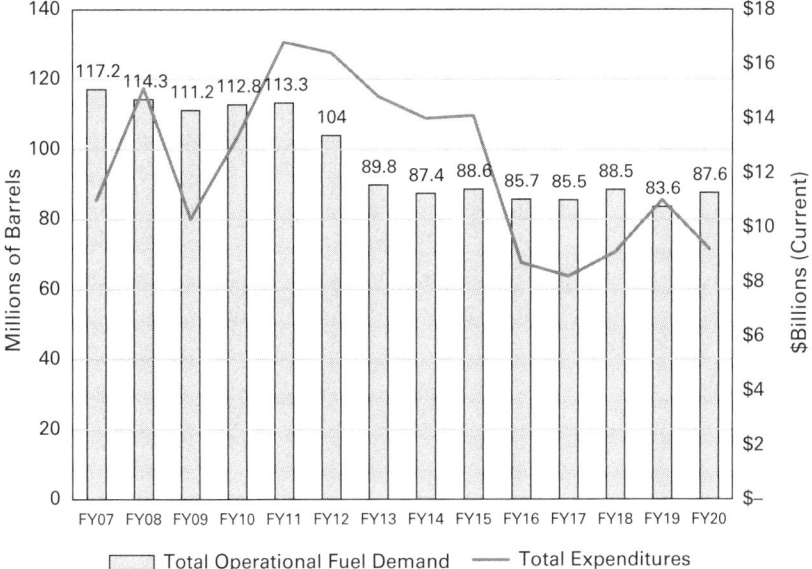

Figure 4.7

Total DOD operational fuel consumption and expenditures, FY2007–2020.

Source: Office of the Under Secretary of Defense for Acquisition and Sustainment, *Fiscal Year 2011 Operational Energy Annual Report* (Washington, DC: DOD, March 2013), 9, https://www.acq.osd.mil/eie/Downloads/OE/FY11%20OE%20Annual%20Report.pdf; Office of the Under Secretary of Defense for Acquisition and Sustainment, *Fiscal Year 2012 Operational Energy Annual Report* (Washington, DC: DOD September 2013), 10, https://www.acq.osd.mil/eie/Downloads/OE/FY12%20OE%20Annual%20Report.pdf; Office of the Under Secretary of Defense for Acquisition and Sustainment, *Fiscal Year 2019 Operational Energy Annual Report* (Washington, DC: DOD, March 2020), 19, https://www.acq.osd.mil/eie/Downloads/OE/FY19%20OE%20Annual%20Report.pdf.

98 million barrels per year from fiscal year 2007 through 2020, and total spending on operational fuel also varied during this period, averaging $12 billion annually.

Unlike the profile of fuel use at installations, where the different branches of the armed services each consume about the same amount of energy, operational energy use varies a great deal by service branch. The U.S. Air Force is consistently the largest operational fuel user, with an average annual consumption between fiscal year 2007 and fiscal year 2020 of about fifty-four million barrels of fuel per year (see figure 4.8). The combined average operational fuel consumption of *all* the *other* military branches and the DOD itself was about fifty-three million barrels for the same period.

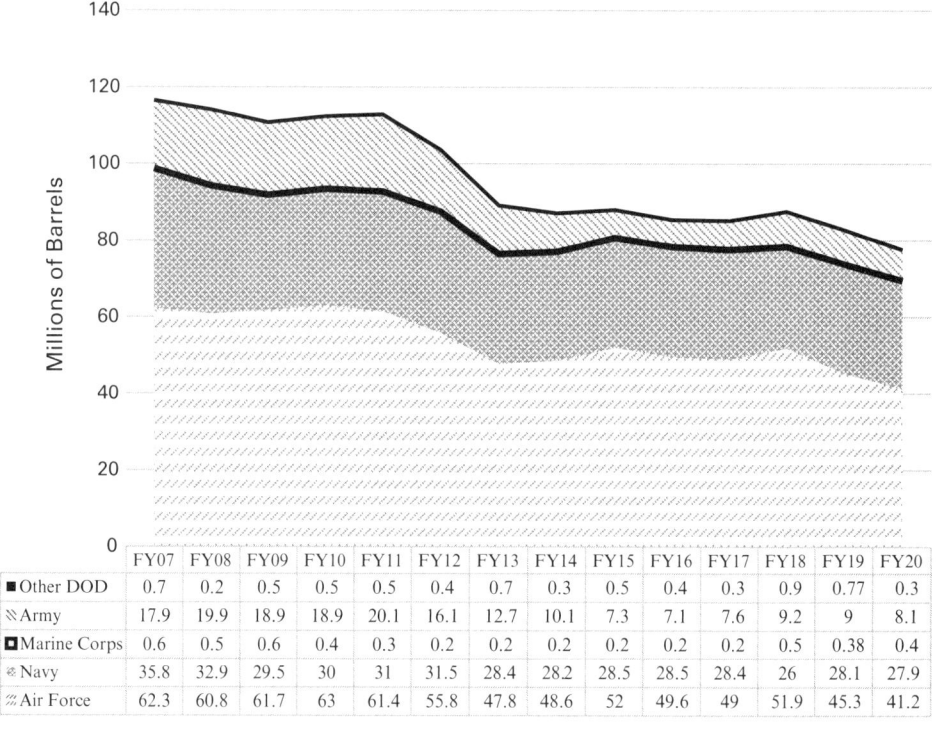

	FY07	FY08	FY09	FY10	FY11	FY12	FY13	FY14	FY15	FY16	FY17	FY18	FY19	FY20
■ Other DOD	0.7	0.2	0.5	0.5	0.5	0.4	0.7	0.3	0.5	0.4	0.3	0.9	0.77	0.3
﹨ Army	17.9	19.9	18.9	18.9	20.1	16.1	12.7	10.1	7.3	7.1	7.6	9.2	9	8.1
❑ Marine Corps	0.6	0.5	0.6	0.4	0.3	0.2	0.2	0.2	0.2	0.2	0.2	0.5	0.38	0.4
❋ Navy	35.8	32.9	29.5	30	31	31.5	28.4	28.2	28.5	28.5	28.4	26	28.1	27.9
╱ Air Force	62.3	60.8	61.7	63	61.4	55.8	47.8	48.6	52	49.6	49	51.9	45.3	41.2

Figure 4.8

Operational energy demand, millions of barrels, FY2007–2020.

Source: Data from the Office of the Under Secretary of Defense for Acquisition and Sustainment, *Operational Energy Annual Reports* (Washington, DC: DOD, March 2013), 9.

Jet fuel consumption accounts for the largest operational fuel use. The U.S. military's heavy reliance on airpower for strategic and tactical purposes is illustrated in fuel consumption statistics. Recall, for example, that the U.S. wars in the Balkans during the 1990s were entirely conducted from the air. Further, as noted earlier, the wars in Afghanistan and Iraq began with days of massive air strikes, and in each case, war material was flown to the war zones. Figure 4.9 shows the consumption in millions of barrels of the main fuels

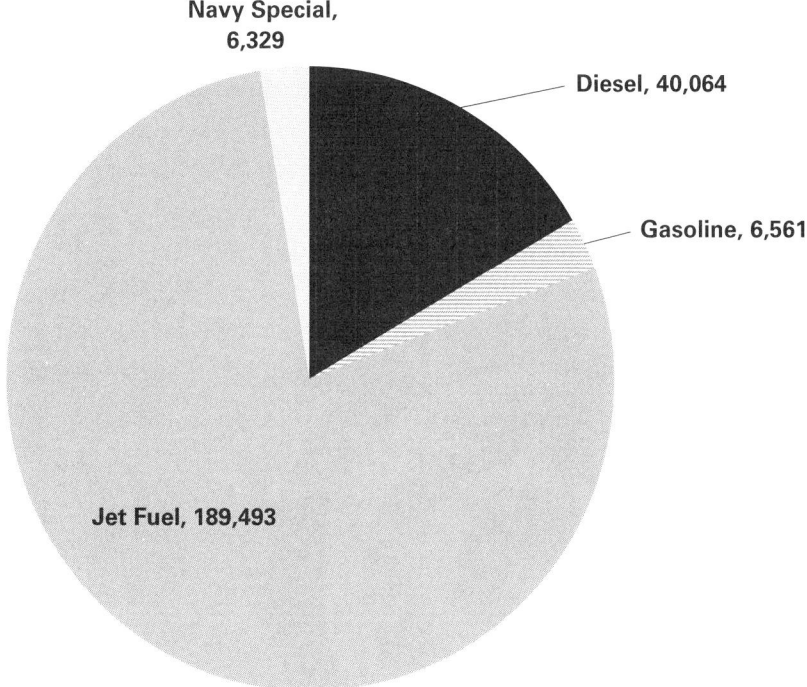

Figure 4.9

DOD vehicle fuel consumption in millions of gallons, by major fuels, FY1975–2020. The U.S. military halted use of Navy Special Fuel for ships in 2010. The U.S. Air Force and Army use JP-8 fuel; the navy uses JP-5 fuel. Commercial fuel has been certified for military jets since 2014. Military vehicles also used 593 million gallons of aviation gas and 14 million gallons of liquid propane gas/propane from FY1975–2020. There were also 77,620 BTUs generated for vehicles by "other" unspecified means.

Source: Data from the Department of Energy, Vehicles and Equipment Energy, https://ctsedwweb.ee.doe.gov/Annual/Report/HistoricalFederalEnergyConsumptionDataByAgencyAndEnergyTypeFY1975ToPresent.aspx.

used by the United States, with jet fuel comprising more than 75 percent of fuel used since 1975.

Energy is consumed differently by the different regional military commands. Figure 4.10 illustrates operational energy use by both service and mission and by regional command for fiscal year 2014. In that year, total DOD operational consumption was 87.4 million barrels of fuel "enterprise-wide to deploy and sustain worldwide missions" at a cost of $14 billion. Aircraft fuel consumption by all services accounted for more than 70 percent of operational energy use. The bar on the right attributes operational fuel use by regional command. NORTHCOM (which is responsible for Continental United States, Mexico, and Canada) consumed the most operational fuel that year, 35 percent. Central Command (CENTCOM) whose area of responsibility includes the Middle East from Egypt through Pakistan, as far north as

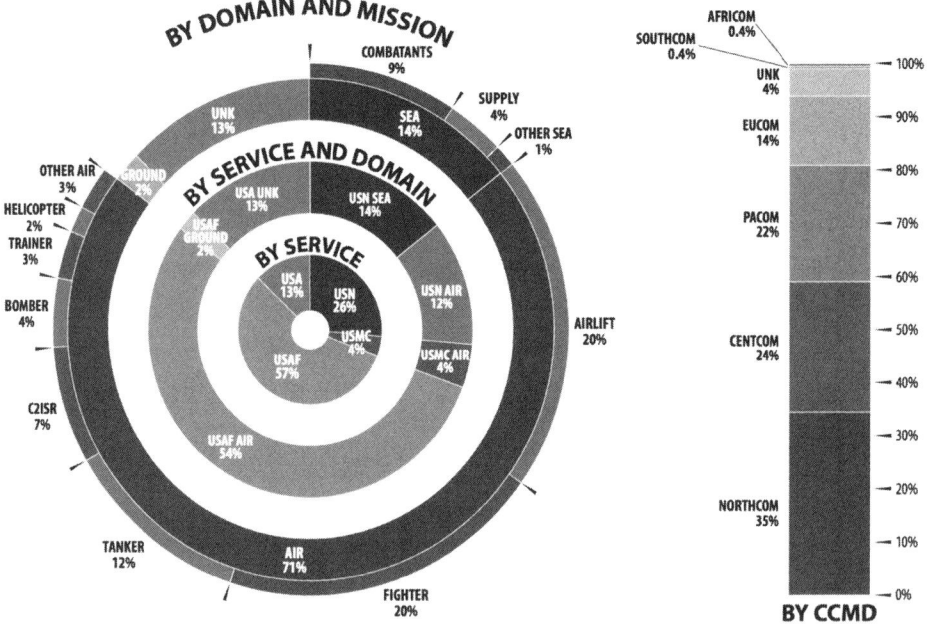

Figure 4.10
Operational energy use by service, domain, and mission, FY2014.
Source: Department of Defense, "2016 Operational Energy Strategy," 4, https://www.acq
.osd.mil/eie/Downloads/OE/2016%20OE%20Strategy_WEBd.pdf.

Kazakhstan, and as far South as Yemen, consumed 24 percent of operational energy that year. In 2014, the United States had reduced its forces in Afghanistan, and begun its war against ISIS in Syria, which started in August 2014 with air strikes. Note that Pacific Command (PACOM, renamed Indo-Pacific Command in 2021) accounted for nearly as much operational energy use as CENTCOM, while Africa Command (AFRICOM) and Southern Command (SOUTHCOM) together were less than 1 percent of operational fuel use.

It is also important to note that the Department of Defense could not account for 13 percent of its total operational energy consumption in fiscal year 2014. In other words, the use of approximately 11.4 million barrels of fuel, worth about $4.2 billion dollars in fiscal year 2014 was unknown. And it is obviously not possible to say with any degree of specificity why the fate of millions of barrels of fuel is "unknown"—portions of it could have been stolen, destroyed in transit in Pakistan, leaked from storage and pipelines, or simply misplaced.

MAJOR TRENDS IN MILITARY EMISSIONS

Pentagon greenhouse gas emissions correlate with fuel use during hot war and cold wars. When the United States is at war or engaged in exercises and war games in preparation for war, emissions are higher than when the U.S. military is relatively less mobilized, in what the military calls "peacetime." In 1975, I estimate that U.S. military greenhouse gas emissions were just over 109 million metric tons of CO_2 equivalent (MMTCO$_2$e).[50] Just one metric ton of CO_2e emissions is equivalent to consuming about 113 gallons of gasoline, which would allow the average passenger vehicle of 2021 to drive 2,513 miles. A million metric tons of CO_2e emissions are equivalent to consuming 112.5 million gallons of gas allowing 217,000 passenger vehicles to drive for one year. Emissions of 109 million metric tons would be the equivalent of the emissions of 23,705,310 passenger vehicles driving a year or the emissions that result from powering 13.12 million homes for a year.[51]

After leaving Vietnam, the total energy consumed by the DOD declined from 1975 until 1979. In the last year of the Carter administration and during the Reagan military buildup of the 1980s, fuel use and therefore greenhouse gas emissions rose. Following the end of the Cold War in 1989, military emissions briefly declined, but I estimate that they reached a peak of about 110 MMTCO$_2$e during the 1991 Gulf War that, as I described in chapter 2, relied heavily on airpower. Thus, U.S. military emissions in 1991 were the equivalent of burning nearly 255 million barrels of oil, roughly equivalent to the emissions of 23.9 million passenger vehicles driving a year, or enough energy to power 13.24 million homes for a year. After the 1991 Gulf War, there was not only a reduction in military spending, but also a significant reduction in greenhouse gas emissions through the 1990s. U.S. military energy consumption and greenhouse gas emissions declined to 62 million metric tons of CO$_2$e in 2000.

U.S. military fuel use and hence greenhouse gas emissions grew again after the terrorist attacks on New York and Washington on September 11, 2001. The United States began major air strikes on October 7, 2001 in Afghanistan and then a ground war and occupation of Afghanistan with its allies. U.S. fuel consumption increased again when the United States invaded Iraq in March 2003, and again, this war began with an air strike campaign, described by its architects as "shock and awe." In 2005, total U.S. military energy consumption hit its highest level in a decade. Only in fiscal year 2014 did DOD energy consumption return to the level it was in 2000. The headline from the Energy Information Administration announcing the transition stated, "Defense Department Energy Use Falls to Lowest Level Since 1975."[52] Total military emissions during the wars in Afghanistan and Iraq occurred over two decades and correlate with the phases of operations in both war zones, tracking troop surges, the use of airpower, and withdrawals.

Figure 4.11 illustrates estimated and reported scope 1 and 2 Pentagon greenhouse gas emissions of fiscal years 1975–2020, where the solid line indicates the Department of Energy and DOD reported for DOD emissions, and the dashed line is my estimate of total emissions, based on DOE fuel consumption data. The estimated total cumulative scope 1 and 2 DOD emissions for the period of fiscal years 1975–2020 are 3,789 MMTCO$_2$e, averaging 82 MMTCO$_2$e annually.

The officially reported and estimated DOD emissions are conservative. In my own estimates, I am not counting all U.S. Department of Defense

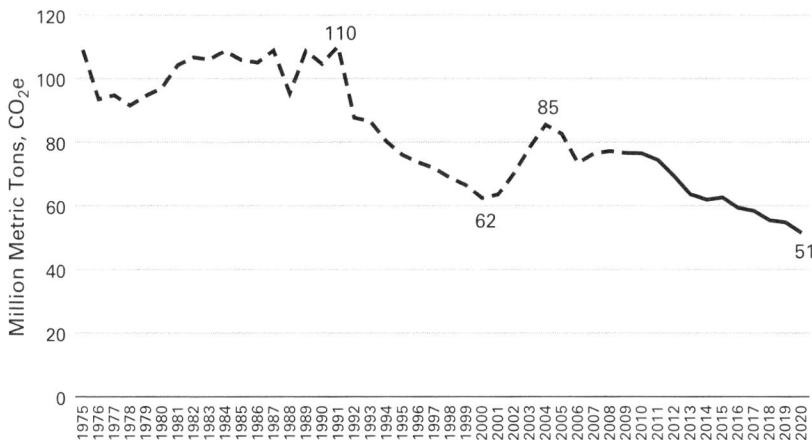

Figure 4.11

Estimated and reported DOD greenhouse gas emissions, million metric tons of CO_2e, FY1975–2020.

Source: For FY2008 and FY2010–2019, see Department of Energy figures, Comprehensive Annual Energy Data and Sustainability Performance annual reports. Emissions estimates for FY1975–2007 and FY2009 are calculated from Department of Energy data for DOD fuel consumption using EPA emissions factors, https://ctsedwweb.ee.doe.gov/Annual/Report /Report.aspx. See the appendix for a discussion of methods.

greenhouse gas emissions because, for example, I do not estimate military biogenic emissions; nor do I estimate the emissions that are associated with destruction and reconstruction. Further, I have not included scope 3 emissions in this estimate. Recall that the DOE, for a time, reported some scope 3 emissions for federal agencies. The categories it included in that accounting were transportation and distribution losses, employee air and ground business travel, employee commuting, and contracted wastewater treatment and municipal solid waste disposal. For the years it enumerated these specific DOD scope 3 emissions, for fiscal years 2008 and 2010–2016, they averaged 7.4 MMTCO$_2$e annually.[53] If the ratio for the scope 1 and 2 emissions to scope 3 emissions holds, then total DOD scope 1, 2, and the partial accounting of scope 3 emissions (which only includes transportation and distribution losses, employee air and ground business travel, employee commuting, and contracted wastewater treatment and municipal solid waste disposal) for each year are perhaps 10–11 percent higher than what I have estimated.

As noted earlier, military vehicle fuel consumption fairly closely tracks operations in war, meaning that greenhouse gas emissions increase and decrease with activities in a war zone. During fiscal years 1975–1990, the end of the Cold War, annual emissions averaged 102 $MMTCO_2e$. When the Cold War ended, and just before the massive mobilization of the 1991 Gulf War, U.S. military fuel consumption declined and thus so did military greenhouse gas emissions. DOD emissions peaked during the Gulf War at 110 $MMTCO_2e$. In the period following the Iraq War, fiscal years 1991–2001, emissions reached their lowest point in 2000, at 62 $MMTCO_2e$ and the annual DOD emissions averaged 74 $MMTCO_2e$ (including the first few weeks of the war in Afghanistan). During the period of the post-9/11 wars, emissions peaked at 84 $MMTCO_2e$ in 2004 and from fiscal years 2002–2020, total DOD emissions annually averaged about 69 $MMTCO_2e$. Table 4. 3 shows the emissions of the military services in metric tons. While installation emissions were consistently higher for the U.S. Army, operational emissions were consistently higher for the U.S. Air Force than the other services, averaging more than 50 percent of total operational emissions for this period.

Of course, the major takeaway from table 4.3 is the fact that the U.S. military's emissions have significantly declined in recent years, as U.S. operations in Afghanistan and Iraq have declined. The U.S. Army has had the greatest decline—54 percent—in operational energy use between fiscal years 2010 and 2019, while overall, DOD operational energy use has declined 28 percent. The reason that total DOD scope 1 and 2 emissions so closely track war is that operational emissions are about 70 percent of total DOD emissions, and jet and diesel fuel consumption comprise the largest components of operational emissions. The United States has, since the end of the Vietnam War, tended to emphasize the use of airpower in comparison to ground troops, and jet fuel emissions more closely track major U.S. military operations than does diesel fuel. This is so even though, of the main fuels used for vehicle and equipment operations, diesel and Navy Special Fuel are the most greenhouse gas intensive, meaning that for every gallon burned, a greater quantity of greenhouse gases are emitted than for other fuels.[54] Diesel fuel emissions fluctuated during the post-9/11 wars, from a low in 2004 of 1.3 million metric tons of CO_2e, to a peak of 11.3 million metric tons in

Table 4.3

U.S. military service branch scope 1 and 2 emissions, $MTCO_2e$, installation and operational emissions for fiscal years 2010–2019

Department of the Air Force Emissions ($MTCO_2e$)	2010	2011	2012	2013	2014	2015	2016	2017	2018	2019
Installation	7,818,302	7,511,312	7,438,053	7,177,727	7,363,524	7,139,002	6,813,348	6,558,246	6,319,535	6,264,132
Operational	26,592,638	25,495,869	23,742,258	20,785,066	20,739,129	22,444,204	21,099,472	20,903,085	19,187,188	18,692,165
Total Air Force Emissions	34,410,939	33,007,182	31,180,311	27,962,794	28,102,652	29,583,206	27,912,821	27,461,331	25,506,723	24,956,296
Department of the Army ($MTCO_2e$)	**2010**	**2011**	**2012**	**2013**	**2014**	**2015**	**2016**	**2017**	**2018**	**2019**
Installation	10,677,212	9,756,749	9,355,221	8,905,694	8,854,049	8,844,495	8,126,767	7,833,821	7,290,688	7,107,834
Operational	7,713,463	8,265,408	6,634,458	5,218,663	4,122,266	2,991,687	2,856,756	3,042,156	3,383,887	3,531,597
Total Army Emissions	18,390,675	18,022,157	15,989,679	14,124,357	12,976,316	11,836,182	10,983,524	10,875,978	10,674,575	10,639,431
Department of the Navy Emissions ($MTCO_2e$)	**2010**	**2011**	**2012**	**2013**	**2014**	**2015**	**2016**	**2017**	**2018**	**2019**
Installation	6,182,905	6,286,391	6,100,351	6,141,043	6,039,954	6,117,342	5,477,751	5,384,375	5,954,879	5,786,730
Operational	14,603,433	14,677,602	14,388,143	13,564,831	13,069,283	13,209,893	13,114,180	13,208,457	11,886,933	11,896,774
Total Navy Emissions	20,786,338	20,963,994	20,488,494	19,705,874	19,109,238	19,327,234	18,591,931	18,592,832	17,841,811	17,683,505

Source: Office of the Assistant Secretary of Defense for Sustainment, "Report on Greenhouse Gas Emission Levels" (Department of Defense, August 2021), appendix A.

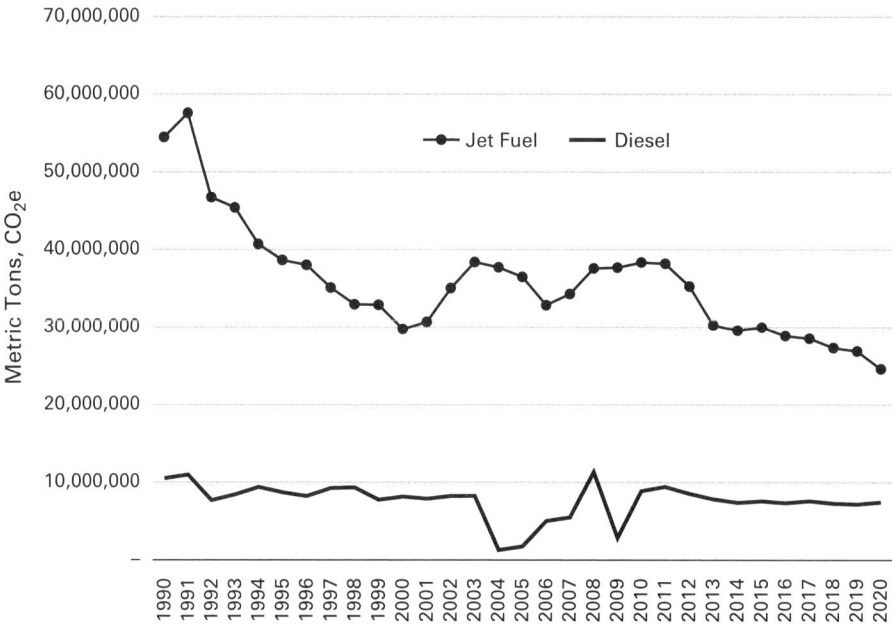

Figure 4.12

Military jet fuel and diesel emissions, 1990–2020, million metric tons of CO_2e.

Source and method: Fuel Consumption Data from the Department of Energy, https://ctsedwweb.ee.doe.gov/Annual/Report/HistoricalFederalEnergyConsumptionDataByAgencyAndEnergyTypeFY1975ToPresent.aspx. Emissions calculated using Department of Energy Emissions Factors, https://www.energy.gov/eere/femp/downloads/annual-energy-management-data-report.

2008. As figure 4.12 illustrates, after a steady decline in the post-Cold War/post–Gulf War period, jet fuel consumption and greenhouse gas emissions increased dramatically after 9/11 and generally tracked surges in troop numbers and operations in the post-9/11 wars.

Jet fuel emissions are also so much higher than diesel fuel emissions because even though the military's diesel vehicles are not what would be considered fuel efficient in the civilian world, they are models of fuel economy in comparison to jet aircraft. The tactical vehicles that use diesel fuel in the military get several miles to the gallon. For example, the approximately sixty thousand HUMVEEs remaining in the U.S. Army fleet in 2017 consumed between four to eight miles per gallon of diesel fuel and those were being phased out in favor of more fuel-efficient vehicles.[55] By contrast, the fuel economy of military jet aircraft is measured in gallons per mile. For purposes of illustration, table 4.4 shows the capacities and estimated relative fuel efficiency, measured in gallons per nautical mile, of several U.S. Air Force aircraft and estimated greenhouse gas emissions without aerial refueling.[56]

However, there is another factor that generally is not included in most discussions of aircraft emissions: the warming potential of burning fossil fuels high in the atmosphere. Specifically, the soot formed when jet fuel is burned at high altitudes produces water vapor condensation trails (contrails) which have a larger radiative forcing effect than the CO_2 alone. Contrails, line-shaped ice clouds, are produced when aircraft fly at altitudes between about 26,000 and 42,000 feet (8–13 km)—altitudes where many U.S. aircraft operate.[57] For

Table 4.4
Estimated U.S. military aircraft jet fuel consumption and CO_2 emissions

Aircraft	Mission	Internal fuel capacity in pounds and gallons*	Range in nautical miles on internal fuel	Fuel consumption in gallons per nautical mile	Maximum estimated metric tons of CO_2e without aerial refueling**
B-2	Bomber	167,000 lbs/ 25,692 gal	6,000	4.28 gallons/mile	251.4 metric tons
F-35A (CTOL)	Fighter bomber	18,499 lbs/ 2,846 gal	1,199	2.37 gallons/mile	27.8 metric tons
A-10	Close air support	11,000 lbs/ 1,692 gal	500	3.38 gallons/mile	17.5 metric tons
KC-135R	Refueling tanker	50,000 lbs/ 7,692 gal	1,500 (loaded with 150,000 lbs of transfer fuel)	4.9 gallons/mile	75.3 metric tons
KC-46A***	Refueling tanker and cargo	*Estimated 16,000 gal*	6,385 (loaded with 210,000 lbs of transfer fuel)	*Estimated 2.9 gallons/mile*	156.5 metric tons

Source: Calculated by the author from data about each aircraft using Department of Energy Emission Factors, https://www.energy.gov/eere/femp/downloads/annual-energy-management-data-report.

*Assuming each pound of jet fuel weighs an average of 6.5 pounds.

**See the appendix. Not including warming effects of water vapor.

***The KC-46A can refuel itself. Boeing has not released data on its internal fuel capacity. The estimate here for fuel capacity and consumption is based on the Boeing 767–400ER range and fuel capacity.

example, B-2 bombers can operate at all altitudes up to 50,000 feet, F-35s up to about 45,000 feet, and B-52s operate at about 35,000 feet.[58]

In sum, it is not surprising that, given the sheer volume of jet fuel consumed in operations, and the United States military's preference for using airpower as much as possible, jet fuel emissions, on average, comprised nearly 50 percent of estimated total military emissions from 1975 to 2019, peaking at 58 million metric tons of CO_2e in 1991, and during the post-9/11 wars at 38 million metric tons in 2003 and 2008 through 2011, the years of most intense air strikes and air lift in those wars.[59] The intense air activity associated with the U.S. withdrawal from Afghanistan in 2021 may also correspond with a spike in military jet greenhouse gas emissions.

Reasons for the Decline in Total Military Fuel Use and Emissions since 1975

While greenhouse gas emissions tend to fluctuate with war and with phases of greater or lesser military activity within wars, overall, U.S. military greenhouse gas emissions have significantly declined as U.S. military fossil fuel uses declined from their post–Cold War peak in 1991. In May 2001, just months before the 9/11 attack, the Defense Science Board noted the downward trend in U.S. military emissions. "Nationally, greenhouse gas emissions have risen from 1,650 million metric tons of carbon in 1990, the base year under the UNFCCC, to 1,835 million metric tons of carbon in 1998, an increase of about 11 percent. Conversely, DoD's emissions have declined by about 20 percent since 1990. There are two primary reasons for this: 1990 was a high consumption year due to Desert Storm and DoD has subsequently downsized its force structure and consolidated a number of installations."[60] The active-duty military has downsized compared to its Vietnam War peak in 1968 of 3.5 million troops. The military moved to an all-volunteer force in 1973, grew during the 1980s, and then significantly declined in the post–Cold War period. Further, there was a decline in the number of troops deployed overseas in the 1990s. In 1989, before the end of the Cold War, there were about 1,600 U.S. military bases all over the world, including a large conventional and nuclear military presence in Europe that was meant to deter the Soviet Union and the Warsaw Pact.[61] The number of U.S. troops deployed in Europe during the Cold War late 1980s, about 340,000, dropped to about 74,000 in 2020.[62]

After the end of the Cold War many U.S. military bases and installations were closed or downsized. Specifically, the number of U.S. military bases were reduced in successive waves of Base Realignment and Closure (BRAC) from 1988 through 2005. The downsizing that the Defense Science Board noted in early 2001 was partially offset by the increased military mobilization and operations following the 9/11 attacks on Washington, DC, and New York.[63] After 9/11, the number of active duty military increased, but then started to decline as the intensity of the major U.S. wars in Afghanistan and Iraq decreased. But, overall, there has been a steady decline in the number of U.S. bases and structures. The most recent BRAC process, from 2005 to 2011, led to an overall decline from over 600,000 individual buildings and structures located on more than 30 million acres of land before the BRAC in fiscal year 2003 to about 585,800 buildings and structures on 26.9 million acres in fiscal year 2018.[64] Thus, energy consumption at military facilities declined more than 50 percent, with the biggest declines in the 1990s. Emissions fell at the remaining installations because the mix of fuels used at facilities changed, toward less greenhouse gas-intensive fuels.

In addition to the overall downsizing of the U.S. armed forces and the number of bases, the decline in emissions also correlates with, and was likely caused by several additional changes. First, as noted, overall, the mix of fuels for U.S. operational forces changed. Since the end of the Cold War, the United States has retired and decommissioned eight fossil fueled nonnuclear aircraft carriers, so that the navy's entire fleet of eleven aircraft carriers is nuclear powered. Yet, as Christopher McMahon notes, the nuclear navy still requires a great deal of petroleum fuel. "Yes, today's nuclear-powered carriers and submarines can steam for decades without refueling, but gas turbine–powered destroyers and cruisers," which travel along with the aircraft carriers, "require fuel nearly as often as coal-burning steamships did, and much more often if they are engaged in combat operations."[65] Each aircraft carrier travels with conventionally powered frigates, destroyers, logistics, and supply ships. Further, as McMahon notes, the aircraft operating from aircraft carriers need to be refueled even if a nuclear aircraft carrier itself does not need refueling

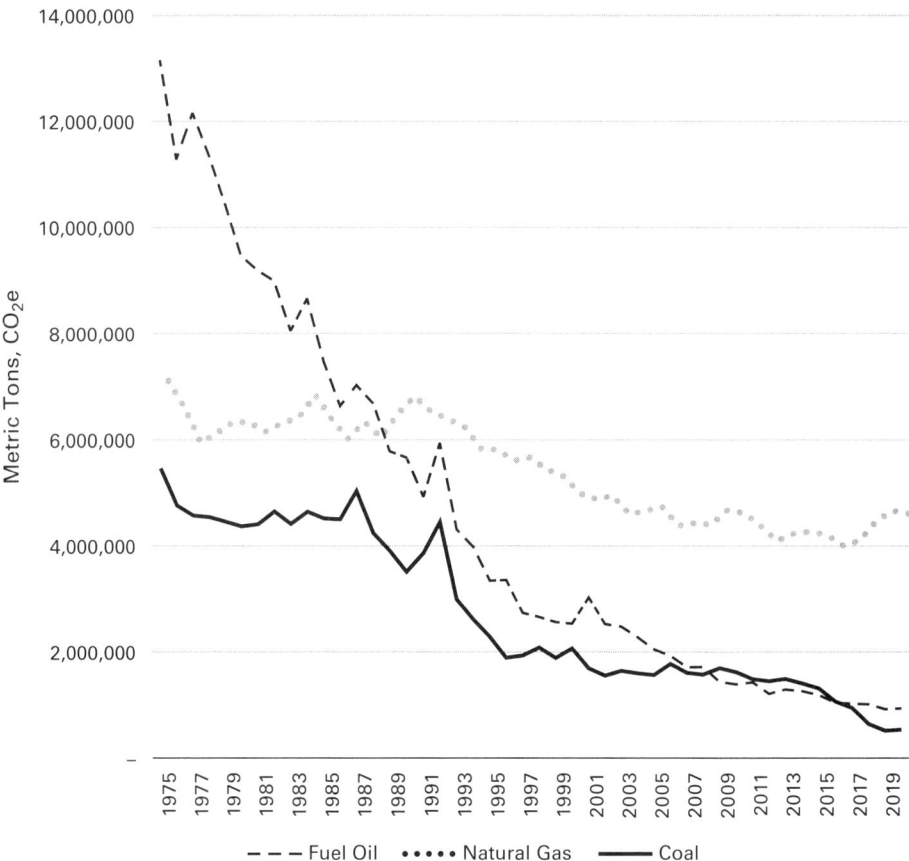

Figure 4.13

Greenhouse gas emissions of fuel oil, natural gas, and coal at military facilities, 1975–2020 in metric tons, CO_2e.

Source: Calculated by the author from Department of Energy, Comprehensive Annual Energy Data and Sustainability Performance, https://ctsedwweb.ee.doe.gov/Annual/Report/Historical FederalEnergyConsumptionDataByAgencyAndEnergyTypeFY1975ToPresent.aspx.

while at sea: "maintaining combat air operations requires a carrier to replenish jet fuel at least every five days." In fact, McMahan observed, there has not been that much improvement since the 1991 Gulf War: "During Operation DESERT STORM, the conventionally powered carriers in the Persian Gulf replenished aviation fuel every 2.7 to 3 days. USS *Roosevelt*, the only nuclear-powered carrier in the DESERT STORM air campaign that was also operating in the Persian Gulf, replenished its aviation fuel about every 3.3 days."[66]

Indeed, he argues, "Navy combatants today are as dependent on logistics ships as their predecessors were during the voyage of the Great White Fleet."[67]

Further, the DOD has become more efficient with the fuels they use. Thus, when weapon modernization occurred, it was sometimes the case that although the same fuel was used, less efficient vehicles and weapons were replaced with those that were more efficient. For example, the newer F-35 and KC-46 aircraft, as described in table 4.2 are much more fuel efficient than previous generations of aircraft. The Pentagon has, even during the wars in Iraq and Afghanistan, as I will discuss in greater detail, started to emphasize operational changes as a way to increase efficiency and thus to decrease their dependence on fuel in war zones. For instance, aircraft can decrease total consumption of fuel through operational efficiencies such as reducing engine idling time and changing flight profiles.[68] Further, the substitution of drones for piloted aircraft has likely led to decreased fuel consumption overall since remotely piloted vehicles are more fuel efficient than piloted aircraft engaged, for example, in the same hours of high-altitude reconnaissance.

In addition, while it is difficult to track the total number of U.S. military exercises with allies, and to gauge their scale, the United States has conducted fewer and smaller military exercises in the post–Cold War era. In 1982, NATO held eighty-two major exercises in West Germany alone. Operation REFORGER, for example was held annually between 1969 and 1993. In 1987, REFORGER moved 115,000 troops from six NATO members to West Germany's border with East Germany. Exercises are generally much smaller in scale than during the Cold War.[69] While the total number of multilateral exercises and their size are difficult to determine, in recent years, the number of NATO exercises seems to have declined. In 2016, NATO planned 240 military exercises. In 2017, NATO conducted 108 military exercises, in 2018, 103 exercises, and NATO planned to conduct 102 exercises in 2019.[70] In 2020, NATO conducted 88 of the 113 military exercises that it had scheduled, with the reduction and modification of exercises due to the COVID-19 pandemic.

MILITARY-INDUSTRIAL GREENHOUSE GAS EMISSIONS

Military industry is, arguably, one of the leading industrial sectors in the United States. The military-industrial sector manufactures real goods such as the specialized vehicles, weapons, and equipment that the armed forces need to do their jobs. The defense industry includes the obvious producers of large equipment and weapons—aircraft, tanks, ships, missiles, and bombs—and components large and small, from satellites to microelectronics, which enable the U.S. military to function as it does. Some of that material can be purchased off the shelf, and it is. But many of the inputs for the military, including software, must be made especially for the military to meet high levels of performance, durability, and reliability, with materials specifically engineered for the task.

A complete accounting of the total greenhouse gas emissions related to war, and preparation for it, would include the GHG emissions of military industry—also called the defense or arms industry.[71] It also includes the private construction industry associated with building bases, runways, and ports. The defense industry is a part of, but also distinct from, military contractors who supply services to the military such as fuel deliveries, equipment maintenance, food preparation, prisoner interrogation, and base security.

As I noted earlier, military industry has, since the nineteenth century, when the manufacturing of arms and vehicles moved from artisans to factories, been increasingly energy intensive. As weapons systems have become more specialized, powerful, and complex the requirements for weapons and weapons platforms have become more rigorous and standardized. Further, the military technology and manufacturing sector has often "spun off" its systems and technologies to the civilian sector after wars (e.g., radar), and military surplus weapons and equipment have been sold or donated to the civilian sector (e.g., airplanes and jeeps). And, while it is true that the size of the arms industry waxes as the United States mobilizes and goes to war, and wanes after wars end, U.S. military industry has often been the world's arms supplier of choice—from small arms to sophisticated aircraft and missiles—and thus the arms industry is a relatively stable feature of the U.S. economy even when the United States is not at war.

Military industry is an essential part of the United States military capability. Each year, the U.S. Department of Defense outlines priorities for the military services and the department as a whole. For example, in fiscal year 2021, the Congress appropriated to the DOD $141.67 billion for military procurement and $106.45 billion for military research, development, test, and evaluation (RDT&E), both increases over in the previous year when the budgets were, respectively, $143.8 billion for procurement and $104.5 billion for RDT&E.[72] The procurement budget includes new advanced aircraft already being built and delivered, such as the F-35, the B-21 Raider Strategic Bomber, which is under development, new nuclear weapons designs, and large ships such as aircraft carriers, destroyers, frigates, and submarines, which take many years to build. Further, $8.5 billion was appropriated for military construction in fiscal year 2021. The construction budget includes both military construction in the United States and abroad, for which, since the 1980s, the DOD is legally obliged, in most instances, to give preference to a U.S. contractor.[73]

Louis Uchitelle writing in the *New York Times* estimated in 2017 that "roughly 10 percent of the $2.2 trillion in factory output in the United States goes into the production of weapons sold mainly to the Defense Department for use by the armed forces."[74] Industrial manufacturing accounted for 373.7 MMTCO$_2$e greenhouse gas emissions in 2019.[75] If military industry is about 10 percent of all U.S. manufacturing output, and 10 percent the emissions associated with the entire U.S. manufacturing sector are attributable to U.S. military industry, then military-industrial emissions would be about 37 MMT CO$_2$e. But using the dollar value of factory output to extrapolate emissions could yield an overestimate of emissions if the military-industrial sector is more capital intensive than other manufacturing.

In this section, I provide a rough estimate of military-industrial emissions using two alternative methods. The first method estimates military-industrial emissions by extrapolation, assuming that military-industrial emissions are correlated with their portion of the U.S. manufacturing economy. In this method, the number of people employed in that sector is used as an indicator of the share of military-industrial employment and it is assumed that military industry has the same greenhouse gas intensity as other manufacturing. Using the employment method, I estimate that in 2019, military industry accounted for between 27 million metric tons of CO$_2$e (based on direct) employment and 64 million metric tons of CO$_2$e (based on direct and indirect employment) that year. The second method totals what is known about the military emissions of the largest military-industrial corporations.

I found that the direct (scope 1 and 2 emissions) for the twelve largest corporations accounted for more than 5.2 million metric tons of CO_2e in 2019, and if estimated supply chain emissions are included, the total is nearly sixty-three million metric tons of CO_2e.

Employment as a Proxy for the Size of the U.S. Military-Industrial Economy and Emissions

One could try to estimate the greenhouse gas emissions associated with military-industrial manufacturing based on the proportion of the total number of people employed in manufacturing who work in military industry. Unsurprisingly, there are different estimates of the number of people working in military industry. In 2017, Deloitte reported that the defense and aerospace sector of the U.S. economy employed directly and indirectly about 4.1 million people in 2016, a decline of about 344,600 workers since 2011. In 2016, Deloitte reported 839,171 people worked directly in U.S. defense industries.[76] The Aerospace Industries Association (AIA) said that out of the 2.5 million jobs "up and down the supply chain" in defense and aerospace in 2018 there were 881,575 direct aerospace and defense industry jobs, and 370,084 of those were in Defense and National Security in 2018.[77]

According to the Bureau of Labor Statistics, there were about 12,839,500 jobs in the U.S. manufacturing sector in 2019. Using data from the Aerospace Industries Association (see table 4.5), this suggests that end-use aerospace and defense industry employment in 2019 was about seven percent of total U.S. manufacturing employment. If one includes both end use and supply chain

Table 4.5
Aerospace and defense industry employment 2016–2019

Year	End-use	Supply chain	Total
2016	851,582	1,175,207	2,027,789
2017	851,488	1,173,191	2,024,679
2018	880,662	1,216,153	2,096,814
2019	922,619	1,275,100	2,197,719

Source: Data from Aerospace Industries Association (AIA), "2020 Facts and Figures, U.S. Aerospace and Defense," 3, http://aiafactstg.wpengine.com/. Total includes aeronautics and aircraft, space, land and sea systems, and cyber.

employment, the aerospace and defense industry jobs account for about 17 percent of U.S. manufacturing employment.[78] Thus, if aerospace and defense industry emissions represented between 7.0 and 17.0 percent in 2019 of U.S. manufacturing emissions, which were 373.7 MMTCO$_2$e in 2019, the defense and aerospace industry contributed between 26 and 63 MMTCO$_2$e emissions that year.[79]

However, there are a couple of caveats to keep in mind. If the defense-related manufacturing has been disproportionately affected by automation, this would yield an underestimation of the size of the military-industrial economy as reflected by the numbers employed in the sector. In making this estimate, I also assumed that U.S. military-industrial emissions are "typical" of other industrial emissions—in other words, that military industry is no more greenhouse gas intensive than, say, automobile or aircraft manufacturing. However, military manufacturing is likely to be more greenhouse gas intensive than other manufacturing because of the scale of the buildings involved in final assembly, the number of components that are often involved, and the other characteristics of producing sophisticated pieces of machinery, using metals and components that are intended to function with high levels of reliability and operate in extreme environments. Stuart Parkinson found that in the United Kingdom, "companies with higher proportions of military sales tend to have significantly higher emissions per employee. This indicates the more capital-intensive nature of military work."[80] Of course, a much more detailed analysis of the greenhouse gas emissions of the military-industrial sector is required. Thus, this estimate may be conservative, since some military-industrial applications (such as armored vehicle and jet aircraft manufacturing) will likely, on average, be more greenhouse gas intensive than other forms of manufacturing.

The Emissions of Large Military-Industrial Corporations

A second way to estimate the scale of U.S. military-industrial greenhouse gas emissions would be to total all the military-industrial manufacturer's emissions and determine which portion of those were military emissions—produced as part of the process of making weapons. There are hundreds of companies that supply the U.S. military with nearly all or some portion of their industrial output and which also export weapons to foreign governments or in commercial transactions. The State Department reports that total "authorized arms exports (commercial and government-managed) rose by 2.8 percent from $170.09 billion to $175.08 billion" from fiscal year 2019 to 2020. The State Department reports that for fiscal year 2019 total government-to-government foreign military sales were about $55 billon and commercial sales were $114.7 billion. In fiscal year 2020 total government-to-government foreign military sales were about $55 billion and commercial sales were $124.3 billion.[81]

I focus here on the emissions of the largest military-industrial companies. Although companies change as they merge and recombine, the United States has about thirty companies in the Stockholm International Peace Research Institute (SIPRI) list of the world's top 100 military-industrial corporations. Table 4.6 lists the top companies in the United States in 2018 and 2019, including recent mergers and the portion of their sales that is military related. According to SIPRI data the twelve largest U.S. arms manufacturers sold $221.2 billion in 2019. The biggest arms corporations in the United States focus on the U.S. market, but depending on the company, some portion of U.S. military-industrial output is exported.

I used the top military-industrial corporations' emissions reports for recent years to estimate the emissions of the top corporations in the U.S. military-industrial sector. Many large military industries have started reporting their scope 1, 2, and 3 emissions in annual statements, and some aerospace and defense companies have made detailed submissions describing their emissions and offsets to the CDP, an independent nonprofit organization that collects and reports city and company self-reports on environment and governance.[82] Recall that scope 1 emissions are those that the company directly produces through its operations; scope 2 emissions are from purchased energy; and scope 3 emissions are from inputs into a company, including all employee business travel, travel to and from work, the material inputs that the company purchases, and its leased assets. Scope 3 emissions document the emissions from the supply chain that in the case of weapons manufacturers can consist of hundreds of small companies that contribute to the larger company's end product. The overall trend among the

Table 4.6

Top U.S. military contractors and arms sales in millions of dollars as a portion of total sales in 2018 and 2019

U.S. rank	World rank 2018	World rank 2019	Company	$ Millions in arms sales (2018)	$ Millions in total sales (2018)	Arms sales as a % of total sales (2018)	$ Millions in arms sales (2019)	$ Millions in total sales (2019)	Arms sales as a % of total sales (2019)
1	1	1	Lockheed Martin Corp.	47,260	53,762	88	53,230	59,812	89
2	2	2	Boeing	29,150	101,126	29	33,580	76,559	44
3	3	3	Northrop Grumman Corp.	26,190	30,095	87	29,220	33,841	86
4	4	4	Raytheon*	23,440	27,058	87	25,320	29,176	87
5	5	5	General Dynamics Corp.	22,000	36,193	61	24,500	39,350	62
6	-	10	L3 Harris Technologies**				13,920	18,074	77
7	14	11	United Technologies Corp.*	9,310	66,501	14	13,100	77,046	17
8	12	-	L3 Technologies	8,250	10,244	81			
9	16	16	Huntington Ingalls Industries	7,200	8,176	88	7,740	8,899	87
10	18	18	Honeywell International	5,430	41,802	13	5,330	36,709	15
11	19	19	Leidos	5,000	10,194	49	5,330	11,094	48
12	17	-	Harris Corp.	4,970	6,801	73			
13	20	22	Booz Allen Hamilton	4,680	6,704	70	5,140	7,464	69
14	28	21	General Electric	3,650	121,615	3	4,760	95,200	5

Source: Data from Stockholm International Peace Research Institute, "SIPRI Arms Industry Database," https://www.sipri.org/databases/armsindustry, accessed July 1, 2021.

*Raytheon and United Technologies Merged in 2019.

**L3 Harris Technologies is the result of a merger between Harris Corp. and L3 Technologies. Its arms sales figure for 2018 is "pro forma," i.e., it is the combined 2018 arms sales of Harris Corp. and L3 Technologies.

top companies in military-industrial greenhouse gas emissions has recently been to reduce their CO_2 emissions as they switch to fuels that produce less greenhouse gas, install their own solar arrays, and achieve improved energy efficiency in their buildings and operation. They have also achieved reductions by purchasing renewable energy credits.

Consider the case of Lockheed Martin. Lockheed's first military ventures were in aircraft used for commercial and military purposes. Lockheed Martin describes itself today as a "global security, innovation, and aerospace" company. It continues to make fixed-wing aircraft, but it also makes helicopters, unmanned "drones," ships, submersibles, armored vehicles, bombs, missiles, radar, and software.[83] The company employed about 110,000 people at 375 locations in 2019, mainly in the United States, and its largest customer is the U.S. DOD and other U.S. federal government agencies. Arms sales totaled 88 and 89 percent of Lockheed Martin's total sales in 2018 and 2019. The other major world arms suppliers based in the United States that derive more than 85 percent of their sales revenue from arms sales are Northrop Grumman, Raytheon, and Huntington Ingalls.

Table 4.7 summarize the emissions profile of Lockheed Martin, to the extent that I could determine, based on public records.[84] Their company-wide scope 1 and 2 emissions were a total of 1,271,358 metric tons of CO_2e from November 2009 through October 2010. That year they purchased

Table 4.7
Total Lockheed Martin emissions and RECs, in metric tons CO_2e (MTCO$_2$e)

Emissions	2010	2016	2017	2018	2019	2020
Scope 1	346,734			291,782	305,362	
Scope 2	1,096,826			521,766	662,659	
Total scope 1 and 2	1,271,358	918,635	844,373	819,548	968,021	790,535
RECs*	*172,202*				*172,292*	
Scope 3		30,644,500				

Source: Lockheed Martin corporate reports: *Climate Change 2020*, CDP, and *2020 Sustainability Report: Propelled by Principle*, https://sustainability.lockheedmartin.com/sustainability/content/Lockheed_Martin_2020_Sustainability_Report.pdf.

*RECs are renewable energy credits that count as a reduction in total emissions. Lockheed Martin used location-based accounting for scope 2 emissions.

renewable energy credits (RECs) allowing them to reduce their total emissions count by 172,292 metric tons of CO_2e.[85] From 2014 though 2020, Lockheed Martin increased the share of renewable energy from 16 to 21 percent of its total electricity use. From 2018 to 2019 the corporation reported a 6 percent reduction in emissions through purchases of renewable energy and initiatives that improved efficiency and reduced consumption.[86]

Figure 4.14 shows the total scope 1 and 2 greenhouse gas emissions reported from 2010 through 2020 by Lockheed Martin, Northrop Grumman, and General Dynamics.[87] In most years, these companies were able to reduce their emissions through a combination of measures including the purchase of renewable energy. And they are not unique: all the major arms manufacturers appear to be reducing their greenhouse gas emissions.

Further, not all of these company's emissions can be attributed to military-related production. Some of these companies, like Honeywell International

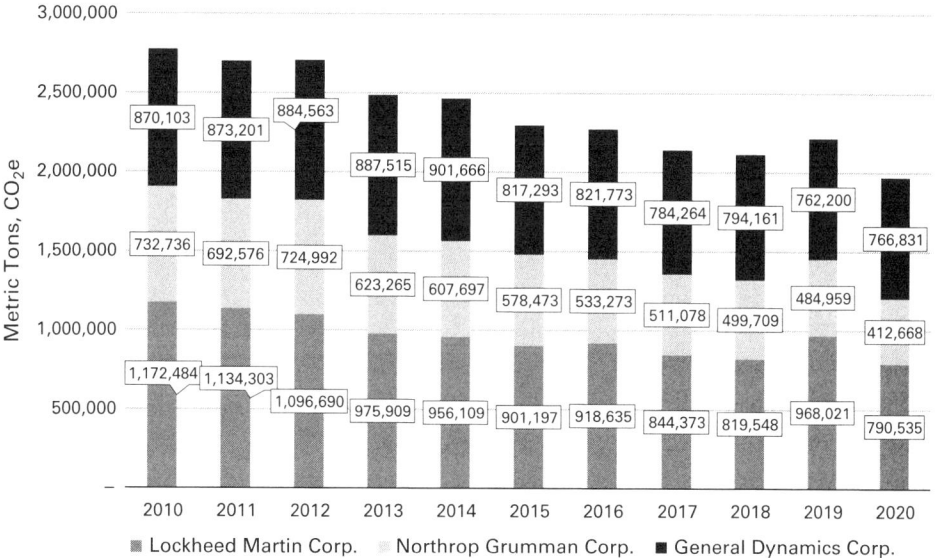

Figure 4.14

Total Scope 1 and 2 metric tons of CO_2e emissions of Lockheed Martin, Northrop Grumman, and General Dynamics, 2010–2020.

Sources: Corporate sustainability reports by Lockheed Martin, Northrop Grumman, and General Dynamics, and climate change submissions to CDP, https://www.cdp.net/en.

and General Electric are so large that even though their military sales are high, their civilian production dwarfs their military production. Lockheed Martin, Northrop Grumman, and Raytheon were the most dependent on their military-related sales. The world's second-largest defense manufacturer, Boeing, with large civilian aircraft sales, is much less dependent on weapons manufacturing. Like the other large weapons manufacturers, Boeing has reduced its greenhouse gas emissions since 2010. For example, Boeing announced in 2020 that through a combination of conservation, the use of renewable energy, and the purchase of offsets, that it had reached net zero emissions in its operations in 2020: "New renewable energy procurements reduced greenhouse gas emissions by 10% in 2020 to significantly lower environmental impact. Boeing procures enough renewable electricity—solar, wind, and hydropower—to power Boeing's factories in Renton, Washington and Charleston, South Carolina, and a large data center in Arizona, as well as meet 97% of the electricity needs for Boeing's Everett factory—the largest building in the world by floor area. Boeing sites in Texas, Illinois, Ohio, Pennsylvania and Indiana operate partially on renewable electricity."[88] But the devil is in the details. Boeing's more detailed environmental report stated that in 2019 in its global operations, the company had 1.976 MMTCO$_2$e in scope 1 and 2 emissions and purchased renewable energy credits (RECs) to offset 134,000 metric tons of CO$_2$e.[89] In other words, not net zero.

Table 4.8 shows the emissions of the top twelve companies, by value of arms sales, of U.S. military industry companies in 2019 and the estimated portion of their emissions that may be attributed to aerospace and military sales.[90] Most of the military equipment—missiles, radar, tanks, control systems, small arms, planes, and ships—produced by these corporations are sold to the U.S. Department of Defense and the other agencies that work closely with the U.S. military, such as Homeland Security and the Coast Guard.

Military greenhouse gas emissions are counted in the national inventories of the country where they are produced. While it does not really matter—from the perspective of total global emissions—where these emissions are produced, most of these emissions are produced on U.S. territory. For example, Lockheed Martin generated 98 percent of its scope 1 and 2 greenhouse gas emissions in 2019 in the United States.[91] I have also included

Table 4.8

Scope 1 and 2 emissions of top 12 U.S. arms companies in 2019, metric tons CO_2e

Company	Total emissions reported in 2019, metric tons CO_2e	Portion of company revenue from arms sales	Estimated military emissions, 2019, metric tons CO_2e
Lockheed Martin	968,021	0.89	861,539
Boeing	1,842,000	0.44	810,480
Northrop Grumman	484959	0.86	417,065
Raytheon	1,963,729	0.87	1,708,444
General Dynamics	762,200	0.62	472,564
L3 Harris Technologies	349,493	0.77	269,110
Huntington Ingalls Industries	55,379	0.87	48,180
Honeywell International	2042631	0.15	306,395
Leidos	55276	0.48	26,532
Booz Allen Hamilton	16983	0.69	11,718
General Electric	2,390,000	0.05	119,500
Textron	603,104	0.25	150,776
Total scope 1 and 2 emissions			5,202,302

Sources: SIPRI and individual corporate sustainability reports, including corporate climate change submissions to CDP, https://www.cdp.net/en.

the emissions for the production of weapons that are exported to the armed forces of other countries. Indeed, the U.S. arms export industry is the largest in the world, accounting for 37 percent of the global total from 2016–2020. Even though not all the U.S.-produced defense industrial products are sold to the U.S. military and other armed elements of the U.S. government, they are mainly sold to countries that are, at least at the moment, U.S. allies, and in that sense they serve U.S. foreign policy objectives, and those exports are approved by the State Department. About half (47 percent) of U.S. arms exports are sold to countries in the Middle East, with Saudi Arabia being the largest importer of U.S. weapons.[92] Incidentally, without these additional sales, and the larger production runs that they require, the cost of military equipment per unit would be somewhat higher. However, the exports sustain a larger military industry than would be the case if the United States were not a major arms exporter.

Estimating Total U.S. Military-Industrial Emissions

A complete inventory of all military-industrial emissions by U.S. companies would include the direct emissions of the other, smaller arms producers and the emissions of the industries in the supply chain that feed the raw materials, supplies, component parts, and capital machinery into the processes of manufacturing weapons.

One way to estimate the scope of the emissions of smaller military-industrial corporations is to estimate the emissions of the larger corporations' supply chains. While some corporations have documented their scope 1 and 2 emissions, there is less documentation of the total emissions of the military industry's supply chain. Although Lockheed Martin does not produce scope 3 emissions statements for each year, in 2016, they conducted a scope 3 analysis of the emissions. Of this, 7,182,300 metric tons of CO_2e were from purchases of capital goods and other goods and services for military-industrial purposes (*not* including other scope 3 emissions, such as transportation to and from Lockheed Martin's vendors).[93] Lockheed Martin reported total scope 1 and 2 emissions in 2016 of 918,635 metric tons CO_2e and of this we can attribute about 817,685 metric tons to military production, or a ratio of 8.78 metric tons of CO_2e emissions to 1.[94] If this ratio of emissions of supply chain to end product is similar for most major military-industrial producers, the supply chain emissions of the twelve largest producers in the U.S. military-industrial sector in 2019 totaled an additional 45,676,211 metric tons of CO_2e. Adding the estimated supply chain emissions to the direct scope 1 and 2 emissions of the twelve largest suppliers in 2019 yields the total 50,878,513 metric tons of CO_2e, or about 51 $MMTCO_2e$.[95]

This estimate of U.S. arms industry emissions in 2019 is likely conservative because it is only based the reported emissions of the largest companies, discounted by the proportion of their emissions that can be attributed to military production, and an estimate of the emissions that come from their larger supply chains. I have not included the emissions from companies operating abroad from which the United States purchases its weapons—these purchases are likely quite small given that during the Trump administration the United States began to implement a "buy American" policy. Finally, I have not included the emissions of the military contractors that supply the U.S.

military with services such as construction, security, food preparation at overseas bases, and so on.

THE MILITARY EMISSIONS OF OTHER COUNTRIES

Of course, although the U.S. military is the focus here, the emissions of some other countries are related to U.S. emissions. Specifically, the military fuel use and greenhouse gas emissions of the formal and informal allies of the United States are linked to U.S. emissions either because, in the case of NATO, there are shared exercises and a commitment to a minimum level of military spending, or because the United States works on an ad hoc basis with other countries to conduct military operations, such as its wars in Afghanistan, Iraq, and Syria.

Stuart Parkinson of Scientists for Global Responsibility based in the UK estimated that the UK military's real estate and operational emissions in 2017–2018 totaled 3.025 million metric tons of CO_2e. Of these, aviation fuel, diesel, and gas/petrol for operational capabilities and equipment accounted for 1.8 $MMTCO_2$e.[96] Parkinson estimates that UK military industry accounts for an additional 6.5 to 11 $MMTCO_2$e. Stuart Parkinson and Linsey Cottrell estimated individual EU country emissions and calculate that total military greenhouse gas emissions for the European Union's twenty-seven members had total reported military and military-industrial emissions of 24.83 $MMTCO_2$e in 2019.[97]

Further, the adversaries of the United States react to U.S. military operations, procurement, and arms sales to its allies, and this may increase their own military greenhouse gas emissions. Using a method based on military spending, Ho-Chih Lin and Deborah Burton have estimated global military greenhouse gas emissions in 2017 to be 445 $MMTCO_2$e. Lin and Burton suggest that if we added the oil consumption of the world's armed forces together, the total fuel consumption of these militaries would rank just ahead of the annual consumption of Belgium or South Africa in 2016.[98] Lin and Burton's work points to the fact that war and arms races can drive the military spending of alliances and regions in a dynamic process, and hence drive regional military greenhouse gas emissions.

OTHER WAR AND MILITARY-RELATED EMISSIONS

Several other sources of emissions are not calculated or estimated here, but they may also be significant sources of war-related greenhouse gases. For example, I mentioned the emissions of military contractors that over the last several decades have taken up many roles the military itself used to perform—but I have not estimated those emissions. This would be important to know if U.S. contractors are doing some of the work that the military would otherwise do, but their emissions are off the DODs books, and so the total DOD emissions reduction in recent years may be offset by the emissions of these contractors. I also have not estimated the emissions that result from the reconstruction of a nation where buildings and infrastructure have been destroyed by the U.S. military.

Further, as I described in chapter 2, belligerents have frequently targeted the petroleum industry and infrastructure of their adversaries, leading to emissions from the burning of oil or the release of methane due to the destruction of oil production equipment. In some instances, countries have sabotaged their own oil infrastructure to deprive an invading army of those resources. The direct and deliberate destruction of fossil fuel infrastructure has also been a part of the strategy of wars in the Persian Gulf.

It is difficult, however, to calculate how much greenhouse gas is released when oil infrastructure is targeted during war. In the 1991 Gulf War, Iraq set oil production facilities in Kuwait alight as they retreated, and in April and May 1991, an estimated three million barrels of oil were burning each day, releasing one or two million tons of carbon dioxide, or about 2 percent of worldwide CO_2 emissions from fossil fuels and biomass. Out of the 82 million barrels released on land and at sea during the 1991 Gulf War, an estimated 11 million barrels of oil spilled into the Persian Gulf, coating the coastlines of not only Kuwait, but also other countries in the Gulf including Saudi Arabia and Iran. More than a decade later, much of that oil remained in coastal areas. In the 2003 invasion of Iraq, oil wells were again set alight by the Iraqi military and burned for several months.[99] And when ISIS retreated, it set oil wells and pipelines on fire in Iraq and Syria.[100] In many cases, these fires burned for several months.

Soon after the United States began its war against ISIS in Syria and Iraq, starting in September 2014, it began to target ISIS-controlled oil production and distribution facilities as a means of cutting off their revenue stream. In 2015, ISIS oil revenue was about $140 million per month and the United States believed that it was critical to reduce that revenue stream. In October 2015, the United States attacked more oil-producing ISIS-controlled infrastructure in what was known as Operation Tidal Wave II, targeting everything related to petroleum, including tanker trucks, oil well heads, oil pump jacks, oil refineries, gas-oil separation plants, and storage sites.[101] The first Operation Tidal Wave, which occurred in August 1943, was the United States attack on Romanian oil facilities. On November 18, 2015, the United States destroyed 300 ISIS tankers; by January 2016, ISIS had lost 90 percent of its oil production capacity.[102] By late 2016, the United States had destroyed thousands of pieces of oil equipment in about 1,600 strikes against oil infrastructure, which included more than 850 strikes against oil tanker trucks and 316 strikes on crude oil well heads.[103] During Operation Tidal Wave II, the United States dropped leaflets warning civilians when oil infrastructure was going to be targeted, allowing civilian truck drivers and other workers to escape.[104] In addition to ISIS control of existing Iraqi or Syria oil production facilities, in some instances ISIS leaders set up makeshift oil wells and refineries to supply themselves with oil and as a source of revenue for their regime. One can assume that these makeshift facilities likely released a great deal of methane since explosions have been reported at some sites.[105]

The United States treats and evaluates strikes on oil infrastructure in the same manner it does the actual and potential consequences of attacking any other infrastructure. The targeting balances the estimated gains of an attack against the *immediate* risk to civilians.[106] When civilians were present, the Joint Chiefs of Staff instructions emphasized, "Joint force targeting during such situations is driven by the principle of proportionality, so that otherwise lawful targets involuntarily shielded with protected civilians may be attacked, and the protected civilians may be considered as collateral damage, provided that the collateral damage is not excessive compared to the concrete and direct military advantage anticipated by the attack. In cases where civilians voluntarily act as human shields, those civilians may be taking a direct part in hostilities and lose protection from attack. Such civilians need not be taken into account when assessing collateral damage under the law, though there may be diplomatic or strategic concerns that affect targeting decisions."[107] While the idea may be to weigh "concrete and direct" military advantage "rather than one that is merely hypothetical or speculative," military advantage is broadly understood.[108] See figure 4.15 illustrating how the U.S. Joint Chiefs of Staff target an adversary's petroleum assets.

The DOD defines advantage as "the advantage anticipated from an attack when considered as a whole, and not only from its isolated or particular parts. The advantage need not be immediate." The concept of military advantage is illustrated by examples: "The military advantage in the attack of an individual bridge may not be seen immediately (particularly if, at the time of the attack, there is no military traffic in the area), but can be established by the overall effort against bridges in order to isolate enemy military forces on the battlefield. Similarly, military advantage is not restricted to tactical gains, but is linked to the full context of war strategy. It may involve a variety of considerations, including the security of the attacking force."[109] If that were not clear enough, the Joint Chiefs of Staff note that there should be an effort to minimize civilian casualties. "Unless otherwise prohibited by ROE, attacks are not prohibited against military targets even if they might cause incidental injury or damage to civilians or civilian objects. In spite of precautions, such incidental casualties are inevitable during armed conflict."[110] Since at least the 1990s, the military tends to avoid targets or may even call off a strike where there is a high likelihood of harming civilians. But the military is less restrictive when it is evaluating a "high value and high payoff targets" such as an adversary's weapons of mass destruction or "target systems" such as their "petroleum, oils, and lubricants industry."[111] In sum, the basic targeting guidance does not mention that planners should consider lasting implications of greenhouse gas emissions—most importantly the uncontrolled release of methane—that will result from attacking oil infrastructure.

Finally, in addition to these direct sources of emissions, one cannot discount the consequences of the loss of carbon gas sinks as they are destroyed to make way for military bases or destroyed because the enemy lives in those

Example of Petroleum, Oils, and Lubricants Target System and Components

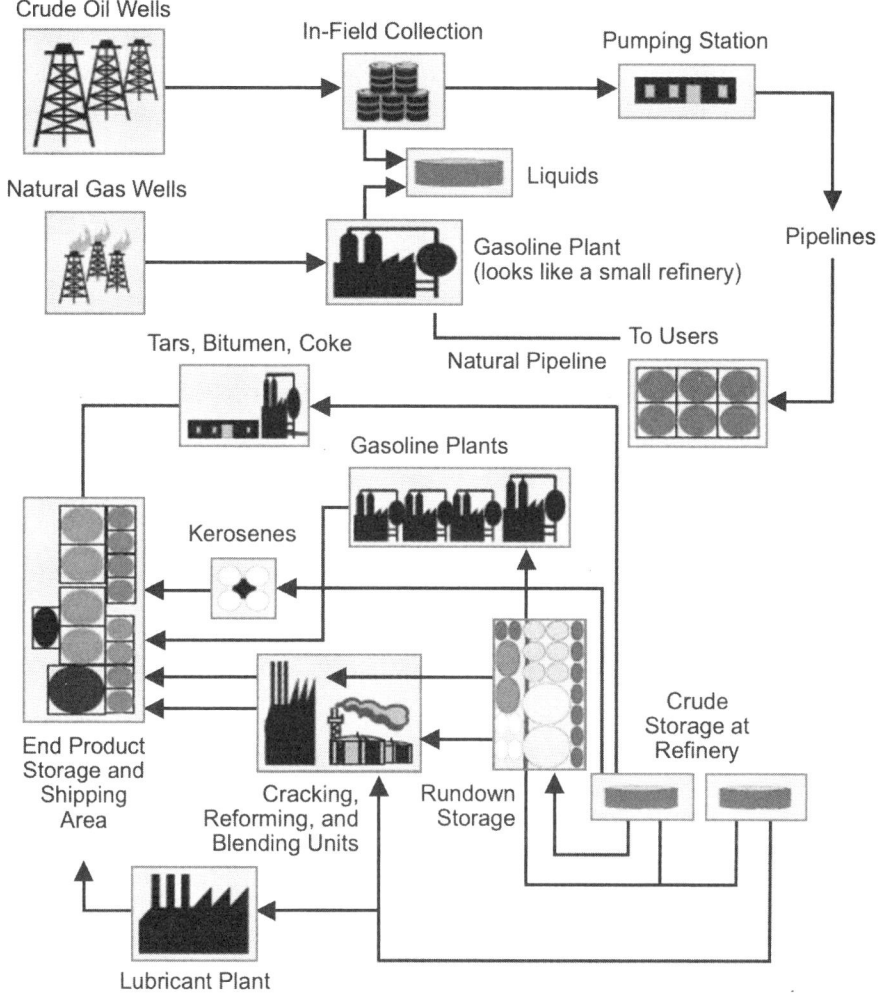

Crude Oil Wells

In-Field Collection

Pumping Station

Natural Gas Wells

Liquids

Gasoline Plant
(looks like a small refinery)

Pipelines

Tars, Bitumen, Coke

Natural Pipeline

To Users

Gasoline Plants

Kerosenes

Crude
Storage at
Refinery

End Product
Storage and
Shipping
Area

Cracking,
Reforming, and
Blending Units

Rundown
Storage

Lubricant Plant

Figure 4.15

The petroleum target system.

Source: Joint Chiefs of Staff, *Joint Targeting*, JP 3–60, January 31, 2013, II-8, https://www
.justsecurity.org/wp-content/uploads/2015/06/Joint_Chiefs-Joint_Targeting_20130131.pdf.

areas. While the United States is not alone in using "scorched earth" tactics, the American military has a long history, since George Washington sent the military against the Iroquois and Kit Carson marched against the Navajo, of destroying Native American forests and orchards. The aim is to eliminate hiding places of the adversary and to destroy the food that could be sustaining adversaries and their civilian supporters. Elimination of trees and forests was also part of U.S. doctrine in the Civil War and the counterinsurgencies in the Philippines and Vietnam. When the United States targeted the forests of South Vietnam in Operation Ranch Hand, the motto was, "Only you can prevent a forest." Forests are also lost inadvertently, due to fires that are sparked by bombing and as displaced people destroy trees to make firewood. An assessment of the climate change consequences of recent U.S. wars would include the loss of carbon sequestration due to war-related deforestation, which may occur as people displaced by war use forests for shelter and fuel. In Afghanistan, war-induced migration and illegal logging were among the chief causes of deforestation. Similarly, the causes of deforestation in Iraq are complex but also include war.[112]

CONCLUSION

I estimate that the U.S. military scope 1 and 2 emissions during the post-9/11 war years from fiscal years 2002–2020 were 1,038 $MMTCO_2e$, an annual average of 69 million metric tons. If the United States had not gone to war with Afghanistan and Iraq, U.S. military emissions might have stabilized at 62 $MMTCO_2e$ annually, the level that I estimate they were in 2001. If that stabilization had occurred, then U.S. military emissions would have been about 140 $MMTCO_2e$ less than they were by the end of fiscal year 2019. But, again, that is a conservative estimate. There are many emissions from war that are difficult to estimate.

Further, without the post-9/11 wars, the downward trend in military emissions already underway following the end of the Cold War would likely have continued. Before the 9/11 attacks, Secretary of Defense Donald Rumsfeld had proposed that the United States was going to be reducing its bomber forces, including cutting all B-1B bombers, and closing some military bases.

It was possible that there would have been further rounds of base realignment and closure. Thus, total emissions were likely to continue declining. Cutting one or two aircraft carriers was also under consideration.[113] If these cuts had occurred—and some eventually did—and emissions averaged 56 MMTCO$_2$e each year for from 2001 through 2019, it is possible that total military emissions for that period could have been 260 MMTCO$_2$e lower than they actually were. And the military and military-industrial emissions of U.S. allies and adversaries would also be likely to have been less as well.

For the last fifty years the United States government—and indeed, the military in particular—has been aware of global warming, its causes, and the consequences of climate change if action to reduce emissions was not taken. U.S. policy makers have not, to date, responded with the same degree of urgency to the threat of climate change as they have acted to defend access to greenhouse gas-producing fossil fuels. Ironically, even as the United States military has led the charge to protect access to Middle East oil, the Pentagon has done the most of any part of the federal U.S. government to prepare for the consequences of climate change for its mission and operations. It has simultaneously worked to become more energy efficient. Fear and worry are driving these policies, including anxiety about ensuring fuel supplies during war and the concern that climate change will harm U.S. military infrastructure, complicate operations, and make armed conflict more likely. I turn to describing these concerns and the way the U.S. military is addressing them in the next three chapters.

III U.S. MILITARY DOCTRINE AND NATIONAL SECURITY STRATEGY

5 ENERGY AND CLIMATE SECURITY: FINDING AND FIXING VULNERABILITIES

On the fuel, it is a significant Achilles heel for us when you have to haul the amounts of fuel that we have to haul around the battlefield for the generators and for the vehicles. . . . I mean, it is an amazingly complex effort to maintain the fuel lines. And it also gives the enemy an ability to choose the time and place of attacking us.
—John Mattis, March 2011[1]

We in the Department of Defense (DOD) know first-hand the national security risk posed by climate change because it affects the work we do every day. Around the world, climate change is a destabilizing force, demanding new missions of us and altering the operational environment. At the same time, climate-related extreme weather affects military readiness and drains our resources. In just the past few years, wildfires have forced evacuation at bases in the western United States, while hurricanes on the East Coast and flooding in the Midwest have inflicted billions of dollars of damage on facilities that are home to key war fighting capabilities.
—Secretary of Defense Lloyd Austin, September 2021[2]

The United States learned the hard way during the Afghanistan war that not only can enemies attack fuel supplies, neutral countries and allies can also complicate fuel delivery. The invasion of Afghanistan in late 2001 and the occupation of that country required enormous amounts of equipment and fuel each day. Much of that fuel and other nonlethal equipment was offloaded at ports in Karachi, Pakistan, and then transported overland in convoys of fuel tankers that then entered Afghanistan through the border crossings at Chaman Pass, near Kandahar, and Torkham Gate near Khyber Pass, close to Kabul. The United States also relied on Pakistani military forces to secure the border with Afghanistan and attack Taliban and al Qaeda militants who were sheltering in the northwest of the country. The United States inadvertently killed Pakistani troops near the Afghan border in early 2011 and then again late November 2011. In response to the second attack, which killed 24 Pakistani soldiers, and injured more than a dozen others, the Pakistani government halted supply convoys for seven months. The convoys resumed in July 2012 after U.S. Secretary of State Hillary Clinton apologized for the deaths. But this was not the first time the Pakistanis had halted the shipments. The route between Pakistan and Afghanistan at Torkham was shut for eleven days in September 2010 after NATO killed three Pakistani troops at the border. The threat of the Pakistani government doing so again was not infrequent. For instance, the Pakistani government threatened to cut off supplies in October 2013 if the United States continued drone strikes against Taliban and al Qaeda militants in Pakistan.[3] This illustrates the need for "energy security," which the DOD defines as "having assured access to reliable supplies of energy and the ability to protect and deliver sufficient energy to meet mission essential requirements."[4]

Around the same time as the Pakistani government halted the supply convoys in 2011, the U.S. Marines were dealing with the aftermath of Hurricane Irene, a Category 1 storm, which hit Camp Lejeune in North Carolina in late August, causing about $2.5 million in physical damage and power outages. Camp Lejeune was hit by Hurricane Florence in September 2018 and suffered $3.6 billion worth of damages. In October 2018, Michael, a Category 5 hurricane, devastated much of Tyndall Air Force Base in Florida, including F-22 aircraft, damaging beyond repair or destroying hundreds of buildings. As Colonel Greg Moseley said, "Every building on the airbase—100% of them spread across 29,000 acres—was either badly damaged or totally destroyed." Estimated to cost a total of 4.9 billion to repair and reconstruct the base, the rebuilding was expected to last as long as five to seven years.[5] In March 2019 one third of Offutt Air Force Base—the home of the U.S. Strategic Command and the Air Force's largest air combat wing, the 55th Wing—located near the Missouri River in Nebraska was flooded after massive rains and melting snow caused the river to overwhelm two levees, leading to part of Offutt's only runway being submerged and damaging about sixty structures beyond repair.[6] The base evacuated nine aircraft and the flooding destroyed about 1.2 million square feet of workspace, including "secure, compartmented information facilities, space for handling highly classified material."[7] In August 2020, smoke and haze from wildfires in California caused Naval Air Weapons Station at China Lake, the U.S. Navy's most important facility for testing its weapons, to cancel several tests and operations.[8]

These incidents illustrate two forms of security the military has become more apprehensive about over the last three decades: energy security and climate security. While energy security has been a growing preoccupation since the 1970s, the DOD has also led the U.S. government in understanding climate change, particularly how U.S. military infrastructure and operations and even its missions already have been and will be affected as global warming continues. In this chapter, I first describe how the military has understood and attempted to provide for military energy security. A concern within the Pentagon that fuel dependency makes the U.S. military vulnerable has prompted the DOD to reduce fuel consumption so that it is less dependent on fossil fuel. Second, I summarize how the DOD believes that climate change will affect U.S. military installations and operations, including how responding to climate disasters will stress operations. Then chapter 6 addresses the U.S. military's understanding of climate change as a national security threat or "threat multiplier."

ENERGY SECURITY, FUEL EFFICIENCY, AND THE "FULLY BURDENED COST OF FUEL"

Because fuel enables and limits the U.S. military's options, the military keeps careful track of the availability and cost of fuel and of the means of delivering it. As we have seen, the U.S. military has been worried about assuring fuel for operations and refueling, and searching for fuel efficiencies, since the nineteenth century, when the U.S. Navy tested its coal steamers to determine which system to use, paddle versus propeller, and which steam engine configurations would allow it to travel the furthest with the least coal. The navy also compared the combustibility, heat, and smoke qualities of specific coals. The greater heat content of liquid fuel compared to coal led the U.S. Navy to switch to oil in the early twentieth century. The range of motorized equipment, which depends on their rate of fuel consumption, has often determined basing requirements and tactical operations.

As weapons became more complex, capable, and larger, they also often became more fuel hungry. As I described in chapter 4, the Defense Logistics Agency and its predecessors have developed a global system to acquire, prepare, and move the right amount of fuel on time to wherever it is needed, spending whatever it takes to get fuel to the fight. During major wars, the U.S. government has even on occasion taken control of some supplies and set the price for fuel. In the era of "cheap" and "easy" oil, and when the United States had thousands of bases all over the world, there was relatively little attention paid to the fuel economy of military vehicles and buildings; the efficiency of military operations and practices has not been a major consideration when compared to other criteria such as speed and lethality. But the relative inattention to fuel efficiency has cost time, money, and lives.

The limits caused by inefficient fuel consumption were evident during the 1991 Gulf War. The fuel requirements of the U.S. Army's Abrams tank in the 1991 invasion of Iraq were so significant—the tank consumed about seven gallons per mile—that General Paul Kern, commander of the U.S. Army's Second Brigade of the 24th Infantry Division on its way to Baghdad, said, "As we considered the route and began planning, our biggest concern was not our ability to fight the Iraqis; it was keeping ourselves from running out of fuel." The tanks required refueling every few hours by fuel trucks. "We also made a decision to never let our tanks get below half full, because we didn't want to refuel in the middle of a fight."[9]

The entire non-nuclear portion of the U.S. fleet also requires frequent refueling, and as Christopher McMahon notes, naval operations still depend on refueling at foreign ports. "There are only six U.S.-flag product tankers to supply fuel for the entire U.S. Navy around the world. There are no other U.S.-flag tankers in international trade, and very few, if any, other product tankers in domestic trade that could be used in an emergency. In a manner similar to its practice during the voyage of the Great White Fleet, the

Navy today frequently relies on foreign-flag tankers and cargo ships to carry Navy fuel and supplies because of the limited number of U.S.-flag merchant ships."[10] Refueling is, clearly, an opportunity for bad things to happen. In October 2000 al Qaeda militants attacked the guided missile destroyer U.S.S. *Cole* while it was refueling in port at Aden, Yemen, on its way to support trade sanctions against Iraq. The attack killed seventeen sailors and injured forty-two others.

In May 2001, months before the United States was attacked by al Qaeda on 9/11 and then responded by going to war in Afghanistan, the Defense Science Board listed the warfighting advantages of increasing fuel efficiency and reducing military fuel use.

> Surprise: Fuel efficiency increases platform stealth by diminishing the platform's heat signatures, exhaust, and/or wakes; and affords less chance of compromising movement by reducing the logistics tail and resupply communications.
>
> Mass: Fuel efficiency decreases the time required to assemble an overwhelming force.
>
> Efficiency: Fuel efficiency increases commander's flexibility in efficiently assembling an overwhelming force.
>
> Maneuver: Platforms will travel faster and farther with reduced weight and smaller logistics tails that improve platform agility, loiter and flexibility.
>
> Security: Fuel efficiency decreases platform vulnerability to attacks on supply lines, and reduces demand for strategic reserves.
>
> Simplicity: Fuel efficiency decreases the complexity and frequency of refueling operations and logistics planning, while reducing vulnerability to the "Fog of War."[11]

The May 2001 report of the Defense Science Board noted that an armored division consumed about six hundred thousand gallons of fuel each day and that "over 70 percent of the tonnage required to position today's U.S. Army into battle is fuel."[12] But the Defense Science Board also found that the way the military thinks about fuel costs did not incentivize fuel economy. The organization within the military then tasked with acquiring and providing fuel, the Defense Energy Supply Center (DESC) within the Defense Logistics Agency charged the services a fixed standard price for fuel "that is supposed to cover both the acquisition costs of the fuel and DESC's operating costs."[13] In fiscal year 2001, the standard price was $1.01 per gallon. But this was, the Defense Science Board argued, an underestimate of what it called the "true cost" of providing fuel, which includes the people, training, delivery, operating costs such as tankers, and other infrastructure. The board calculated that once all the other expenses were included, it cost the U.S. Air Force $17.50 to deliver a single gallon of fuel by aerial refueling tanker; delivering a gallon of fuel to the army to the forward edge of battle cost $400 a gallon; delivering a gallon by truck would cost about $10 per gallon over "modest distances," and up to $40–50 per gallon over larger distances.[14] The argument the Defense Science Board was making suggested that if the military understood the total "true" costs of supplying fuel to the forces, they would see an incentive to increase efficiency and that "improving the fuel efficiency of weapons platform has large and unrecognized potential to strengthen warfighting capability and free resources for other high priority military needs."[15] Until it was nudged by the Defense Science Board, the military had not tended to account for the "true" cost of delivering a gallon of fuel to the front.

The Defense Science Board weighed in again in 2008 with the report of a task force cochaired by James Schlesinger, the former secretary of energy and former secretary of defense, called *"More Fight—Less Fuel."*[16] Here again the challenge was framed in terms of both the cost of energy and the growing demand for fuel in war zones as well as the need for more resilient infrastructure within the United States itself. But global warming was mentioned, as "an important and growing issue."[17] The Defense Science Board recognized it was important that the DOD not be "oblivious" to the global movement to constrain carbon emissions and it recommended that "if DoD decides to provide financial backing to synthetic fuel production plants, it should avoid investing in processes that exceed the carbon footprint of petroleum."[18] Thus within a decade of the DOD taking the view that the United States not be constrained by international agreements, the Defense Science Board was recommending a lower emission solution to DOD energy problems. Global warming was not a major focus of the report, but it was a key acknowledgment and reflection of a changing understanding of the conditions the Pentagon faced.

The question of fuel supply in war zones was vital in 2008; about 50 percent of the load carried by supply convoys in war zones was fuel.[19] Returning to the questions raised in its earlier report, in 2008 the Defense Science Board found that "little progress" had been made in implementing its recommendation to account for the total or "fully burdened cost of fuel." Indeed, the DOD had not yet developed the analytical capability to do so.[20] Rather, the DOD Defense Energy Support Center was still setting a standard price for fuel that was much lower than if it included the rest of the costs associated with delivery, including the cost of transporting the fuel, delivery equipment, and force protection. For example, the "standard" price of the jet fuel JP-8 that DESC sold to the services was $2.14 per gallon in April 2007 and $3.04 in December 2007. But estimates of the true, fully burdened cost of fuel ranged from "a low of $4 per gallon for ships on the open ocean to $42 per gallon for in-flight refueling to several hundred dollars per gallon for combat forces and FOBs [forward operating bases] deep within a battlespace."[21] And while the Defense Science Board was encouraging of the DOD's efforts to develop better estimates, it did not support the DOD's methods for calculating the fully burdened cost of fuel. The DOD was proposing to base its analysis of the fully burdened cost of fuel on peacetime operations, not including combat scenarios. The Defense Science Board suggested that such an estimate of total costs would be incomplete.[22] The board also highlighted something that had become increasingly obvious during the wars in Afghanistan and Iraq: troops were put at risk every time they transported fuel to and within a war zone. Thus, the Defense Science Board argued, the costs of protecting fuel deliveries to the services in war zones were "difficult to measure and often not monetary costs. They include reduced combat effectiveness, risk to missions, and casualties."[23]

The concept of "energy security" not only for the United States as a country, but also the military as an organization was legislated in the 2007 Energy Independence and Security Act, which set goals for reducing energy consumption.[24] Further, the fiscal year 2009 National Defense Authorization Act stipulated that the secretary of defense was to "require that analyses and force planning processes consider the requirements for, and vulnerability of, fuel logistics." The National Defense Authorization Act also defined the fully burdened cost of fuel as "the commodity price for fuel plus the total cost of all personnel and assets required to move and, when necessary, protect the fuel from the point at which the fuel is received from the commercial supplier to the point of use."[25] This law made energy security a goal, requiring the services to study solar and wind energy for use in "expeditionary forces" and to study alternative and synthetic fuels.

In 2009, Deloitte, one of the world's largest accounting firms, published a report called *Energy Security: America's Best Defense*, which made the correlation between fuel use and casualties explicit. While the "U.S. Department of Defense (DoD) is the single largest institutional consumer of oil in the United States of America," they argued that the United States should reduce its reliance on fossil fuels as a way to decrease the military's vulnerability in war zones. The "increasing number of convoys required to transport an ever increasing requirement for fossil fuels is itself a root cause of casualties, both wounded and killed in action. The [adversary's] use of IEDs [improvised explosive devices] and roadside bombs has been an especially effective means to disable friendly fighting forces by disrupting their supply of energy."[26] In sum, if the United States reduced fossil fuel consumption, it would reduce the need for fuel convoys, which in turn could lower the vulnerability of the troops who move the fuel and guard it. Deloitte suggested many ways to cut consumption including "widespread and aggressive conservation techniques; the use of renewable resources, in particular, solar and wind energy within the theater; renewable carbon-based fuels generated in theater, such as algae, biomass, and other alternative fuels; the use of highly efficient electric vehicles; nuclear fission; hot/ cold fusion; fuel cell technology."[27]

The Urgent Incentive to Reduce Fossil Fuel Use

The human and tactical costs and consequences of high levels of fuel use were illustrated quite dramatically during the U.S. occupation of Afghanistan. As the then secretary of the U.S. Navy Ray Mabus said in October 2010, "Fossil fuel is the No. 1 thing we import to Afghanistan and getting that fuel is keeping the troops doing what they were sent there to do, to fight or engage the local people."[28] U.S. and NATO forces were dependent on the transit of supplies by road through Pakistan and fuel comprised between 30 and 80 percent of each convoy. Each fuel truck, which carried thousands of gallons on a more than 435-mile journey, traveled in convoys that had to be protected in transit from the port in Karachi in southern Pakistan to Afghanistan. But the fuel was particularly vulnerable in transit, and convoys of NATO oil tankers were often attacked by militants and burned during their passage through Pakistan to NATO bases in Afghanistan during the U.S. war there.

The height of the attacks on convoys coming from Pakistan came between 2010 and 2011. At this point, the United States was surging troops and operations in Afghanistan and the need for fuel was thus greater than before. At the peak of the U.S. troop presence in Afghanistan, when the United States had nearly one hundred thousand troops in the country, it depended on moving about 50–90 percent of its supplies to Afghanistan through Pakistan.[29] In 2010, 27,073 trucks passed through the border post at Chaman, with about 25 percent of those vehicles carrying fuel. The Pakistani inspector general of the Frontier Corps, in Balochistan, General Obaidullah Kahn told a reporter that, in 2010, 194 trucks were destroyed in 159 separate attacks at the Chaman crossing.[30] Despite the risks, in early 2011 about a hundred truckloads of fuel per day were crossing the border. In one particularly spectacular series of attacks, which occurred in late September 2010 at the Torkham border crossing, dozens of fuel tankers were ambushed and burned. In another attack on February 25, 2011, on the Ring Road in the Khyber Pakhtunkhwa, fifteen oil tankers in a NATO convoy were attacked and blown up; more than twenty tankers were burned in early December 2011, in another attack.[31]

When the fuel and equipment convoys were halted in late 2011 as a result of the U.S. strikes that injured and killed dozens of Pakistani soldiers at the border, the United States moved its fuel and other supplies through different countries, via the Northern Distribution Network, and increasingly supplied the war by air shipments. When the border between Afghanistan and Pakistan reopened after seven months, and the convoys resumed in July 2012, so did the militant attacks. In sum, from 2008 through 2020, there were, according to one source, about 450 attacks on NATO supply convoys

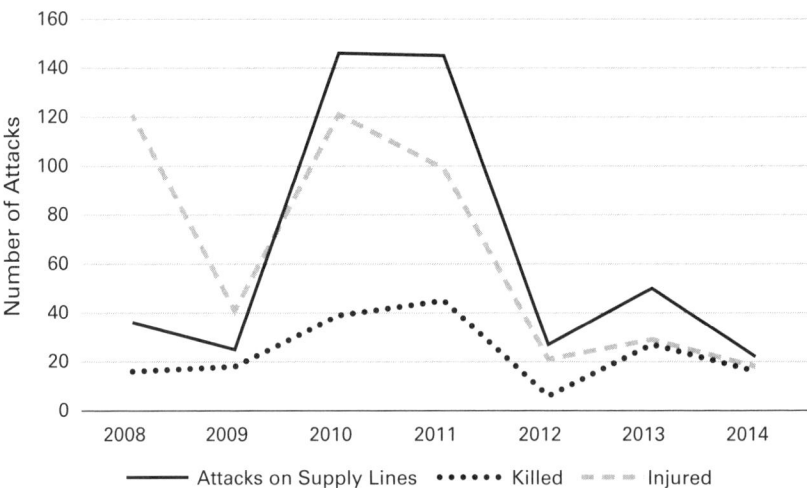

Figure 5.1

Attacks on NATO supply convoys through Pakistan, 2008–2014.

Source: Pak Institute for Peace Studies, annual *Pakistan Security Report*, https://www
.pakpips.com/publications#1512730923805-d52fde57-07fa.

moving equipment and fuel through Pakistan into Afghanistan, causing 167 deaths and 450 injuries.[32]

In sum, the military's concept of energy security—reliability, resiliency, and efficiency—at installations and in operations has become more sophisticated and integrated into the organization at every level. Nevertheless, while the U.S. military takes great care to supply fuel to the services, it has not traditionally done such a good job of accounting for the total, "fully burdened cost of fuel." When pressed in the early 2000s, the United States began to take account of the total, fully burdened cost of military fuel—including the cost to purchase the fuel from refineries, the cost the Defense Logistics Agency would charge military services, and the cost of delivering the fuel. This became explicit and actually extended to decisions about what weapons systems to purchase in the 2009 National Defense Authorization Act, passed in late 2008, requiring the military to consider the logistics support requirements and lifecycle cost of fuel in analyzing and evaluating alternatives for purchasing military equipment.[33]

Yet, a number of costs and consequences of a heavy reliance on fossil fuels, namely the political and environmental consequences of the U.S. dependence on oil, are still generally considered externalities and thus left off the books. This also includes the costs of defending any oil imports to the United States that are used to power the U.S. military, the fuel used by contractors, and the fuel used to produce the weapons.

CLIMATE CHANGE SECURITY AT INSTALLATIONS AND BASES

With the end of the Cold War, and the dissolution of the Warsaw Pact, the United States, the United Kingdom, and France became the countries with the largest overseas operations and bases. And with its about 750 bases, the United States has by far the most extensive network of bases and operations in the world compared to other countries, including Russia and China. All told, the Department of Defense manages a reported 26.9 million acres of land worldwide on which sit about 279,000 buildings, 184,000 structures, and 122,000 of what the DOD calls "linear" structures, e.g. runways, roads, rail lines, pipelines, and electrical distribution lines.[34] The DOD reported energy consumption at about 120 overseas bases in fiscal year 2019, not including bases of operation in Iraq and Afghanistan.[35]

The military lives and works on the same planet as the rest of us, and the armed forces face the same problems all of us face as carbon dioxide and other greenhouse gas emissions accumulate in the atmosphere and as deforestation and the loss of natural habitats like wetlands and natural grasslands eliminate important mechanisms for pulling carbon out of the atmosphere. We all face higher surface temperatures, greater acidification of the oceans, increasingly ferocious and numerous storms, too much rain in some places on the earth, and too little rain elsewhere. Sea levels will continue to rise as glaciers melt and warmer water expands.[36] Extreme weather events will also increase. The DOD understands all of this and has begun to prepare.

The DOD began to track and publish the number of "utility outages" to its infrastructure by cause in 2014. Most of the disruptions are to electrical service, and in any one year, most were caused by equipment failure and or planned maintenance. But there are many events that the DOD says are caused by an "act of nature"—weather or storms. For example, in fiscal year 2019, there were 2,572 unplanned utility outages at DOD facilities, with

542 of the outages lasting eight or more hours. Of these, over 90 percent were electrical outages. The rest were disruptions to the water supply and other utilities. That year, 47.6 percent of the outages were caused by equipment failure and 29.2 percent were "caused by acts of nature." As figure 5.2 suggests, while the percentage of these incidents varies, the number of outages lasting more than eight hours at DOD installations that were caused by "acts of nature" seem to be trending up.

Climate change can obviously make operations more difficult when bases are impacted to the point where they are barely functional or nonfunctional. For decades, the U.S. Navy has understood the most urgent threat to military infrastructure is rising sea levels and major storms that inundate coastal infrastructure and limit the use of naval bases.[37] In 2008, the National Intelligence Council indicated that more than thirty military facilities were already facing risks from rising seas.[38] In early 2018, the DOD published the result of a survey conducted in 2015 of its more than 3,500 primary installations and sites and identified flooding due to storm surge, flooding due to rain or other non-storm surge events, extreme temperatures, wind, drought,

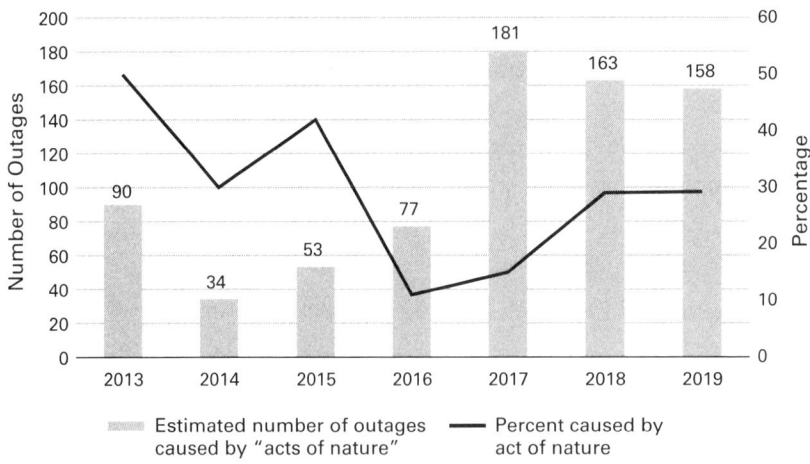

Figure 5.2

Outages lasting eight or more hours at DOD installations caused by "acts of nature."

Source: Calculated from data in the Office of the Assistant Secretary of Defense for Sustainment, *Department of Defense Annual Energy Management and Resilience Report* (AEMRR), fiscal years 2013–2019.

and wildfire as the major threats to its infrastructure.[39] They said that about half of all Department of Defense installations had already experienced climate change-related effects and that many installations experience more than one of these effects.[40] Figure 5.3 shows that the bases experiencing multiple vulnerabilities are located in every part of the United States. In 2019, the DOD survey of seventy-nine important installations found climate change effects at dozens of installations, including recurrent flooding (fifty-three installations), drought (forty-three installations), wildfires (thirty-six installations), and desertification (six installations).[41]

This included the facilities located at Hampton Roads, situated at the confluence of the James, Elizabeth, and Nansemond Rivers and the

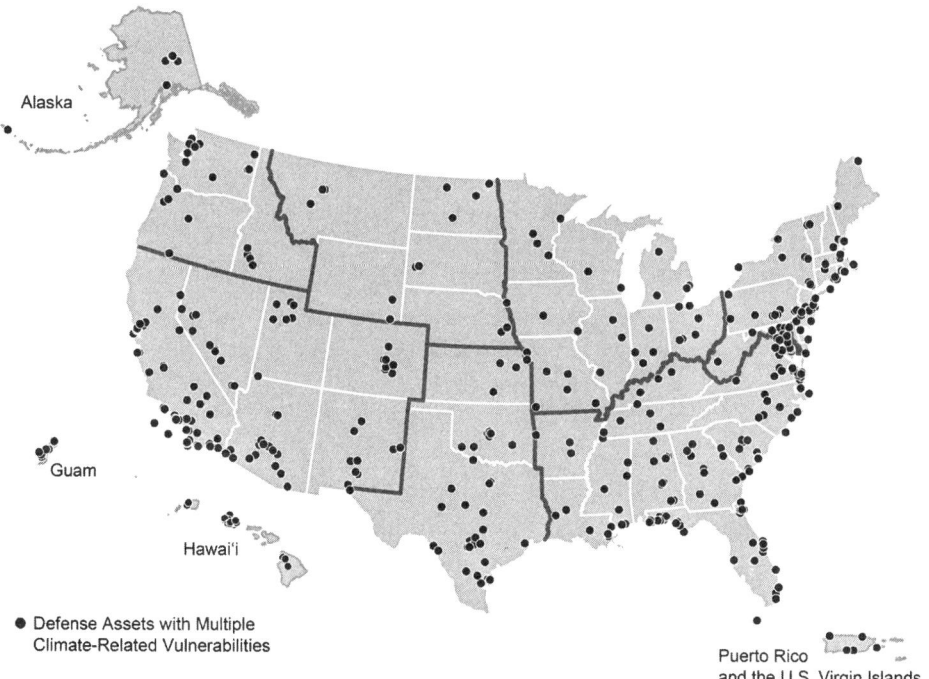

Figure 5.3
U.S. military assets with climate-related vulnerabilities.
Source: Office of the Undersecretary of Defense for Acquisition, Technology and Logistics, *Department of Defense, Climate-Related Risk to DOD Infrastructure Initial Vulnerability Survey (SLVAS) Assessment Report*, January 2018.

Chesapeake Bay, an area with features favorable to navigation that has been site of continuous occupation for thousands of years. English colonists landed there in the early seventeenth century and over the next three hundred years the colonists and later the U.S. military established military bases there including the United States' largest naval facility. The Great White Fleet left Hampton Roads in December 1907, refueling every two weeks or so over the course of the fourteen-month voyage until it returned to Hampton Roads in 1909. Although the fleet resupplied in California, foreign ports and coaling ships were essential throughout the voyage.[42]

Hampton Roads is greatly changed since the Great White Fleet returned in early 1909. Military and military-industrial infrastructure dramatically expanded in World War II, and today the area is home to twenty-nine military sites including the 3,600-acre naval station at Norfolk. With fourteen piers and eleven aircraft hangars, Norfolk Naval Base, the world's largest, can support 75 ships and 134 aircraft and it is the home port for four aircraft carrier strike groups, U.S. naval submarines, and Military Sealift Command. The area is also the home of a naval supply center, the Craney Island Fuel Depot, and a U.S. Coast Guard base in Portsmouth and Joint Base Langley-Eustis. All told, the region is the home to 100,000 military personnel. Further, nearby Newport News is the site of Huntington-Ingalls Industries owned Newport News shipbuilding, the manufacturer of the new Ford Class aircraft carriers and two classes of submarines.

The entire Hampton Roads area, where much of the land is at or just a few meters above sea level is vulnerable to climate change. And sea level is rising—between a projected 1.61 feet (.49 meters) and 7.55 feet (2.3 meters) by 2100. High tides already flood much of the Hampton Roads area and over the next several decades, the region will likely experience extreme heat and more frequent and intense storms, including snowstorms.[43]

The map shown in figure 5.3 also includes Guam, Puerto Rico, and Hawaii. Pearl Harbor and Wahiawa Annex in Hawaii have already experienced flooding and drought. The DOD reported that "recurrent flooding limits capacity for a number of operations and activities including Navy Expeditionary Forces Command Pacific, submarine squadrons, telecommunications, and a number of other specific tasks supporting mission execution" at the U.S. naval base on

Guam.[44] But the map doesn't include most of the United States' other island bases although number of island bases are particularly vulnerable.

The U.S. facilities at Diego Garcia in the Indian Ocean, south of the Maldives and East of the Seychelles are among the most vulnerable U.S. military installations. Diego Garcia is a fourteen-mile-long, U-shaped coral atoll with an average elevation of four feet, and a maximum elevation of twenty-two feet above sea level. Most areas are no more than 6.5 feet above sea level. While formally controlled by the United Kingdom, the United States has operated on the atoll by agreement since 1966 and in 2016, it renewed its fifty-year lease for another twenty years, until 2036. Diego Garcia is the home of a naval support and prepositioning facility, nuclear submarines, air mobility command, telecommunications, and a detachment of the Pacific Air Force. The prepositioned ships hold tanks, helicopters, ammunition, and fuel. The facility has a 12,000-foot runway, long enough to accommodate the B-52, B-1, and B-2 strategic bombers and the special hangars required to protect the sensitive B-2 bomber's special stealth surface, as well as a 2,000-foot-long wharf, which can support submarines and surface ships. Diego Garcia's location allows it to support operations in Africa Command, Pacific Command, and Central Command and it has done so, as the launching point for the air strikes in Afghanistan and Iraq in the post-9/11 wars. It is also comparatively close to important "choke points" for the transit of oil—the Strait of Hormuz to its northwest and the Strait of Malacca to its east.[45]

All of this capacity built up over decades makes Diego Garcia extremely valuable to the United States. As John Pike, the founder of Global Security.org told David Vine, Diego Garcia is "the single most important military facility we've got. It's the base from which we control half of Africa and the southern side of Asia, the southern side of Eurasia." Further, Diego Garcia is "the facility that at the end of the day gives us some say-so in the Persian Gulf region." And, Pike said, "Even if the entire Eastern Hemisphere has drop-kicked us" out of every other base, the U.S. would be able to "run the planet from Guam and Diego Garcia."[46] But the U.S. position in Diego Garcia is precarious.

The United States leases Diego Garcia from Britain, which arguably took it from Mauritius illegally. The island's former inhabitants, the Chagossians, say they were removed illegally in the 1960s and 1970s and without just

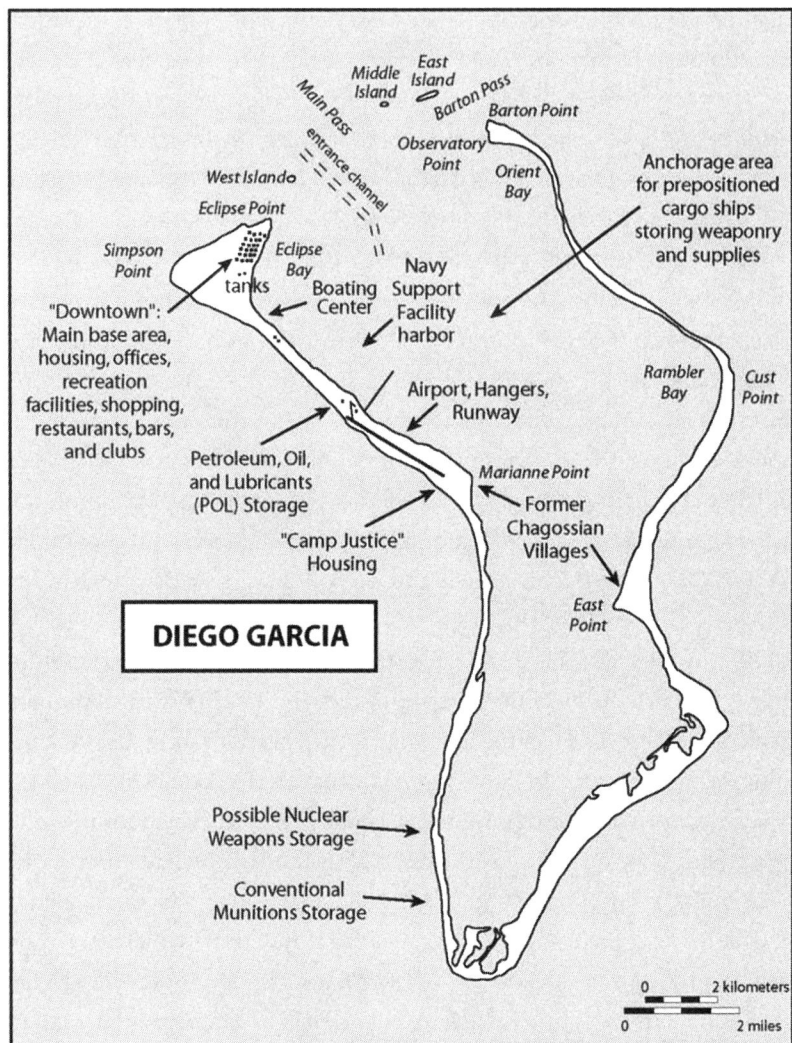

Figure 5.4

U.S. installation at Diego Garcia.

Source: David Vine, *Island of Shame: The Secret History of the U.S. Military Base on Diego Garcia* (Princeton, NJ: Princeton University Press, 2009), 11.

compensation.[47] In February 2019, the International Court of Justice in fact found that the entire Chagos archipelago is part of Mauritius territory and that the continued occupation of the islands including Diego Garcia was illegal. In May 2019, the UN General Assembly voted 116 to 6 (with 56 abstentions) that the territory should be decolonized as soon as possible.[48] The no votes were Australia, Hungary, Israel, Maldives, United Kingdom, and United States. In late 2020, Mauritius offered to lease the Diego Garcia installation to the incoming Biden administration for ninety-nine years—a proposition that bypassed and essentially ignored the British.[49]

Over the long run, the atoll is extremely vulnerable to sea level rise. Indeed, a 2018 study funded by the U.S. Department of Defense found that most atolls will be uninhabitable by the middle of this century as their ground water is contaminated by sea level rise and wave-driven flooding. Even in the cases where an atoll is not limited by its ground water, "the tipping point at which most of those atoll islands would be flooded annually" is projected to occur in the 2060s.[50] The U.S. Army has active facilities on seven islands in the Kwajalein atoll of the Marshall Islands, with its largest facility consisting of 517 buildings located on Kwajalein Island. The Marshall Islands are also vulnerable to sea level rise.[51]

Other installations in Guam, Bahrain, and Eglin Air Force base in Florida, on the Gulf of Mexico, are also vulnerable to sea level rise and storms. At the other extreme, melting permafrost in the Arctic threatens the physical stability of the U.S. Army's facilities in the Arctic, including Fort Greeley. "Military bases located where climate change is acute (e.g., polar regions, coastal areas) are likely to face unanticipated local effects, such as the exposure of pollution previously thought to be buried or sequestered, problems with sewage flows, and the structural weakening of buildings or infrastructure."[52] This may include exposing radioactive waste in Greenland buried at the bases that the United States left in the 1960s.[53]

CLIMATE CHANGE AND MILITARY OPERATIONS

In 2011 the National Research Council published a report, *National Security Implications of Climate Change for U.S. Naval Forces*, that emphasized the operational challenges the U.S. Navy could expect as a consequence of climate change.[54] On the one hand, climate change-caused natural catastrophes could overstretch military capabilities as the military would be called to react to the direct and indirect effects of climate change—natural disasters and the migration that results. On the other hand, global warming will cause conditions on the planet that make operations more difficult or unpredictable for equipment and for military personnel. The DOD "Climate Change Adaptation Roadmap" in 2014 outlined the now familiar list of challenges climate change posed to DOD plans and operations. It stressed the necessity of preparing for and adapting to climate change, including more clearly a focus on supporting civilian authorities as they respond to climate emergencies.[55] In September 2016, President Obama issued a National Security Memorandum that said, "Climate change and associated impacts on US military and other national security-related missions and operations could adversely affect readiness, negatively affect military facilities and training, increase demands for Federal support to non-federal civil authorities, and increase response."[56]

National security officials anticipating a growing role supporting civil authorities in disaster relief missions are concerned that natural disasters, made worse as a consequence of climate change, will stress the operational capacities of the U.S. military. As sea levels rise and prolonged droughts intensify and increase the number and intensity of wildfires, critical civilian infrastructure will be at risk. The U.S. active-duty military and National Guard are increasingly used to respond to climate-related emergencies in part because, after decades of building their capacity, they are in some ways the best resourced, flexible, and most capable organizations in the country.

For example, in 2012 the National Guard and military were called to assist in the aftermath of Superstorm Sandy, deploying over fourteen thousand people. In late summer of 2017, the DOD was called to assist in the response to three hurricanes—Harvey, Irma, and Maria. National Guard and active-duty forces were mobilized to respond in Texas, Florida, the U.S. Virgin Islands, and Puerto Rico. It was, as Michael Klare called it, a "triple punch" that "made the nightmare more palpable." Fed by warmer ocean temperatures, the storms were huge and powerful, dumping enormous quantities of rain on the areas they hit. As Klare notes, "with electricity out and many roads impassable, the guard and military services were called upon to rescue stranded citizens and perform a wide variety of other relief operations. An estimated 20,700 DoD personnel were committed to these activities in the Southeastern United States, while another 10,000 were performing similar tasks in Puerto Rico and the US Virgin Islands." Puerto Rico was devastated. "With all of Puerto Rico's ports obstructed by storm damage, the Marines sent to provide disaster relief supplies were forced to use their amphibious assault vessels to disembark."[57] In August 2018, the National Guard was called out to fight wildfires in California. Two years later, the U.S. Marines and Navy were called out to join the National Guard in the far west during an even worse fire season to fight some of the more than one hundred fires that were burning at one time.

Richard Kidd, the deputy assistant defense secretary for environment and energy resilience, acknowledged that the trade-off in time and effort between training and responding to climate change was significant. Kidd told *Defense One* in 2021 that "In terms of current operations, we have National Guardsmen, we have active-duty soldiers, we have active-duty airmen right now participating in firefighting support efforts. So these are . . . folks who are not doing a primary job. So right now we are experiencing climate change and effects. Right now, we know that these are going to only increase over time." Kidd projected that the military's other work in reaction to climate change, such as the "U.S. Army Corps of Engineers response for hurricanes and droughts in support of the national response through FEMA," would increase.[58] As Deputy Secretary of Defense Kathleen Hicks said in late October 2021, "Climate change is really increasing the number and frequency of missions that we're executing here at DOD. Let's look at firefighting. Severe drought has led to increasing fire seasons, lengthening of fire seasons. It's to the point where our National Guard bureau chief has started to talk about fire season becoming fire year. And in fact, we have in the last five years gone from about 14,000 personnel days for U.S. National Guard members to, in 2021, about 176,000 person days spent just on firefighting."[59]

Military personnel are also concerned that the operational environment will become more challenging for training and warfighting. The army says it must prepare for "imperiled Soldier health through exposure to airborne irritants like smoke and dust, disease vectors, and temperature extremes."[60] For example, as the season and range for mosquitos carrying malaria and Dengue fever expand, so will the risk to troops of contracting those and similar illnesses. Extreme heat and humidity have already been a growing concern in training and operations, "especially under conditions typical of military activity, including wearing heavy clothing and carrying heavy packs."[61] During the period of 2016 to 2020, 341 heat illnesses were documented in Afghanistan and Iraq, with 7 percent of those being the most serious form, heat stroke. The DOD expects an increased number of "black flag" days, where heat causes training to be suspended. "In 2008, 1,766 cases of heat stroke or heat exhaustion were diagnosed among active-duty service members, according to military data. By 2018, that figure had climbed to 2,792, an increase of almost 60 percent over the decade."[62] Although the method of reporting was changed, rates appeared to have fallen in mid 2020 from their peak in 2018 and 2019.[63] After a number of deaths, and a sharp rise in incidents of heat exhaustion and heat stroke, the military responded by changing the treatment of heat illness and heat stroke.

Some of the more subtle potential effects of climate change on operations were already noted in 1990, for instance in Terry Kelley's speculations about the effects of climate change on sonar.[64] The U.S. Navy is specifically concerned that melting sea ice in the Arctic will make its submarines more detectable and vulnerable to anti-submarine warfare. This goes both ways: each side's submarines may be more detectable. In 2000, the United States Arctic Research Commission suggested that "melting of sea ice in the Arctic will turn it into a conventional open-ocean ASW environment, with none of the advantages it now affords to an adversary strategic submarine."[65] However, a 2011 report by the National Research Council of Sciences suggested that if the United States monitored changes in temperature and salinity, ASW would not be affected because the changes were "nothing outside the operating scope of current systems."[66]

In contrast, air operations are affected by climate change. Warmer air, for example, means that aircraft will require longer runways to reach takeoff speed and aircraft engines may more easily overheat. As the climate becomes more arid in some regions, the dust generated during more routine training may require more and more frequent maintenance and repairs. Further, when U.S. operations occur in arid regions that are prone to sandstorms and dust storms, the fine debris may harm turbofan engines and diminish visibility. Finally, as storms increase or intensify, this may cause catastrophic equipment failures. In fact, a dust storm was one of a series of issues that caused the April 1980 U.S. hostage rescue attempt in Iran to fail. Specifically, the sandstorm limited visibility during the helicopter flights that were part of the rescue attempt, caused one helicopter to fail, and another to become lost and abandon the mission.[67]

CONCLUSION

The Pentagon has been quite attentive to the vulnerabilities of its installations to climate change-caused fires, drought, sea level rise, and other "acts of nature." As I will show in chapter 7, this has sparked both adaptation and innovation at installations. The adaptation effort is Herculean. However, the DOD's consideration of the larger question of vulnerability to climate change of the United States' global presence—and the military's careful, and expensive, exertions to maintain and repair facilities—is not matched by a consideration of whether the missions that the bases serve are necessary.

The Pentagon has been interested in reducing fuel usage for tactical purposes quite apart from the consequences for greenhouse gas emissions and climate change. The idea that the military should consider other costs for fuel—besides the cost per barrel or gallon paid to the companies that provide the fuel—such as the cost of transportation to the war zone was an important advance in DOD thinking about the costs of fossil fuels. In addition, it was important that the DOD paid greater attention to high fuel consumption rates during war—for weapons, equipment, and heating and cooling tents. High fuel consumption rates obviously decreased mobility while it increased the need to transport fuel to conflict zones, and therefore the vulnerability of the troops and contractors who provided the fuel to the war zone. Some argued that the "non-monetary" cost in lives should also be part of the fully burdened cost or true costs of fuel.

Recall the quote that opened this chapter. In early 2011 Marine Corps General John Mattis was asked by Representative Adam Smith during the House Armed Services Committee hearings about his widely reported 2003 comment that the U.S. military had to be "unleashed from the tether of fuel." Mattis was then the commander of Central Command, and thus in charge of the U.S. military presence in Afghanistan and Iraq. Mattis elaborated: "On the fuel, it is a significant Achilles heel for us when you have to haul the amounts of fuel that we have to haul around the battlefield for the generators and for the vehicles. . . . I mean, it is an amazingly complex effort to maintain the fuel lines. And it also gives the enemy an ability to choose the time and place of attacking us."[68]

In conclusion, then, the Department of Defense has been concerned that by requiring enormous quantities of fuel for its operations, the U.S. military is too dependent on fossil fuels at a tactical level. But the DOD and the foreign policy establishment do not seem to have factored in the larger costs of patrolling the Persian Gulf in peacetime—the annual and cumulative costs of protecting access to Persian Gulf oil. Nor have the budgetary and political costs and consequences of arming U.S. allies that DOD hopes will ensure access to oil apparently been factored into the fully burdened costs of fuel.

In chapter 7 I return to the question of efficiency and greening of the Pentagon, and discuss the choice between adaptation and mitigation. But before that, in chapter 6, I describe the other way that the Pentagon thinks about climate change: as the cause of future instability and conflict—a "threat multiplier."

6 CLIMATE CHANGE AS "THREAT MULTIPLIER"

As climate change converges with other drivers, especially geostrategic competition, emerging technology, and global-demographic trends, it is reshaping the risk landscape. An increasingly accessible Arctic has already emerged as a competitive space with national security implications, including competition over resources such as fish, water, and minerals that will intensify. The global increase in natural disasters will also drive up the number and severity of humanitarian crises.

The corrosive impact of these trends will make nations increasingly vulnerable to domestic instability, with sweeping implications for regional and border security and core national security interests. Simultaneously, revisionist nations will seek to exploit climate change induced instability to erode the rules-based order that is central to the security and prosperity of the United States and its allies.

—Department of Homeland Security, October 2021[1]

In 2010, the Department of Defense *Quadrennial Defense Review (QDR)* put climate change in a long list of "other powerful trends that are likely to add complexity to the security environment."[2] Since then, the people we regard as national security experts—the generals, admirals, defense department leaders, intelligence experts, and academics who study the causes of war and peace—appear to have come to a consensus: climate change has gone from being one among many issues, in the category of "other" trends or concerns, to a top priority. In October 2021, Secretary of Defense Lloyd Austin said, "To keep the nation secure, we must tackle the existential threat of climate change."[3]

Many national security experts now argue that climate change poses not only a threat to the environment, and the ability of people in the most affected regions of the world to make a living, but also a threat to political order and national security. Indeed, some strategists paint nightmare scenarios where climate change leads to armed conflict—such as when extreme heat and drought crop failures produce famine, and water shortages lead to unwelcome mass migration and to conflicts over water and other natural resources. Unless carbon dioxide, methane, and nitrous oxide emissions are dramatically reduced, and carbon is sequestered—most quickly in trees—this is our fate. As I noted earlier, in 2013 U.S. Navy Admiral Samuel J. Locklear II argued that climate change-caused instability "is probably the most likely thing that is going to happen . . . that will cripple the security environment, probably more likely than the other scenarios we all often talk about."[4] A year later the DOD called climate change a "threat multiplier" in its *QDR*: "Climate change may exacerbate water scarcity and lead to sharp increases in food costs. The pressures caused by climate change will influence resource competition while placing additional burdens on economies, societies, and governance institutions around the world. These effects are threat multipliers that will aggravate stressors abroad such as poverty, environmental degradation, political instability, and social tensions—conditions that can enable terrorist activity and other forms of violence."[5]

How did climate change move to the center of the national security debate? Through much of this period, Congress was controlled by people who denied that climate change was occurring, or who said that if it was happening it was not caused by greenhouse gas emissions made by the burning of fossil fuels. Thus, the former members of the military and government officials who were concerned about climate change formed nongovernmental organizations to sound the alarm.[6] Their message was relatively simple and quite plausible: climate change is a "threat multiplier." And then, gradually, government officials and members of Congress took up the argument that climate change was a risk factor in national security. These experts argued that climate change will lead to a more chaotic and dangerous world. Or more simply put, climate change-caused chaos is coming to a country and perhaps even a neighborhood near you.

The fear that climate change will lead to war has become, among many, a truism and part of the landscape of the popular imagination. As I noted in the introduction, a whole raft of recent books portray climate change as an imminent security threat. The discourse tends to be alarming and titles of several recent books seem to say it all: Gwynn Dyer, *Climate Wars: The Fight for Survival as the World Overheats*; Todd Miller, *Storming the Wall: Climate Change, Migration, and Homeland Security*; Harald Welzer, *Climate Wars: What People Will Be Killed for in the 21st Century*; Anatol Lieven, *Climate Change and the Nation State: The Case for Nationalism in a Warming World*; David Wallace-Wells, *The Uninhabitable Earth: Life after Warming*; and Michael T. Klare, *All Hell Breaking Loose: The Pentagon's Perspective on Climate Change*.[7] A review of Klare's book in *Sierra*, the magazine of the Sierra Club, seems to use this climate change–conflict and catastrophe nexus as a validation. Environmentalists appear to be glad that, after years of being alone in the wilderness on these issues, that someone is listening. As the reviewer of Klare's book, Carol Polsgrove said, "The military takes climate change very seriously." And also: "When the US goes to war in 'the national interest,' we need to ask ourselves what narrative defines that interest? What stories do national leaders tell themselves to justify the wars the US military fights? What stories do we tell ourselves to ease acceptance of the destructive human systems in which we live?"[8]

All of this alarm seems to have been, well, alarming. A YouGov poll of 30,000 people in twenty-eight countries in July 2019 found that the majority of people believed it likely that climate change would cause small wars, and around one-third thought that it would cause a new world war. Furthermore, many of the people polled thought that climate change would result in the "extinction of the human race."[9] And this was, of course, before two IPCC reports were published (in 2019 and 2021) stating that the world had only a few years remaining to act to avoid a global temperature rise of 1.5 degrees Celsius.

The Biden administration has taken the "climate change as threat multiplier" view on board, and in some ways amplified it. In this chapter, I describe the emergence of a consensus among national security experts that climate change is a national security concern *and* potentially a threat multiplier. The process of taking an issue that was not directly understood as an international and national security concern and transforming it into a national security threat is "securitization."[10] As I will show, this has been a conscious process: through the efforts of a number of national security experts, climate change was first understood as an environmental issue, then gradually an economic issue, and then an urgent national security concern. Through this analysis, climate change became perceived as a threat multiplier. However, not all of the scholars who study the connections between climate-related stress and conflict are convinced that the fear of climate change-caused conflict is supported by the evidence. Put another way, as Jon Barnett has argued, "predictions of international conflict arising from climate change are premature."[11] The questions that arise are whether climate change is a threat multiplier, and if so, whether that relationship is necessary, or if national security strategy can make that causal connection more or less likely.

THE MILITARY AND NATIONAL SECURITY ESTABLISHMENT TRACK CLIMATE CHANGE

Although the urgency among national security analysts may seem newly emergent, climate change was mentioned in President George H. W. Bush's 1991 *National Security Strategy* as among the new threats the United States faced in the wake of the end of the Cold War and the Gulf War. "The environmental depredations of Saddam Hussein have underscored that protecting the global ecology is a top priority on the agenda of international cooperation— from extinguishing oil fires in Kuwait to preserving rain forests to solving water disputes to assessing climate change. The upheavals of this era are also giving rise to human migrations on an unprecedented scale, raising a host of social, economic, political and moral challenges to the world's nations."[12] Though it was not the first time that a national security strategy mentioned the need to reduce greenhouse gas emissions, it was the first time a national security strategy explicitly mentioned climate change, albeit among a list of "global environmental concerns" that also included food security, water supply, deforestation, and biodiversity. The *National Security Strategy* also argued that "the stress from these environmental challenges is already contributing to political conflict."[13]

At the 1992 Earth Summit in Rio, President Bush signed the agreed upon text of the United Nations Framework Convention on Climate Change, which said that the "parties should take precautionary measures to anticipate, prevent or minimize the causes of climate change" and would work to reduce their emissions to their 1990 levels by 2000. Article 4 of the treaty committed all parties to publish national inventories of their greenhouse gas emissions.[14] The UNFCCC was ratified by the U.S. Congress in December 1992. Senator Al Gore, a student of Roger Revelle's at Harvard, who chaired the U.S. Senate delegation to the Rio Earth Summit, had just been elected vice president.[15]

In 1993, President Clinton's Secretary of Defense Les Aspin took the opportunity of the end of the Cold War to engage in a "Bottom Up Review" of the DOD and its missions. Intended to provide the "direction for shifting America's focus away from a strategy designed to meet a global Soviet threat to one oriented toward the new dangers of the post–Cold War era" the "Bottom Up Review" focused on "new dangers" which it saw as nuclear weapons and other weapons of mass destruction, potential aggression by regional powers, the potential failure of democratic reform in the former Soviet Union, and the "potential failure to build a strong and growing U.S. economy."[16] In sum, there was a lot of *potential* danger on the horizon and the *Report on the Bottom Up Review* argues that the United States needed a significant military presence overseas, even in peacetime. "The peacetime overseas presence of our forces is the single most visible demonstration of our commitment to defend U.S. and allied interests in Europe, Asia, and elsewhere around the world."[17] As happened in the immediate aftermath of the end of the Cold War, the chance for a new administration to rethink U.S. military priorities was largely lost as the old priorities were reaffirmed in the face of "potential" threats. As Jonathan Schell said later, "Mere opportunity was mistaken for accomplishment, and nothing was done."[18]

Only late in the *Bottom Up Review* is global warming mentioned: the "most notable environmental threats to U.S. security to which the Department must respond are: *global threats*, such as warming, ozone depletion, loss of biodiversity, and nuclear proliferation." Other *"regional threats"* were "environmental terrorism, accidents or disasters, regional conflicts caused by scarcity or denial of resources, and cross-border and global contamination." On the list of national environmental threats were "risks to public health and the environment from DoD activities" and, oddly, concerns that seem not to be related at all to the environment, specifically, "increasing restrictions on military operations, inefficient use of DoD resources, reduced weapon system performance, and erosion of public trust."[19] The DOD should also, Aspin's report said, establish energy and resource conservation guidelines and incentives to reduce energy consumption.[20] Secretary of Defense Aspin does not link fossil fuel consumption to emissions and climate change. Rather, the *Bottom Up Review* is clear that protecting oil is a major priority. "The ambitions of Iraq or Iran to dominate Southwest Asia, which continue to threaten our friends and allies in the Persian Gulf region . . . could endanger global economic stability through limiting access to oil supplies."[21]

But, as we have seen, the environment was on the radar of national security officials in the 1990s. Some concerns were recurrent, specifically, that once the Arctic Sea was open year-round it would raise questions about the need to patrol it and the Russians perhaps coming to dominate the Arctic and the resources there.[22] But the climate change–conflict nexus got a boost during the Clinton administration when Sherri W. Goodman, a protégé of Senator Sam Nunn, was named the first deputy undersecretary of defense for environmental security in 1993. Goodman, who served in that position through the Clinton administration, has been a leader since then in the effort to institutionalize concern about climate change-caused insecurity. Her efforts, and the urgent interest in climate change in the U.S. Navy, gradually elevated the issue.

For example, the Clinton administration's 1994 *National Security Strategy* worried that a "range of environmental risks" could "jeopardize international stability."[23] In his State of the Union Address in 1996, President Clinton also included "environmental degradation," albeit last, on his list of threats that "respect no nation's borders."[24] Goodman argued in 1996 in remarks to the National Defense University that the U.S. 1994 intervention in Haiti was in part driven by environmental conditions: "In Haiti, environmental conditions, such as deforestation, soil erosion, and water pollution, along with demographic, socioeconomic, and political problems including poverty, urban overpopulation, and a highly centralized government combined to create the societal decay that led to the involvement of American troops." Her analysis did not name climate change, but rather a mix of factors, including deforestation. "Haiti's deforestation is its most severe environmental concern, one that world relief agencies have explicitly tied to the country's refugee crisis which brought in American troops. One need only look at satellite photos of Haiti and its island neighbor, the Dominican Republic: on the Dominican side lie vast, forested areas; on the Haiti side, the land has been stripped bare by rampant clear-cutting. The disappearance of Haiti's forests and its consequent soil erosion are so extreme that coral reefs have been damaged, resulting in devastating reductions in fish stock." The argument was that environmental degradation led directly to poverty. "Economic deprivation has driven people from their land, which in turn has deepened the country's political crisis and intensified the outpouring of people seeking refuge in the United States."[25]

Goodman's office and her work were a high-profile acknowledgment of the importance of the environment and climate change. But there was continuity as well. The George H. W. Bush, Bill Clinton, and George W. Bush administrations sought to ensure that the United States would not be bound by any treaty language that had mandatory limits on emissions. These administrations believed that the only acceptable climate treaty that the United States should agree to would bind all countries—industrial and developing nations—and that any emissions limits would be voluntary and accomplished by "market mechanisms."[26] Comprehensive and sophisticated articulations of the issues linked to global warming occurred in high-level strategy documents across the intelligence community and Department of Defense during the Bush and Obama administrations spurred on by the work of security scholars and analysts working outside the Pentagon. Former fleet commanders and general officers were sounding the alarm.

The DOD's Office of Net Assessment commissioned a study by Peter Schwartz and Doug Randall from the consulting firm Global Business Network, to evaluate the impact for the DOD and national security of climate change. Their study, finished in October 2003, *An Abrupt Climate Change Scenario and Its Implications for United States National Security*, was meant to be a follow-up to the National Research Council's National Academy of Sciences 2002 report, *Abrupt Climate Change: Inevitable Surprises*.[27] Schwartz and Randall's study, which they described as "extreme in two fundamental ways"—the scale and likelihood of the effects they posit—argued that climate change could cause a "significant drop in the human carrying capacity of the Earth's environment."[28] Further, "an abrupt climate change scenario could potentially de-stabilize the geo-political environment, leading to skirmishes, battles, and even war due to resource constraints," including food shortages, declining availability of fresh water, and disruptions to energy supplies.[29] They suggested that, in response to these stresses, nations could respond defensively or offensively: "nations with the resources to do so may build virtual fortresses around their countries preserving resources for themselves. Less fortunate nations especially those with ancient enmities with their neighbors, may initiate . . . struggles for access to food, clean water, or energy. Unlikely alliances could be formed as defense priorities shift and the goal is resources for survival rather than religion, ideology, or national honor."[30] And while on the one hand, they say that the scenarios they suggest are extreme, on the other Schwartz and Randall suggest that there are "indications" that their scenarios may be more likely. But perhaps because the report attracted negative publicity, it was not apparently taken up in the Pentagon.[31] The cause of climate change is not described in Schwartz and Randall's analysis. Nor did their report include in the several steps it suggested any measures to actually prevent climate change by reducing emissions.

> Improve predictive climate models to allow investigation of a wider range of scenarios and to anticipate how and where changes could occur.
> Assemble comprehensive predictive models of the potential impacts of abrupt climate change to improve projections of how climate could influence food, water, and energy.
> Create vulnerability metrics to anticipate which countries are most vulnerable to climate change and therefore, could contribute materially to an increasingly disorderly and potentially violent world.
> Identify no-regrets strategies such as enhancing capabilities for water management.

Rehearse adaptive responses.

Explore local implications.

Explore geo-engineering options that control the climate.[32]

While the DOD continued to explore the science and implications of climate change, climate change was becoming politicized. In March 2007, Senators Richard Durbin and Chuck Hagel, Democrat and Republican, respectively, introduced a bill requesting that the *National Intelligence Estimate* assess the potential for climate change to pose a national security threat. But perhaps more effective was the analysis going on outside the Pentagon. After she left the DOD, Sherri W. Goodman continued her work framing climate change as a national security concern and Goodman led several projects intended to make climate change more visible. She and other former officials and retired officers were likely forced into taking a more direct advocacy role because, as Michael Thomas shows, "the armed forces and wider security institutions were prone to downplaying, or publicly avoiding, the security threats posed by climate change on account of keeping their non-partisan status."[33] Goodman was at the center of several key conversations in 2006 and 2007 that set the agenda and resulted in 2007 in the publication of several reports that were pivotal in framing climate change as a national security problem. In April 2007, the CNA Corporation published *National Security and the Threat of Climate Change*. This report was equal part analysis and testimonials, featuring statements by retired generals and admirals from every service and the participation of many former national security elites who were then out of government.

National Security and the Threat of Climate Change was the outgrowth of the formation of the Military Advisory Board formed in at CNA in 2006 and coordinated by Goodman, the executive director, and a March 2007 conference at the U.S. Army War College, "The National Security Implications of Climate Change." Retired Vice Admiral Lee Gunn was one of those involved in these conversations who worried a great deal about the potential for climate change-caused crises. In an address to the American Association for the Advancement of Science, Gunn explained the forces that led him to join the CNA effort, particularly that since 2006, the retired admirals and generals who joined CNA's Military Advisory Board had been assessing the impact of climate change on national security. Gunn said, "I have been there. And during the course of my Navy Career, I was in many places where the causes and consequences of environmental collapse were evident every day." But Gunn suggests that the DOD knows these crises are already happening. "Men and women in our armed forces are on the front lines, contending not only with direct military threats and confrontations but also with the consequences of crop failures, pollution, disease spread, mass human migrations, failed or failing states, and natural disasters, in many cases related in some way to the heating of the planet." Gunn argued that the DOD is ready to act and is planning for things the civilian world is not yet aware of. "Because men and women in uniform are so heavily engaged in dealing with the results of climate change, the Department of the Navy and Department of Defense have long been focused on helping people and nations avoid and contend with the resulting stresses. Americans expect leaders in defense to plan for every eventuality and to deal with every threat to our security. It should come as no surprise, then, that the Department of Defense and all the armed services are, in my view very far ahead of many other parts of the federal government in planning for and considering the causes and consequences of climate change, energy dependence and related issues."[34] Gunn even suggests that climate is a breeding ground for terrorism. "My last war was in Somalia. Climate change and agricultural conflict contributed to massive displacement of Somalis. I have visited Somali refugee camps in northern Kenya and seen the desperation and the fertile recruiting ground these are for terrorist organizations. When young people, primarily men, have no future, no prospects, no hope, and are offered an alternative to living in the refugee camps and looking at that kind of bleak future, terrorist alternatives are seductive."[35]

The key findings of CNA's *National Security and the Threat of Climate Change* were that projected climate change "poses a serious threat to America's national security," "acts as a threat multiplier for instability in some of the most volatile regions of the world," "will add to tensions even in stable regions of the world," and that "climate change, national security, and energy dependence are a related set of global challenges."[36] Their updated report of 2014, again featuring the statements of former military leaders, was similarly sweeping and alarming: "We are dismayed that discussions of climate change have become so polarizing and have receded from the arena of informed public discourse and debate. Political posturing and budgetary woes cannot be allowed to inhibit discussion and debate over what so many believe to be a salient national security concern for our nation."[37]

If the debate over climate change had become polarized, the narrative of climate change as national security threat was gaining traction among national security experts. For example, when former vice president Al Gore and the IPCC were given a Nobel Prize for their work on climate change in October 2007, the Nobel Committee announcement of the prize touched on the argument that climate change was a national security issue: "Indications of changes in the earth's future climate must be treated with the utmost seriousness, and with the precautionary principle uppermost in our minds. Extensive climate changes may alter and threaten the living conditions of much of mankind. They may induce large-scale migration and lead to greater competition for the earth's resources. Such changes will place particularly heavy burdens on the world's most vulnerable countries. There may be increased danger of violent conflicts and wars, within and between states."[38]

In November 2007, the Center for Strategic & International Studies and the Center for New American Security jointly released a major report, *The Age of Consequences: The Foreign Policy and National Security Implications of Global Climate Change.*[39] The report's authors included a former national security adviser, a former director of the CIA, and a deputy assistant secretary of defense. It was a sobering analysis, focused on exploring the implications of three scenarios—expected, severe, and catastrophic increases in global temperatures and the likely climate and political changes that would result. *The Age of Consequences* argued that climate change had begun and the most likely consequences of all the scenarios the report explored would be widespread and largely negative. There will be no "winners": "Unchecked global climate change will disrupt a dynamic ecological equilibrium in ways that are difficult to predict. The new ecosystem is likely to be unstable and in continual flux for decades or longer. Today's 'winner' could be tomorrow's big-time loser."[40] The report stated that the threat was unprecedented, but it was still possible to learn from previous disasters: "In the past, natural disasters generally have been either localized, abrupt, or both, making it difficult to directly compare the worldwide effects of prolonged climate change to historical case studies. No precedent exists for a disaster of this magnitude—one that affects entire civilizations in multiple ways simultaneously." The authors argued that "human beings have reacted to crisis in fairly consistent ways. Natural disasters have tended to be divisive and sometimes unifying, provoke social and even international conflict, inflame religious turbulence, focus anger against migrants or minorities, and direct wrath toward governments for their actions or inaction. People have reacted with strategies of resistance and resilience—from flood control to simply moving away. Droughts and epidemic disease have generally exacted the heaviest toll—both in demographic and economic terms—and both are expected effects of future climate change." They said that even though the scale and effects of global warming would be unprecedented in some respects, "past human behavior may well be predictive of the future."[41]

Not to be left out, in November 2007, the elite Council on Foreign Relations released a special report by Joshua Busby, *Climate Change and National Security: An Agenda for Action.*[42] Perhaps because its author was a mere academic, and not a collection of former national security officials, *Climate Change and National Security* did not receive as much attention as the other reports of that year. However, listed within it was a prestigious Advisory Committee for Climate Change and National Security that included retired military officers and experts. Busby made what were now increasingly familiar arguments about the potential effects of climate change on national security. He said that "climate change does not pose an existential risk for a country as large as the United States," but it did pose enormous risks to it and many other countries, including some that already posed a national security risk to the United States.[43] But his analysis stressed solutions, what he called "no-regrets policies, those that it would not regret having pursued even if the consequences of climate change prove less severe than feared."[44] Busby called for risk reduction and prevention, adaptation, and mitigation. It was a bold call to action.

The defense intellectuals and Pentagon officials were paying attention to the former military officers and officials. Even though the DOD was starting to grapple with adaptation, global warming remained on the back burner of national security concerns until the Obama administration. It is in this context that the 2010 Department of Defense *Quadrennial Defense Review* put climate change and energy security on the list of security concerns. It stressed two consequences, saying that "climate change will shape the operating environment, roles, and missions that we undertake" potentially serving as "an accelerant of instability or conflict." It stressed that the DOD will "need to adjust to the impacts of climate change on our facilities and military capabilities."[45] In an acknowledgment that "climate change, energy security, and economic stability are inextricably linked" the 2010 *QDR* highlighted programs that the armed services had already undertaken to reduce their fuel use and noted that the U.S. military was preparing for "prolonged outages caused by natural disasters, accidents or attacks."[46] But the *QDR* was clear that "climate change alone does not cause conflict." This distinction—between climate change as a primary cause of conflict as opposed to being an element in a larger context that *could* lead to conflict—becomes less clear in subsequent years.

In late 2011, the Defense Science Board released a report on the *Trends and Implications of Climate Change for National and International Security.* There was little if anything new in the report. The important element here, as with earlier analyses, was who was saying it. The Defense Science Board highlighted "observable trends over multiple decades," including: Increasing land and sea surface temperatures; changing ocean temperature; changing ocean chemistry (acidity and salinity); declining mass of Greenland and Antarctic ice sheets; declining glaciers and snow cover; decreasing Arctic sea ice; more frequent and longer droughts; increased frequency of heavy precipitation events; increased cyclones intensity; rising sea level.[47] The report said that "climate change is likely to have the greatest impact on security through its indirect effects on conflict and vulnerability. Many developing countries are unable to provide basic services and improvements, much less cope with repeated, sudden onset shocks and accumulating, slow onset stresses. These effects span the spectrum from the basic necessities of livelihood to social conflict, including protests, strikes, riots, intercommunal violence, and conflict between nations." But the Defense Science Board was careful to say that climate change is "more likely to be an exacerbating factor for failure to meet basic human needs and for social conflict, rather than the root cause. Climate change is already intensifying environmental and resource problems that communities are facing."[48]

The Director of National Intelligence James Clapper was explicit when he told Congress in 2014, "Extreme weather will increasingly disrupt food and energy markets, exacerbating state weakness, forcing human migrations, and triggering riots, civil disobedience, and vandalism."[49] That same year, the DOD's 2014 "Climate Change Adaptation Roadmap" underscored the potential for increased conflict: "the changing climate will affect operating environments and may aggravate existing or trigger new risks to U.S. interests."[50] In the 2015 *National Security Strategy*, the Obama administration said, "Climate change is an urgent and growing threat to our national security, contributing to increased natural disasters, refugee flows, and conflicts over basic resources like food and water. The present-day effects of climate change are being felt from the Arctic to the Midwest. Increased sea levels and storm surges threaten coastal regions, infrastructure, and property. In turn, the global economy suffers, compounding the growing costs of preparing and restoring infrastructure."[51] The White House said in 2016, "The national security implications of climate change impacts are far-reaching, as they may exacerbate existing stressors, contributing to poverty, environmental degradation, and political instability, providing enabling environments for terrorist activity abroad. For example, the impacts of climate change on key economic sectors, such as agriculture and water, can have profound effects on food security, posing threats to overall stability."[52] In 2016, the National Intelligence Council underscored the threat in a report, *Implications for U.S. National Security of Anticipated Climate Change*.[53] The National Intelligence Council listed a range of concerns from increased migration, to food shortages, to greater conflict and war caused by shortages of fresh water and access to arable land.[54]

In sum, Congress had been hearing the argument from former and current national security leaders that climate change was a national security threat for a decade, and so when the Trump administration ignored climate change in the 2017 *National Security Strategy*, more than one hundred members of Congress wrote to the president in January 2018 to underscore the risks and to urge Trump to include climate change in the *National Security*

Strategy. They said, "We have heard from scientists, military leaders, and civilian personnel who believe that climate change is indeed a direct threat to America's national security and to the stability of the world at large. As global temperatures become more volatile, sea levels rise, and landscapes change, our military installations and our communities are increasingly at risk of devastation. It is imperative that the United States addresses this growing geopolitical threat."[55] And at the same time, Retired Admiral James Stavridis argued that climate change was arguably the most pressing national security challenge the United States faced. Stavridis said, "What makes climate change so pernicious is that while the effects will only become catastrophic far down the road, the only opportunity to fix the problem rests in the present. In other words, waiting 'to be sure climate change is real' condemns us to a highly insecure future if we make the wrong bet. We are in danger of missing not only the vast forest of looming climate change, but the ability to see some of the specific trees that will cause us the most problems."[56]

The intelligence community kept its eyes on climate change even as President Trump denied global warming and had members of his administration scrub it from administration publications and policies. In January 2019, Daniel R. Coats, director of national intelligence, told the Senate Select Committee on Intelligence, "Global environmental and ecological degradation, as well as climate change, are likely to fuel competition for resources, economic distress, and social discontent through 2019 and beyond. Climate hazards such as extreme weather, higher temperatures, droughts, floods, wildfires, storms, sea level rise, soil degradation, and acidifying oceans are intensifying, threatening infrastructure, health, and water and food security. Irreversible damage to ecosystems and habitats will undermine the economic benefits they provide, worsened by air, soil, water, and marine pollution."[57]

National security intellectuals and former military in Europe and the United States formed the International Military Council on Climate and Security (IMCCS), with Sherri Goodman, again, at the top of the organizational chart. IMCCS, which is comprised of the U.S.-based Center for Climate and Security, The Council on Strategic Risks, and The Hague Center for Strategic Studies, includes national security experts from Europe, the United States and the UK. They constitute what they call an Expert Group, which in 2020 and 2021 produced reports aimed at influencing expert understanding of the links between climate change and security. Their *World Climate and Security Report, 2021* offers analysis of the interrelationships reinforcing relationships between climate change and climate-related events, such as heat waves and risk factors, and adverse impacts. The report also identifies areas in the world that are prone to climate change-caused risks.[58] Although there are sections focused on solutions to local and regional security problems, like the earlier "*Age of Consequences*" report, the predictions are dire.

After a four-year lull during the Trump administration and at least the appearance of a lack of careful official consideration of climate change as a national security issue, the Biden administration made climate change a major focus. In a March 2021 memo to Senior Pentagon leadership, Biden's Secretary of Defense Lloyd Austin said climate change "presents a growing threat to U.S. national security interests and defense objectives . . . altering the global security and operating environments, impacting our missions, plans and installations." Austin said the DOD would "incorporate climate risk analysis into all of our work from installation planning; to modeling, simulation, and wargaming; to the National Defense Strategy and all other relevant strategy." To that end, he established a Department of Defense Climate Working Group to coordinate DOD work on climate and "energy related" objectives and track implementation of any directives.[59] In May 2021, Avril Haines, the Biden administration's director of national intelligence, stated to Congress: "To address climate change properly it must be at the center of a country's national security and foreign policy."[60] The DOD's new "Climate Adaptation Roadmap," published in September 2021, threw agnosticism out the window. The plan stressed "climate informed decision-making" and argued that "climate literacy" in the department would help it achieve its goals.[61] A few weeks later, the DOD, National Intelligence Council, Department of Homeland Security, and the White House released four reports in October 2021, in advance of the Glasgow Conference of the Parties scheduled to begin later that month. The DOD released its *Department of Defense Climate Risk Analysis*; the White House produced a *Report on the Impact of Climate Change on Migration*; the Department of Homeland Security produced a "Strategic Framework for Addressing Climate Change";

and the National Intelligence Council published the *National Intelligence Estimate: Climate Change And International Responses Facing Challenges to US National Security through 2040*.[62] The analysis in these documents was consistent with previous reports in asserting that climate change is a major threat to U.S. national security and the United States had better be prepared.

But, in a way, these documents and the interviews Biden administration officials were giving to publicize them were the most alarming official assessments of the threat posed by climate change yet. For instance, the observation that climate change was causing the U.S. military to spend time away from its core missions potentially diminishing military readiness was explicit. According to Deputy Secretary of Defense Kathleen Hicks, "Frequent fire, loss of power through frequent storms like we saw in the deep freeze in Texas. That's very costly and it takes us away. Those forces that are located in those locations, they aren't focusing on mission. They're not flying on their training days, perhaps, or they're not out to sea or getting prepared to go out to sea. Rather, they're moving in and out for storm purposes. All of those are ways that we both are reduced in our ability to do our main mission, and it costs us money to repair."[63]

The Hicks interview did not include some of the things the DOD had recently been asked to do that were also, in part, a response to climate change. For several years, the U.S. military had been positioned at the U.S. border to keep back the human migration that is in part caused by the drought and flooding caused by climate change. While many of the people attempting to enter the United States through the southern border were political asylum seekers, others were fleeing the conditions that made it increasingly difficult to make a living in drought-, storm-, or flood-damaged regions of the world. In April 2018, President Donald Trump ordered troops—members of the National Guard from thirty-four states or territories—to the southern border to support the Department of Homeland Security and U.S. Customs and Border Protection in preventing illegal border crossings. In November 2018, the DOD sent 5,815 active duty personnel to the Southwest border to join in the enforcement of the border between the United States and Mexico. The DOD spent at least $841 billion on those operations.[64] The DOD report included a map of "representative climate change hazards and potential impacts on DOD missions." But it also did not include operations at the U.S.-Mexico border.

Climate change *will* likely increase migration to the countries that have the capacity to take in refugees. This may fuel anti-immigrant sentiment and heavy-handed policing to find, detain, and deport refugees. Christian Parenti worries that in response to climate change, "another type of political adaptation is already under way, one that might be called the *politics of the armed lifeboat*: responding to climate change by arming, excluding, forgetting, repressing, policing, and killing."[65] He suggests that the U.S. response to people coming to the United States through Mexico is a preview of this kind of adaptation.[66]

The DOD's *Climate Risk Analysis* was the familiar list of concerns. "As the frequency and intensity of these hazards increase, impacts are likely to expand competition over regions and resources, affect the demands on and functionality of military operations, and increase the number and severity of humanitarian crises, at times threatening stability and security." The DOD acknowledged that the worse impacts were not inevitable: "Climate change is one of many factors that contribute to instability and conflict; resilience and strong governance responses can reduce the likelihood of climate hazards having security implications." But it seemed to be bracing for the worst, stating, "However, in worst-case scenarios, climate change related impacts could stress economic and social conditions that contribute to mass migration events or political crises, civil unrest, shifts in the regional balance of power, or even state failure. This may affect U.S. national interests directly or indirectly, and U.S. allies or partners may request U.S. assistance."[67]

The most sobering of these four October 2021 reports was perhaps the *National Intelligence Estimate* (NIE). The National Intelligence Council predicted increasingly severe impacts as global average temperatures climb over the next twenty years. Like previous assessments it argued that there might be competition over natural resources such as water. But the report was frank in its assessment that countries were unlikely to meet the emissions reduction goals they set in Paris and that the pace of transition to low-carbon technologies was too slow to reduce the likelihood of the worst effects of climate change. This, they argued, would lead to growing tension

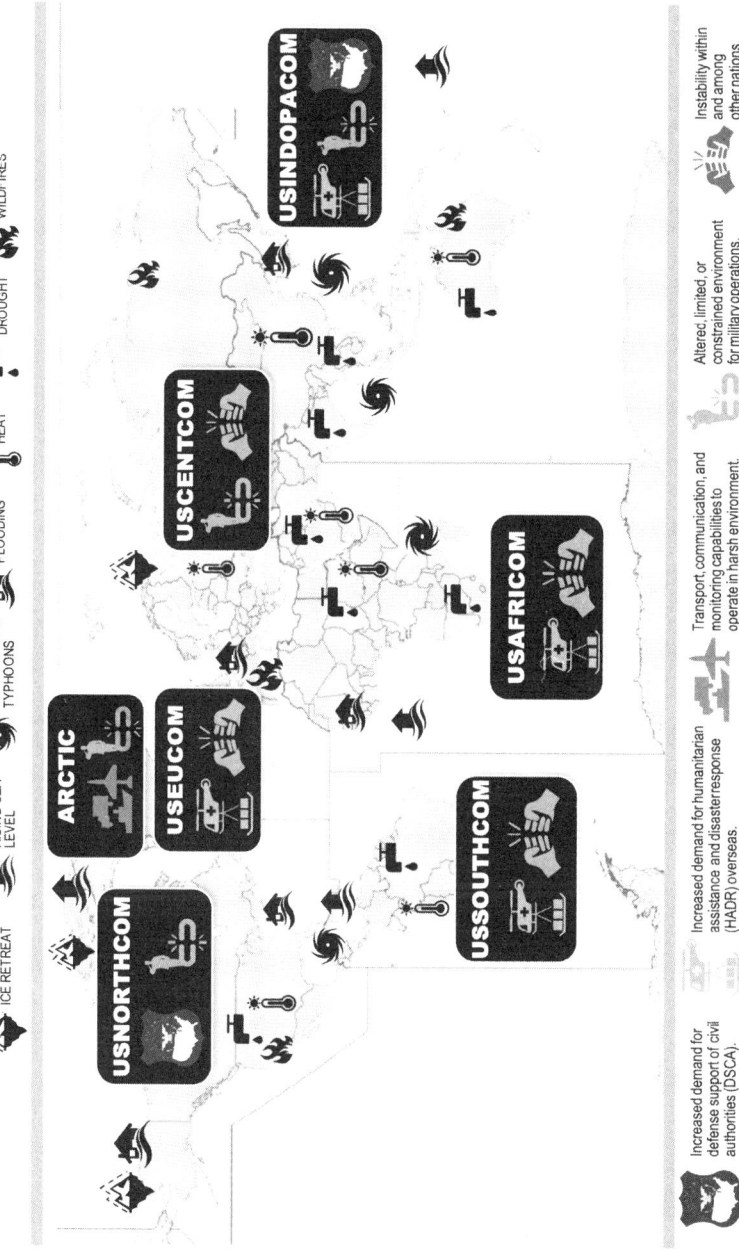

Figure 6.1

DOD map of representative climate change hazards and potential impacts, 2021.

Source: Department of Defense, Office of the Undersecretary for Policy (Strategy, Plans, and Capabilities), *Department of Defense Climate Risk Analysis*, report to the National Security Council, October 2021, 10, figure 2, https://media.defense.gov/2021/Oct/21/2002877353/-1/-1/0/DOD-CLIMATE-RISK-ANALYSIS-FINAL.PDF.

among countries. Further, military activity in the Arctic would increase and multilateral institutions would not be able to manage it. Most alarming, the NIE said that there was "a growing risk that a country would unilaterally test and possibly deploy large-scale solar geoengineering technologies as a way to counter intensifying climate effects if it perceived other efforts to limit warming to 1.5 degrees Celsius had failed. Without an international agreement on these technologies, we assess that such a unilateral effort probably would cause blowback." Not to mention the potential this would unleash for uncontrolled consequences to the environment.[68] The NIE map (figure 6.2) differed from the Pentagon map (figure 6.1) in emphasizing the countries in the world that it felt were most vulnerable to climate change effects, including the risk of instability.

The DOD argued that climate change makes hazards and risks worse. "The majority of climate hazards are not new; however, climate change is altering the frequency, intensity, and location of the hazards, contributing to vulnerability and compounding risks. Additionally, when climate change intersects with other forms of environmental degradation, such as deforestation and erosion, the impact can be even greater." For the DOD, climate change is clearly linked to potential conflict. "Climate impacts, such as increased competition over scarce resources, are likely to contribute to internal tensions within countries, as well as external tensions between countries. As the likelihood of multiple converging extreme events increases with climate change, risks can compound and put enormous pressure on any government's capacity to respond, increasing the possibility of cascading security impacts."[69]

While the Biden administration was struggling to get climate change legislation through the Congress in the fall of 2021, the kind of legislation that would make greenhouse gas emissions reductions at a pace and a scale possible to avert the disasters it predicted in those four reports, the administration itself was not emphasizing military emissions reduction, known as mitigation. Rather, the DOD seemed to assume that the worst-case scenarios it outlined could not be averted through mitigation.

The best, it seemed, the United States military could do was reduce its emissions, but the emphasis was on the need to adapt and prepare for

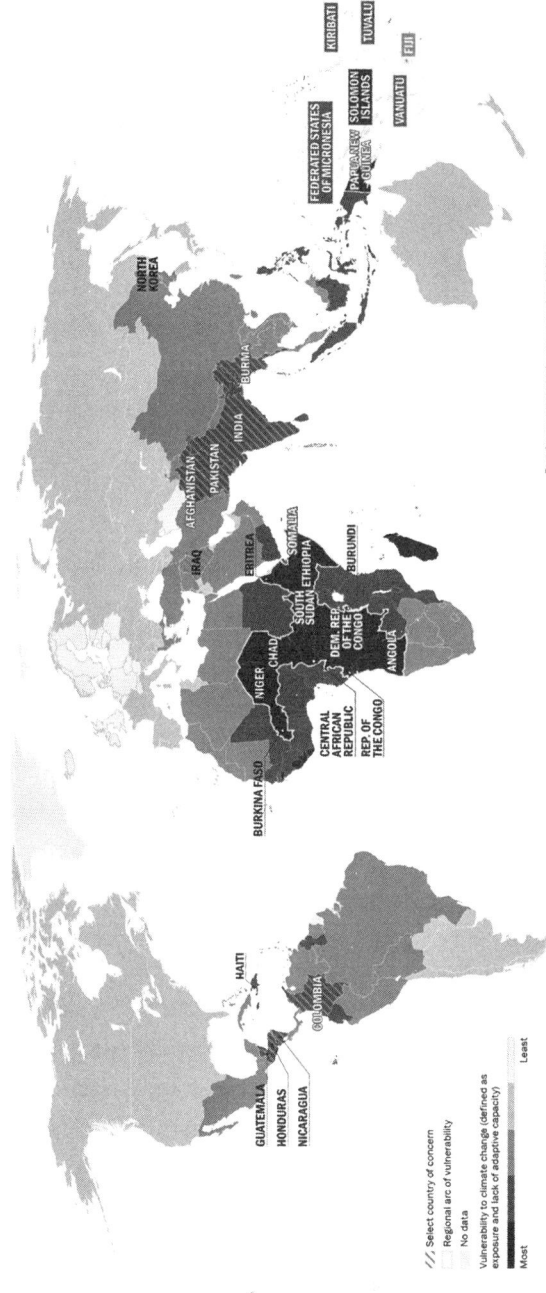

Climate Change in Select Highly Vulnerable Countries of Concern

The IC identified 11 countries and two regions of great concern from the threat of climate change. Building resilience in these countries and regions would probably be especially helpful in mitigating future risks to US interests. Two regional arcs also stand out because these groups of countries are clustered together, are relatively poor, and have little capacity to assist their neighbors.

Boundary representation is not necessarily authoritative.

Figure 6.2

National Intelligence Council map of vulnerable countries, 2021.

Source: National Intelligence Council, *National Intelligence Estimate: Climate Change and International Responses Facing Challenges to US National Security through 2040,* NIC-NIE-2021-10030-A, October 2021, 12, https://www.dni.gov/files/ODNI/documents/assessments/NIE _Climate_Change_and_National_Security.pdf.

whatever might happen. The implication is that the military will need more resources to deal with the threats.

When the U.S. Army released its own *Climate Strategy*, in February 2022, it described mitigation in some detail; it suggested that it could increase capability and prepare for new hazards and decrease energy demand, "all of which in turn will reduce greenhouse gas (GHG) emissions." But the army not only underscored the institutional concerns over increased vulnerability and "competition for scarce resources," it also emphasized the potential for armed conflict due to climate change.[70] The army argued, in fact, that these risks were "even more dangerous" than the institutional "hazards" climate change posed. "Chief among them is an increased risk of armed conflict in places where established social orders and populations are disrupted. The risk will rise even more where climate effects compound social instability, reduce access to basic necessities, undermine fragile governments and economies, damage vital infrastructure, and lower agricultural production. Adversaries and other malign actors may seize dwindling resources while seeking new opportunities to threaten U.S. national interests. Taken together, climate hazards will result in less economic and social stability, fewer goods to meet basic needs, and a less secure world."[71]

DOES CLIMATE CHANGE LEAD TO CONFLICT?

While current and former U.S. government officials have increasingly asserted that climate change could lead to war or that it already has, scholars have been more cautious, asking *if* climate change *might* directly cause war or cause environmental stress that exacerbates preexisting tensions and lead to war. Scholars like Thomas F. Homer-Dixon, who argued in several papers in the early 1990s that environmental hardship and resource scarcity could cause violent conflict, were early advocates of taking the climate change–conflict nexus seriously.[72] Michael Klare joined Homer-Dixon in raising the alarm in the early 2000s.[73]

For example, national security analysts sometimes suggest that a prolonged drought in northeast Syria from 2007 to 2010 triggered subsequent mass migration to cities, where overcrowding and lack of services created conditions that contributed to the emergence of the civil war there in 2011. The Syrian civil war itself then triggered mass internal displacement and refugee flows of about half the population and gave rise to ISIS, which the United States went to war to defeat in 2014.[74] Other candidates for a causal relationship between climate change and conflict include the Ogaden War on the Horn of Africa in 1977 and the Rwandan genocide in 1994.[75] The fact of a long drought in Afghanistan made drought-resistant poppy the crop of choice for some farmers, where the sale of opium was part of the income stream of the Taliban.[76]

The academic literature has explored these and other conflicts where climate change may have played a part in exacerbating existing tension and sparking armed conflict. Climate change and conflict literature often focuses on the Middle East, South Asia, and Africa. For instance, India is vulnerable to water shortages in already arid regions including Bihar, Uttar Pradesh, and Rajasthan. In the south, Andhra Pradesh, Haryana, and Karnataka may suffer water shortages that lead to agricultural failures and ultimately food shortages. Further, as the monsoon rains become less dependable, agriculture may fail. If sea levels rise in neighboring Bangladesh, millions could be displaced into India, adding further stress. In addition, water shortages could stress relationships with neighboring Pakistan. The same tensions over water exist in North Africa and the Middle East. Sub-Saharan Africa is also extremely vulnerable to climate change disruption, and it has more recently become highly militarized as governments fight Islamic extremists.[77]

The trouble is, while the connections have often been painted in vivid terms, social science has not reached a consensus about the relationship between climate change and war. Indeed, while some national security experts and journalists have been arguing that climate wars are coming, scholars of the environment and security say we should not be so fast to jump to the conclusion that climate changed caused conflict is inevitable.[78] In a recent assessment, Nils Petter Gleditsch argues that the academic evidence for a causal connection going from climate change to war is ambiguous.[79] For example, Marwa Daoudy disputes a connection between climate change and the Syrian civil war, arguing that the initial spark for that war was protests in solidarity with the larger Arab Spring movement then underway.[80] In any case, the relationships are much more complicated. For example, in Syria, researchers found that crop scarcity, in part due to drought, was related to violence to capture or destroy agriculture in the war.[81] Kathryn Mach and her colleagues argue that governmental capacity and science policy will determine whether governments will be able to cope with the stresses induced by drought, sea level rise, flooding, fires, and famine.[82] In sum, the politics of conflict and peace is harder, more complex in some ways, than the science.

Climate change has and will cause people to be miserable and insecure, but it is not inevitable that global warming will lead to violence and war.[83] There is no doubt that climate change will stress some regions very severely, and that it has and will cause people to move. On the one hand, it is true that social and economic stressors such as mass migration and tensions over access to fresh water that might lead to or exacerbate conflicts will increase as temperatures increase because of climate change. On the other hand, communities could and already have used existing institutions and developed mechanisms to cooperate in the face of climate change stressors. International agreements can make conflict over scarce resources less likely.

For instance, water resources are going to become increasingly important as global warming progresses. There are hundreds of transboundary aquifers and surface freshwater basins, as well as regions that depend on snow melt from mountains located in other countries. But there are interstate agreements on many transboundary water issues. Cooperation on water is much more the norm than conflict, and even during periods of conflict, these agreements are usually still observed.[84] "In the vast majority of cases water resources are shared in a cooperative fashion and conflicts are worked out via treaties."[85]

Even though India and Pakistan have clashed over other issues, and fought two wars, the Indus Water Treaty of 1960 between India and Pakistan has been successful.[86] The management of the Nile and the 1959 Nile Waters Treaty is another example of cooperation. The dispute between Egypt and Ethiopia over the Grand Ethiopian Renaissance Dam (GERD) on the Nile River has been contentious, because Ethiopia is damming Egypt's only source of freshwater. But so far, despite Egypt's much greater military power, the two states have been able to use negotiations and international law as they resolve the dispute.[87] In sum, other tools besides armed conflict are available. The international community could support dispute resolution in cases such as these by providing financial support to states that lack the economic resources necessary to build desalination plants.

Similarly, national security experts who are concerned about the Arctic sea becoming ice free year-round worry a great deal about the potential for conflict there between states rushing to exploit economic resources and claim a strategic advantage. However, the Arctic Council, formed in 1996, its members countries that border the Arctic, provides a forum that can alleviate and regulate competition there. The council has already been a forum for agreements on scientific cooperation, oil pollution, and search and rescue. Of course, its work will be more difficult if states continue to militarize their presence in the Arctic. That is not to say that coming to cooperative agreements about how to use resources in the Arctic is easy. But it is also true, as James Lee argues, that environmental "cooperation is far cheaper than violence."[88]

CONCLUSION

There is a growing belief among national security experts that climate change will cause not only stress but also potentially violent conflict as nations scramble to meet their resource needs or large numbers of people flee natural disasters. At least some of these experts seem to be preparing us for climate change–induced war. State and community failure *could* be caused by climate change; climate change–caused conflict is also possible.

However, *possible* does not mean *inevitable*. The research on climate change and conflict shows that cooperation over resources, including water, is not only possible, it is common. Preventing climate change stress by reducing emissions (mitigation) will ultimately be less conflictual and over the long term likely less expensive than responding after the fact to stress caused by climate change, including the collapse and failure of states incapable of reacting to the stresses of climate change. Acting now to reduce the most dangerous and disruptive effects of climate change by reducing emissions of carbon dioxide, methane, nitrous oxide, and other greenhouse gases; and pulling carbon dioxide out of the atmosphere by sequestering carbon in forests and soil will reduce human misery and decrease the likelihood of conflict brought on by climate change.

While those who want to encourage urgent action on climate change are eager to see the national security community validate their concerns about global warming, focusing on the risk of war can backfire if it undermines emissions reduction efforts. As Franziskus von Lucke argues, "The securitisation of climate change resulted in a climatisation of the security, defense and intelligence sector in the US and refocused the debate on adaptation instead of mitigation measures."[89] It may be that it is easier to get consensus on adaptation, especially among those who assume that it is too late to prevent the worst effects of climate change. But emissions reduction must still be a priority. The climate change–conflict nexus is taken for granted, perhaps because it seems intuitive, but like most of what we have seen so far in this book, policy makers can make events in the social world more or less likely depending on the policies they pursue. It *is* prudent to prepare for climate change by adaptation. But to the extent that emissions reductions are not the priority, and emissions remain high, we make the necessity for adaptation much more urgent and the risks of the worst outcomes greater. In other words, to the extent that adaptation and preparing for war caused by climate change fail to reduce emissions, or even increase them, adaptation may be counterproductive when those preparations increase the likelihood of the worst possible outcomes, including war. As I argue in the next chapter, the Pentagon and national security experts have focused more on adapting to climate change than on diminishing greenhouse gases.

IV THE WAY AHEAD

7 A LEAN GREEN FIGHTING MACHINE? MITIGATION VERSUS ADAPTATION

Unleash us from the tether of fuel.
—General Jim Mattis, 2003[1]

Our tag line is expeditionary energy. We don't do green.
—Colonel Brian Magnuson, 2016[2]

Whether they call it green energy or not, the DOD's work on sustainability is impressive, and perhaps no more so than the "Great Green Fleet." As part of a larger effort to decrease fuel dependency begun in 2009, the U.S. Navy collaborated with the Department of Agriculture to develop a 50–50 fossil fuel/nonfood biofuel mixture, substituting alternative fuels for military applications in military vehicles, including jets.[3] The navy had already "commissioned the USS *Makin Island*, its first electric-drive surface combatant, and tested an F/A-18 engine on camelina-based biofuel in 2009—two key steps toward the vision of deploying a 'green' carrier strike group using biofuel and nuclear power by 2016."[4]

Called mitigation, emissions reduction can be accomplished by reducing fossil fuel use through reducing demand and more efficient energy consumption to meet the remaining demand, or through switching to alternative forms of energy that do not use fossil fuels. As noted earlier, the United States demonstrated the Great Green Fleet's capacity to both conserve energy by using measures such as LED lighting and dashboards, and to use biofuel to power ships and aircraft in January 2016. The ships in the Great Green Fleet ran on a 10–90 blend of 10 percent biofuel, sourced from beef fat, and 90 percent conventional petroleum. In the ceremony launching the fleet, Secretary of the Navy Ray Mabus said, "It gives us an edge tactically, it gives us an edge strategically. It keeps . . . fuel from being used as a weapon against us." In fact, Mabus said, the U.S. Navy had cut its oil consumption by 15 percent and the Marine Corps by 60 percent since 2009.[5] Mabus noted that there had initially been skepticism about the concept of biofuels because they ostensibly were much more expensive than conventional fossil fuel. "We announced in 2009 that we would demonstrate the Great Green Fleet in 2012 and sail in 2016. And when we demonstrated the new fuels in 2012, we bought small quantities of experimental fuels. And we paid $26 per gallon, and some people said 'you'll never make this affordable.'" Mabus was proud to say that the navy had paid much less than that. "So here we are, in 2016, sailing the Great Green Fleet, and we paid $2.05 per gallon for the largest renewable fuel contract ever made. Let me repeat that. $2.05 per gallon. Even in these times of low oil prices, that's cost-competitive. And that's 13 times less expensive than just four years ago."[6]

There was both more and less to the promise of the Great Green Fleet. The Great Green Fleet also participated in Rim of the Pacific Exercises (RIMPAC) exercises in early 2017.[7] However, Mabus was skipping the fact that the U.S. Department of Agriculture was subsidizing the cost of the fuel. Secretary of Agriculture Tom Vilsack, also at the launch ceremony, said, "Today's deployment proves that America is on its way to a secure, clean energy future, where both defense and commercial transportation can be fueled by our own hardworking farmers and ranchers, reduce landfill waste, and bring manufacturing jobs back to rural America."[8] And while energy conservation measures and hybrid fuel drives have been installed in some of the U.S. Navy's new ships, the entire fleet is far from green.

As I have shown, the military has been convinced for more than a decade that climate change is indeed likely to cause a range of negative consequences to the military itself—persistent disruptions to military training, the loss of power and water at installations, a major distraction from the military's core missions as they help governments respond after disasters, and potential competition for federal budget share. The military's concern with energy security and the effects of climate change on bases and operations in part motivates its efforts to be more resilient. In other words, gains in greater efficiency and renewable energy at bases and installations have, on balance, been motivated by a mix of concerns about cost, resilience, and reliability as well as the desire to reduce greenhouse gas emissions. Fear of the potential for major wars arising from the climate crisis lurks in the background. Because the armed services and the Department of Defense have taken climate change as a fait accompli, and are increasingly stressing the inevitability of climate change caused conflict, adaptation is the priority.

The DOD *has* been adapting to climate change and has turned its attention to emissions reductions, framing its policies as resilience. Are these reductions likely to be substantial enough to address the climate crisis? The best way to know whether the DOD is striking the right balance is to examine what it has already done to reduce emissions and to evaluate its plans for future emissions reductions. For example, the Army's "Climate Strategy," released first among the armed service branches in February 2022, is a case study in future plans. It acknowledges the importance of emissions reductions and says that "climate change will only make [the army's] mission more challenging, and the Army must proactively reduce the risks that climate change imposes."[9]

I have framed the possible response to climate change as mitigation *versus* adaptation, but they are not necessarily opposed. Mitigation can be adaptive— that is, microgrids powered by solar energy at bases reduce emissions but are also adaptive when they make local power generation more resilient in the face of storms. Adaptation can be accomplished in a way that may or may not reduce emissions. For example, on the one hand, moving a power plant to higher elevation to prevent flooding will require energy consumption that may be offset by a switch to a different source of energy. On the other hand, rebuilding a base that has been damaged or destroyed in the same location so that it is more resilient, and then rebuilding it again after the next storm may be energy intensive, and in some cases, futile. Adaptation and mitigation are related, and the DOD sees it that way, although it treats the two differently at an organizational level. The Pentagon's response to the infrastructural and operational climate change-related challenges at installations has been to urge preparation through adaptation and increasing resilience. The DOD's operational efficiencies are creative and, in some instances, high tech. Some might argue that the innovations can be transferred to the civilian realm. In fact, on the basis of the military's overall record on climate change some analysts, such as Michael Klare, suggest that the military should be leading the way on climate change.[10]

What is the right mix and balance between mitigation and adaptation? Are the actions the Pentagon has taken so far, and its future plans for action to reduce emissions, significant and on the appropriate scale? As I argue here and in chapter 8, as effective as these efficiency and adaptation measures are, they do not make the deep reductions of greenhouse gas emissions that are needed, nor do these measures reflect a fundamental rethinking of the beliefs that have led to high military and military-industrial emissions. But, before I make that argument, it is important to appreciate scale and the scope of what the DOD has already done, and the steps that it is likely to take in the next several years based on the plans and promises that the military articulated in the first year of the Biden administration.

MITIGATION: REDUCING FOSSIL FUEL ENERGY USE

In 2000, the DOD said that "conserving energy is important to the Department, because it saves money and reduces greenhouse gas emissions harmful to the environment."[11] Specifically, the military thinks in terms of "operational" and "installation" energy consumption, and reductions in fuel use and emissions track this scheme. As I discussed earlier, strides in reducing military operational fuel use have largely been motivated by energy security concerns, a desire to reduce tactical vulnerability and increase tactical flexibility and reduce the costs of fuel. Lower fuel consumption rates increase the resilience of the military in war zones.

As General Jim Mattis told members of Congress when he testified in 2011, the United States was already working to reduce operational vulnerabilities associated with fossil fuel use. "There are efforts under way to make more expeditionary bases which would actually generate some of their own energy requirements using, for example, solar power. In many of these places, there is a lot of sunshine. If we can get expeditionary capability to capture that and then basically recharge our batteries. . . . We are engaged with Science and Technology, we are engaged with DARPA [Defense Advanced Research Projects Agency], and we are looking at very pragmatic ways of doing this. We are also looking at what we can do to actually change how we distribute fuel, to reduce the enemy's opportunities to come after us."[12] The quest for operational "energy security" led the U.S. military to make improvements in fuel efficiency, and as these fuel economies scale, they amount to significant savings.

The Pentagon became more fuel efficient and focused on "sustainability" for several reasons. First, as I showed in chapter 5, there was a growing realization during the "highly energy-intensive operations in Afghanistan and Iraq" that Pentagon fuel use was a problem in and of itself.[13] The DOD thus had an incentive to reduce the cost and the vulnerability of fuel in war zones.[14] The DOD wanted "energy resilience," or "the ability to avoid, prepare for, minimize, adapt to, and recover from anticipated and unanticipated energy disruptions in order to ensure energy availability and reliability sufficient to provide for mission assurance and readiness, including mission essential operations related to readiness, and to execute or rapidly reestablish mission essential requirements."[15] Some leaders in the DOD and the military services became convinced that not only greater fuel efficiency but also renewable energy itself made good sense—renewable energy provided backup power, reduced or replaced fuel consumption in the front line, and were even capable of providing "spin-offs" for the civilian sector. The focus on "resilience" made it possible to see renewable energy as a win-win. "Solving military challenges—through such innovations as more efficient generators, better batteries, lighter materials, and tactically deployed energy sources—has the potential to yield spin-off technologies that benefit the civilian community as well. DoD will partner with academia, other U.S. agencies, and international partners to research, develop, test, and evaluate new sustainable energy technologies."[16] While the military was cautious not to be seen as advocates of renewables for renewables' sake, some outsiders who were former insiders were enthusiastic about "clean energy" even if they were still reluctant to stress the importance of greenhouse gas emissions reductions. In early 2021, the Atlantic Council published a blog that argued, "Military and national security leaders from across the political spectrum know that climate change is a threat, and the clean energy transition is critical to a secure future. Now it is time for action to meet the magnitude of the challenge." Further, the authors, former advisers to the U.S. Army and the U.S. Air Force, said, "Clean energy can be a weapon, enabling forces to deploy more quickly and for longer, and helping keep troops and the homeland safe. A new DoD clean energy innovation agenda will build military capabilities and resilience, strengthen US leadership in the race to develop breakthrough clean energy technologies, and create thousands of American jobs to accelerate the economic recovery from the pandemic."[17]

Second, as the 2008 Defense Science Board report indicates, there was external political pressure to be more fuel efficient as the world became increasingly focused on emissions from all sources. This is part of a larger trend—with the Trump administration excepted—in the U.S. government since the late 1990s, accelerated during the Obama administration, toward energy conservation and efficiency.

However, the military often went out of its way to stress the tactical reasons for reducing fuel consumption and was careful to say, at least until recently, that its fuel efficiency is about operational gains, and emissions reductions are a nice though largely unintended consequence. As former assistant secretary for installations, environment, and logistics of the U.S. Air Force Bill Anderson said, "We're concerned about climate change . . . but the first mission is bombs on target." Anderson, who worked for President George W. Bush told *Scientific American*, "Where renewable sources and efficiency can help us there, I have seen the military grasp it like nobody's business, but you've got to stay on target. The Department of Defense is not the Department of Energy."[18]

Part of the Pentagon's response to concerns about climate change has been to create new organizations within the DOD and to study the issue.[19] For instance, the U.S. Navy created the "Task Force Climate Change" (TFCC) in 2009.[20] The idea that the military should be more sustainable was institutionalized in 2010 when the DOD established a Senior Sustainability Council and sustainability and resilience were articulated as an explicit element of the DOD's mission in the 2010 *Quadrennial Defense Review*. The motive for increasing renewables in the energy mix and decreasing overall demand was explicitly readiness and cost, as well as the need to reduce emissions, in other words, "to improve operational effectiveness, reduce greenhouse gas emissions in support of U.S. climate change initiatives, and protect the Department from energy price fluctuations." At this point, the investment in "noncarbon" sources was worth noting. "The Military Departments have invested in non-carbon power sources such as solar, wind, geothermal, and biomass energy at domestic installations and in vehicles powered by alternative fuels, including hybrid power, electricity, hydrogen, and compressed national gas." The DOD pointed to projects that demonstrated emphasis on sustainability including the Rucksack Enhanced Portable Power System (REPPS), which included a solar panel "blanket" for portable battery recharging.

The 2010 *QDR* also mentioned the Marine Corps Expeditionary Energy Office, created in 2009 to address operational energy risk. The Marine's Energy Assessment Team "identified ways to achieve efficiencies in today's highly energy-intensive operations in Afghanistan and Iraq in order to reduce logis-tics and related force protection requirements."[21] In 2009, the office began to collect data on Marine Corp energy usage from bulk distribution to the unit level and made its first study of fuel use in Afghanistan.[22] In 2010 the Marine Corps sent solar panels and chargers to its forces there.[23] A few months later, the solar panels deployed in the Ground Renewable Expeditionary Energy System (Greens) had reportedly cut generator fuel consumption by nearly 90 percent, from 20 to 2.5 gallons per day.[24] In 2011, the Marine Corps India Company deployed solar power at a Marine Corps post in Sangin, Afghani-stan, that cut fuel use by the diesel generator there by 60 percent.[25]

The 2010 *QDR* established ambitious targets including these: "By 2016, the Air Force will be postured to cost-competitively acquire 50 percent of its domestic aviation fuel via an alternative fuel blend that is greener than conventional petroleum fuel. Further, Air Force testing and standard-setting in this arena paves the way for the much larger commercial aviation sector to follow. The Army is in the midst of a significant transformation of its fleet of 70,000 non-tactical vehicles (NTVs), including the current deployment of more than 500 hybrids and the acquisition of 4,000 low-speed electric vehicles at domestic installations to help cut fossil fuel usage."[26]

Energy for the Warfighter: The Operational Energy Strategy, published by the DOD in May 2011, and its implementation plan, set other ambitious goals for the U.S. military.[27] These were summarized as "more fight, less fuel";

"more options, less risk," and "more capability, less cost."[28] It listed several positive outcomes of using less fuel, diversifying fuel sources, and considering the cost of energy in all its operations.

Saving lives now lost moving and protecting fuel on the battlefield;

Improving the range, endurance, and reliability of ground, air, and naval forces and information assets;

Lightening the logistics load and reducing the vulnerability of fuel supply lines;

Refocusing some combat forces and capabilities from supply lines and fuel logistics to operational missions;

Strengthening the Department's resilience to energy price and supply volatility and disruption;

Posturing the future force for success in meeting 21st century challenges by better aligning resources to tactical, operational, and strategic goals;

Building capacity and stability in and good relations with partner nations by sharing improved operational energy capabilities, including in civilian applications; and

Contributing to national goals, such as reducing reliance on fossil fuels, cutting greenhouse gas emissions, and stimulating innovation in the civilian sector.[29]

However, the military did not in this or subsequent analysis factor the costs of climate change in the cost of fuel use. The closest the DOD has come to factoring in some of the costs of climate change is their accounting of power outages at military bases, as I noted in chapter 5, which it began to report in fiscal year 2013.

In 2017, the military's push to reduce energy consumption ran into politics when the Trump administration decided to deny climate change. The Pentagon was put in an awkward position. A headline in *Fortune* magazine framed the issue nicely: "The Military Is Getting Greener, but That Clashes with Trump's Promises."[30] During the Trump administration, the DOD phrased its sustainability efforts at installations as *not* about greenhouse gas emission reduction because the president was explicit in his rejection of climate change, and perhaps because some conservatives believed renewable energy was wasteful. Rachel Zissimos of the Heritage Foundation told a reporter in the first year of the Trump administration, "The administration right now needs to focus specifically on combat power. Investing money on optional initiatives right now I think is problematic."[31] In May 2018 the Trump administration rescinded the Obama administration's federal energy efficiency goals (Executive Order 13653).[32] But the military continued its work anyway for tactical reasons, including the fact, for instance, that electric vehicles would be much quieter than conventionally powered vehicles. Colonel Brian Magnuson who led the Marine Corps Expeditionary Energy Office said green energy was a means to an end: "These technologies are a way to become more effective in combat. This is about war-fighting capability."[33]

The rhetoric shifted during the Biden administration. The U.S. Army *Climate Strategy*, released in February 2022, set a goal of reducing army net greenhouse gas emissions 50 percent by 2030, compared to 2005 levels.[34] The army does not provide a number for its emissions in 2005. However, DOD data, reproduced in table 4.3 in chapter 4, show that in fiscal year 2019, total U.S. Army emissions were about 10.6 $MMTCO_2e$, already a 42 percent reduction from its emissions levels in fiscal year 2010, which were about 18.4 $MMTCO_2e$. These were deeper emission reductions than the other services, which averaged 27 percent for the air force and 14 percent for the navy; the average emissions decrease for the entire DOD was 28 percent for the period from 2010–2019. Assuming that army emissions levels in 2005 were about the same or greater than emissions in 2010, it seems that the army may already be very close to its target for reductions. The army's goal is also to reduce emissions from all buildings 50 percent by 2032, from a 2005 baseline; they have already reduced installation emissions 33 percent from fiscal years 2010 to 2019. The army also set a goal of being net zero at installations by 2045. This is not to say that the changes are insignificant. The army's emissions reductions will be accomplished by electrifying all non-tactical vehicles, purchasing or producing carbon-pollution-free electricity at installations, and increasing building energy efficiency.

Mitigation can also be accomplished by carbon sequestration, and the army has over 13 million acres of land, which it says it intends to manage with an eye to sustainability. Typically, the army emphasizes the how land use that protects the environment also meets its mission. "Stewardship of Army lands can also help mitigate climate change threats by safeguarding forests and other beneficial environments alongside Army RDTE and training."[35] The Army Compatible Use Buffer (ACUB) program also preserves 420,000 acres of private land near army installations as buffers between its installations and its neighbors, limiting development and increasing safety near training areas. "Camp Shelby, Mississippi, recently used ACUB to sequester the equivalent of 120,000 metric tons of carbon dioxide annually, or roughly 2,500 average households' carbon emissions every year."[36]

Renewable Energy: Geothermal, Solar, Wind, Microgrids

The military has increased use of off-site renewable energy and its investment in producing renewable energy on site. Figure 7.1 shows changes in the Department of Defense's mix of electric and nonelectric renewable energy sources, on- and off-site over a ten-year period. Figure 7.2 is a snapshot of

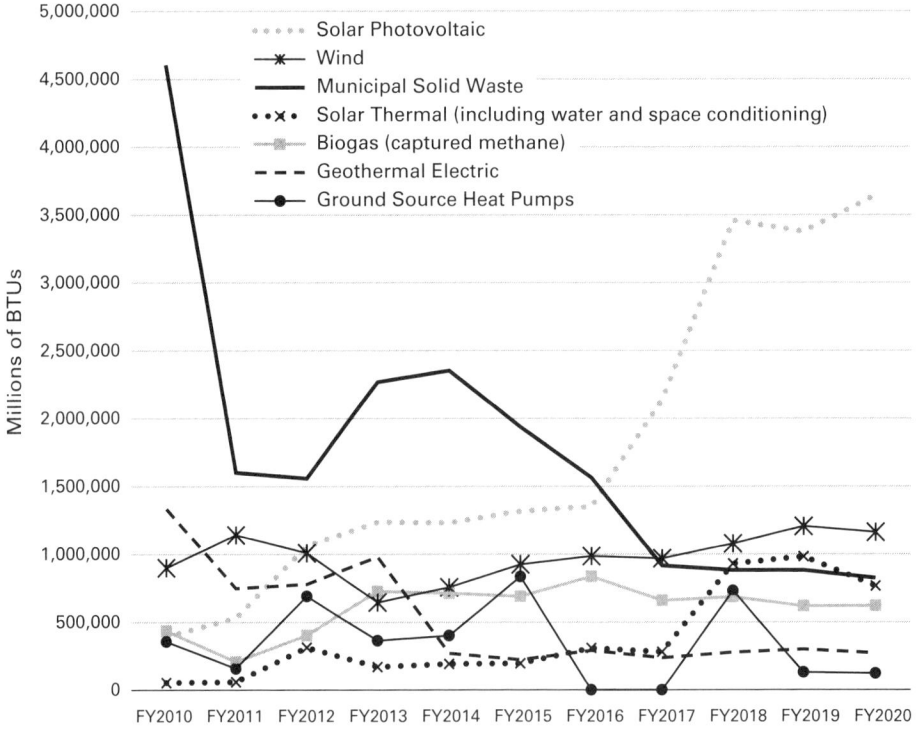

Figure 7.1

DOD electric and non-electric renewable energy sources, millions of BTUs, FY2010–2020.
Source: Department of Energy, "Federal Government Energy and Water Use in 2020," Federal Facility Reporting and Data, Comprehensive Annual Energy Data and Sustainability Performance, https://ctsedwweb.ee.doe.gov/Annual/Report/Report.aspx.

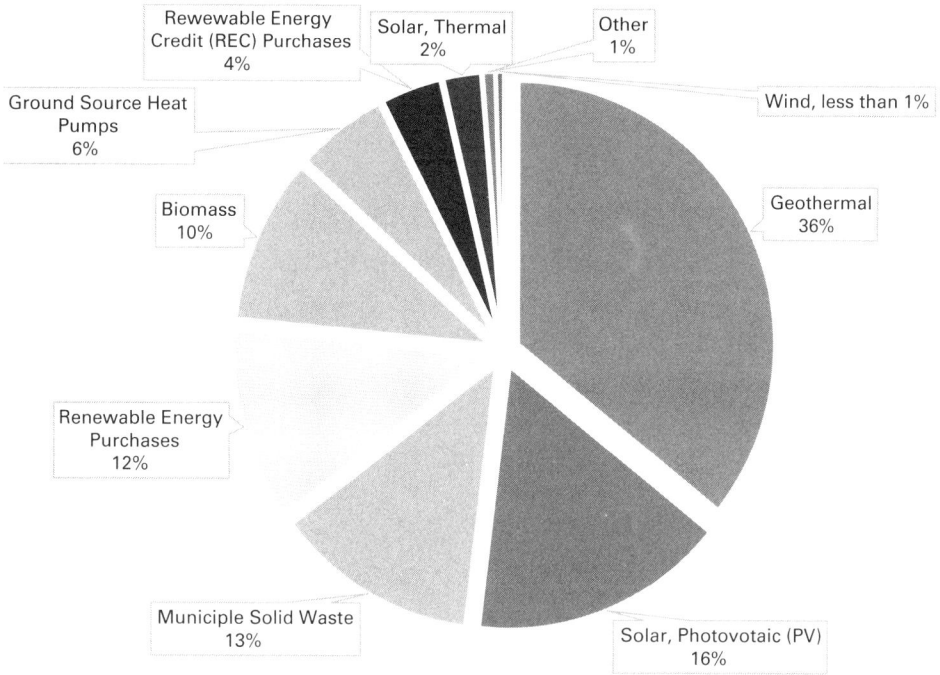

Figure 7.2

Alternative energy production and purchases at DOD installations in FY2016.

Source: Data from Office of the Assistant Secretary of Defense for Sustainment, *Department of Defense Annual Energy Management and Resilience Report (AEMRR) Fiscal Year 2016*, 35, https://www.acq.osd.mil/eie/Downloads/IE/FY%202016%20AEMR.pdf.

on-site production of renewable energy and purchases at DOD installations in fiscal year 2016.

While the DOD purchases most of its electricity for installations from outside its bases, it has increased resilience at its bases by adding renewable energy. The data show that the military has steadily invested in solar generation and other renewable energy by installing its own capacity at installations or purchasing it from off site.[37] An increasing number of facilities are public/private partnerships where a private company installs solar panels, for example, at a military installation and the energy produced there is used on the base and also distributed to the local community in a power purchase agreement (PPA).[38] For example, the army has on-site solar generation and off-site wind production at Fort Hood; solar on site at Fort Detrick, and biomass on site at Fort Drum, and the Air Force has solar on site at Vandenberg Air Force Base.[39]

The DOD has, on occasion, underscored the larger context, such as when it said in 2020 that "reduced electricity use at DoD installations, along with increased use of renewable energy, contributed to a reduction of more than 150,000 MTCO$_2$e in FY 2019 across the Department. This is the equivalent of removing over 32,000 passenger vehicles from the road for one year."[40] But in general, the Department of Defense is careful to say that efficiency and effectiveness are the reasons for its energy supply choices. The fiscal year 2019 *Annual Energy Management and Resilience Report*, which was published in June 2020, described its policies as "agnostic" in five locations in the document. "Increasing energy resilience on military installations remains a top priority for the Department. To this end, the DoD is technology-agnostic when deciding on solutions that mitigate energy resilience vulnerabilities and fill gaps in critical energy load requirements. More important than the technology type are the life-cycle cost-effectiveness of the system and the amount of critical load served. Renewable energy systems are among many technological options available to the Department and will continue to be implemented when they are the most feasible solution."[41] In 2020, the DOD also said, "As there are no Department initiatives to directly reduce GHG emissions, reductions are the result of other initiatives."[42]

At the same time that moving to renewable energy enabled some installations to become more resilient to power outages, some of the renewable sources that generate excess energy were supplying it to the regional grid. For example, the U.S. Navy's geothermal power plant at the Naval Air Weapons Station at China Lake, California, the largest source of on-site renewable energy for the military, not only powers the base, but it also sells electricity to the western power grid. The DOD also notes that the China Lake geothermal facility's annual electricity generation of 180 megawatts is equivalent to saving approximately 4.16 million barrels of oil each year.[43]

In addition, microgrids using renewable sources have also been incorporated into the system to increase the resilience of installations. The Redstone Arsenal in Huntsville, Alabama, has had a solar and battery storage backup on a microgrid since 2011 and Fort Hood in Texas had a solar and wind microgrid installed in 2017. Otis Air National Guard Base on Cape Cod, Massachusetts, moved to a microgrid in 2019, using a combination of wind and diesel power generation with battery storage, allowing the base to run off the grid if necessary. The microgrid at Otis designed by Raytheon is also capable of providing electricity to the grid in Massachusetts. Other microgrids were planned for Fort Hunter Liggett in California, where solar would provide the renewable power, as well as for Parris Island, Fort Bragg, and the Portsmouth Naval Shipyard in Maine.[44]

Operational Biofuels

Installations are about 30 percent of DOD energy consumption, and thus fossil fuel use reductions also have to occur on the operational side, which accounts for about 70 percent of total DOD energy consumption. Here, the experiment of the Great Green Fleet is relevant; it proved that it is possible to operate naval surface ships with biofuel. When first launched, the Great Green Fleet ran on a 10–90 blend of renewables to fossil fuels, but the plan was to ultimately use a 50–50 blend of renewables to fossil fuels in the surface ships that were made to use a biofuel blend.

Aircraft are the largest fuel users among all the services. In March 2011, an F-22 Raptor flew a test flight on a 50–50 blend of camelina-based biofuel and conventional JP-8 jet fuel, but biofuels have not made a large contribution to fueling airpower. The test showed that the F-22 was able to reach speeds of Mach 1.5 and it maneuvered at 40,000 feet. The C-17 Globemaster transport fleet has also been certified for using a jet–biofuel blend since February 2011.[45] It is possible, that all aircraft, including commercial aircraft, can run on food waste-derived biofuels. Jet fuel using "wet" food waste has enormous potential for two reasons. When used as tested in a 70–30 blend of biofuel to conventional jet fuel, biofuel can replace billions of gallons of fuel with a cleaner mixture. Biodiesel and other alternative fuels such as ethanol and hydrogen are also part of the mix for the military's nontactical fleet vehicles (e.g., the automobiles that are used on bases). In fiscal year 2005, 2.2 percent of that fleet was powered by alternative fuels. In fiscal year 2016, the share had increased to 8.8 percent and the aircraft on the Great Green Fleet were running on a biofuel–jet fuel mixture.[46]

The advantages of biofuels may be worth their price if they can perform as well fossil fuels and it seems they can. They can be engineered to be "drop-in" replacements for fossil fuels, that do not require any modifications of the equipment in terms of storage or engines. Further, the soot formation of biofuels is much lower than conventional jet fuel, which has the effect of reducing the formation of water vapor contrails, which, as discussed in chapter 4, are more greenhouse gas intensive than carbon dioxide. In addition, if the DOD were to use food waste to produce biofuels, the effect would be to reduce food waste and its associated greenhouse gas footprint. Food waste produces the greenhouse gas methane when it decomposes in landfills; therefore when the food waste is diverted from landfill, this leads to a further reduction in greenhouse gas emissions.[47]

Nevertheless, although this technology works in principle, research and testing of biofuels has not received a large amount of funding and the DOD has not made a major shift toward using them.[48] Biofuels have not been manufactured in large amounts. The construction for Red Rock Biofuels and Fulcrum Sierra Biofuels biorefineries only began in 2018. Further, there are questions about how much the fuel will ultimately cost to produce and whether its production will crowd out food crops.[49] The Biden administration plan to convert all U.S. aviation fuel by 2050 to a biofuel blend, if implemented, could eventually bring the cost of all biofuels down if it increases total biofuel supply.[50]

ADAPTATION

The Pentagon assumes that the most severe effects of climate change are inevitable, and it emphasizes adaptation and resilience in the face of the threats to its operations and infrastructure. Climate change has begun, and adaptation is prudent. The only question, if one assumes that too little mitigation will occur over the next several years, is how much and how fast the DOD will be required to act.

In its 2014 "Climate Change Adaptation Roadmap," the DOD said that it "is responding to climate change in two ways: adaptation, or efforts to plan for the changes that are occurring or expected to occur; and mitigation, or efforts that reduce greenhouse gas emissions." The adaptation goals are to first "identify and assess the effects of climate change" on the DOD, and then build climate change into its programs, operations, and plans so that it can "manage associated risks."[51] The DOD defined adaptation in 2016 as "Adjustment in natural or human systems in anticipation of or response to a changing environment in a way that effectively uses beneficial opportunities or reduces negative efforts" and resilience as the "Ability to anticipate, prepare for, and adapt to changing conditions and withstand, respond to, and recover rapidly from disruptions."[52] The Biden administration's 2021 *Department of Defense Climate Adaptation Plan* leaves discussion of mitigation of emissions largely to another report.[53]

When faced with climate change-related damage at bases, the response has often been to rebuild, albeit with better construction that is more likely able to withstand climate challenges, and often with greener designs. The way the military thinks about the vulnerability of its installations to climate change is illustrated in the response to the major climate damage two air bases suffered. As I described in chapter 5, Tyndall Air Force Base in Florida was nearly completely destroyed by the Category 5 Hurricane Michael in October 2018. Not only planes but also entire buildings were totaled. The military chose to rebuild the base, but in a way that was both more advanced and resilient, making it possible for the structures there to withstand another Category 5 hurricane, with up to 180-mile-per-hour sustained winds. At a total cost of between $4.3–4.9 billion Tyndall would become "the base of the future."[54]

A few months after Tyndall was hit, Offutt Air Force Base in Nebraska experienced major flooding with water reaching up to seven feet high in March 2019; one-third of Offutt was flooded after massive rains and melting snow caused the Missouri River to overwhelm two levees. In the hours before the flood overwhelmed the base, about seven hundred pilots filled 235,000 sandbags and 460 flood barriers in an effort to hold back the water. Nevertheless, Offutt's only runway was partially submerged and about sixty structures were damaged beyond repair.[55] See figure 7.3. The U.S. Air Force values Offutt Air Force Base assets, including Strategic Air Command, and its location in the middle of the country, traditionally far from most threats, so it decided to repair the base. Offutt is also near both the Missouri and Platte Rivers and

Figure 7.3
Flooding at Offutt Air Force Base on March 17, 2019.
Source: U.S. Air Force.

there are fourteen wetlands on the base itself. The base had a close call in 2011 when a flood reached within fifty feet of its runway. And in 2015, a land-use study plan conducted by the base and the Nebraska Metropolitan Area Planning Agency stressed the vulnerability of Offutt Air Force Base to flooding.[56] A year and a half after the flood, *Air Force Times* ran the headline, "After Massive Flood, Offutt Looks to Build a Better Base."[57] Kiran S. Jivnani and Inkoo Kang of the Atlantic Council argued that it would have been better to fully fund adaptation at Offutt than to react to the disaster after occurred: at a cost of $650 billion to repair, they say, "estimates indicate the disaster will likely cost much more to repair than it would have cost to prevent; preventative action would have cost only $22.7 million."[58] Indeed, each year, responding to climate change-related disasters requires billions in funding. In fiscal year 2020 alone, the air force had $3.5 billion in disaster recovery funding.[59]

The post-9/11 wars in Afghanistan and Iraq underscored the urgency of reducing fossil fuel use for tactical reasons, and the Pentagon undertook remarkable investments in technology and made operational innovations that resulted in fuel use reductions during those wars. In this sense, the Pentagon has been more innovative than other sectors of the U.S. economy. The military services have switched light bulbs to LEDs, put advanced thermostats in their tents, and developed new technologies including biofuels. Thankfully, incentives to increase fuel efficiency have dovetailed nicely with the DOD's environmental efforts and increased investment in renewable energy. However, while the efforts to reduce fossil fuel consumption have paid dividends over the last twenty years, and, while the DOD has led in adaptation, they have not made mitigation of emissions a clear priority until quite recently. There is, in other words, a lack of a sense of urgency and agency that is astounding for an institution that believes it can do almost anything. While the Department of Defense has reduced its emissions in ways that are impressive, it is still the single largest institutional energy consumer and greenhouse gas emitter in the United States.

Motivations may not matter if the outcome is the essentially same. But if the magnitude of the effects of climate change scale in proportion to the amount of greenhouse gases that the military and other institutions emit, and if climate change is more likely to harm more people than the conflicts and wars it is still possible to prevent by other means, ought climate change prevention rise to the top of the list of priorities? If the vulnerabilities and risks that most concern the military will only increase over the next twenty years unless the world begins to dramatically reduce all greenhouse gas emissions and sequester carbon dioxide—remove it from the atmosphere, for instance, by storing it in the soil or in trees—one might think that the military, itself a large emitter, should be eager to engage in large-scale emissions reduction. After all, an ounce of prevention is worth a pound of cure.

When Congress overrode President Trump's veto of the fiscal year 2021 National Defense Authorization Act in January 2021 they funded a new Climate Change Adaptation Roadmap. The law also provided for the National

Academies of Science to create a new National Academies Climate Security Roundtable to support the Climate Security Advisory Council established in the previous year's Defense Authorization. In early 2021 the Biden administration made emissions reduction a centerpiece of its climate initiatives for the entire government, including the DOD, which is both remarkable and builds on about fifteen years of work begun late in the George W. Bush administration and accelerated under the Obama administration. But, unlike previous administrations, the DOD is making the link to climate change much more explicit. As Biden's Defense Secretary Lloyd Austin said in March 2021, "We face a growing climate crisis that is impacting our missions, plans and capabilities and must be met by ambitious, immediate action." [60] Deputy Assistant Secretary of Defense for Environment and Energy Resilience Richard Kidd said, "Going forward, we're going to have to embed climate change as a consideration in all that we do. All of our resource allocation decision-making activities, our policies, and our strategies must include a climate change dimension."[61] The DOD started a Climate Change Action Team in March 2021 to make the work more effective and to assess installations.

President Biden unveiled his administration's "whole-of-government" approach to climate change in April 2021 which is intended to "promote safety and security at home and abroad." Specifically, the administration launched, with Canada, the "Greening Government Initiative" that would "lead by example in developing and implementing climate action plans that increase the resilience of and mitigate emissions from national government operations and real property." The White House also said, "Climate change has been identified by the Department of Defense (DoD) as a critical national security threat and threat multiplier. As a result, DoD has undertaken assessments of the impacts that the climate crisis has on American military installations."[62] The administration announced that it would do so by conducting climate exposure assessments on all major continental U.S. installations within twelve months and on all major overseas military installations within twenty-four months using a Defense Climate Assessment Tool (DCAT).[63] "The DCAT helps identify the climate hazards to which DoD installations are most exposed, which is the first step in addressing the potential physical harm, security impacts, and degradation in readiness resulting from global climate change."[64] The United States would also share that tool with its allies.

Unmentioned at the launch of the Greening Government Initiative was the DOD's contribution to U.S. emissions—specifically that DOD emissions are the bulk of U.S. government emissions. The administration seemed wary of saying that emissions should be reduced for their own sake. The political context still required giving additional reasons for action beyond a concern about climate change. For example, when in September 2021, the DOD and the Department of Homeland Security targeted hydrofluorocarbons, the greenhouse gases with greatest warming potential, for reduction the economic benefits of the transition were underscored. In announcing the change, administrator of the General Services Administration Robin Carnahan said, "Confronting the climate crisis is a top priority for the Biden Administration, and presents a historic opportunity to create tens of thousands of good-paying jobs for the American people. By working with fellow federal agencies to reduce harmful pollutants, GSA is fighting against a changing climate and improving air quality, while bolstering our economy here at home."[65]

The U.S. government and the Pentagon are complex institutions, often negotiating internal tensions and even sometimes living with contradictions. And though it sometimes moves at Mach speed and turns on a dime, as a whole, the Pentagon is an institution where, as they say in the navy, you can't quickly turn an aircraft carrier going full speed—it takes time to change course. Reading the Pentagon's statements about sustainability and resilience, it would be easy to believe that the U.S. military is a leader in emissions reduction. In fact, in some ways, these changes have been made at remarkable speed and the armed forces should get credit where credit is due. The military and the defense intellectual community slowly turned over the course of twenty years from a tactical concern about fuel efficiency, to department-wide conservation and energy resilience. Further, the Pentagon has also acknowledged that it is a greenhouse gas emitter and turned from downplaying climate change in favor of military superiority, to touting the virtues of clean energy.

Yet, the DOD has not, until quite recently, acknowledged that its own fuel use is a substantial contributor to global greenhouse gas emissions. Nor does it appear that decision makers within the Department of Defense believe that significant reductions in Pentagon fuel use, or indeed overall U.S. petroleum consumption, are a potentially significant way to reduce the risks of climate change-caused operational vulnerabilities and national security risks. Rather, while the Pentagon is keen to emphasize the advantages of greater efficiency, this has usually been decoupled from the climate emergency and the Pentagon's possible role in averting the risks altogether. The most recent *Department of Defense Climate Risk Analysis* published in October 2021 in advance of the Glasgow United Nations climate change conference, stresses adaptation and preparing for the risks of increased conflict over the need to reduce emissions. As J. E. Surash, a senior official in the army, told the Association of the United States Army in October 2021, "Climate change and its effects obviously pose a very serious threat to the U.S. national security interest. But I want to stress that . . . climate change does not alter the Army's overall mission, which is to deploy, fight and win."[66] The implicit assumption here is that reducing emissions "too much" could weaken the U.S. military.

Further, the DODs greenhouse gas emission reduction targets can appear impressive, as when the army says it will reduce emissions 50 percent by 2030 from its emission levels in 2005. As I showed in chapter 4, U.S. military emissions track U.S. wars. Total U.S. military emissions rose to an estimated 85 $MMTCO_2e$, a post–Cold War peak during the post-9/11 wars, in fiscal year 2004, and declined 40 percent, to 51 $MMTCO_2e$ in fiscal year 2020, as the U.S. reduced operations in Afghanistan and Iraq. Of these, operational emissions declined nearly 41 percent and installation emissions declined about 38 percent between 2004 and 2020. Thus, while the United States has found efficiencies and invested in renewable energy, most of the reductions in DOD emissions in recent years, and those that are likely to get it to its near term targets, are due to the fact that the U.S. military was engaged in less fighting as it ended its major wars in Afghanistan and Iraq.

Thus, the federal government's policies appear to be ambivalent if not contradictory regarding whether it treats climate change as an existential threat. On the one hand, Congress mandated the DOD to produce a plan to reduce military emissions. Specifically, the FY2022 National Defense

Authorization Act, signed in late December 2021, required that the "Secretary of Defense submit to Congress a plan to reduce the greenhouse gas emissions of the Department of Defense" no later than September 30, 2022.[67] The law does not require that the DOD reach net zero or even set a goal for emissions reductions. On the other hand, while the Biden administration issued an executive order mandating federal agencies to be "leading the Nation on a firm path to net-zero emissions by 2050," it exempted national security agencies from the Federal government's net zero requirement on December 2021.[68] Specifically, the order stated:

> Sec. 602. Exemption Authority. (a) The head of an agency may exempt particular agency activities and related personnel, resources, and facilities from the provisions of this order when it is in the interest of national security, to protect intelligence sources and methods from unauthorized disclosure, or where necessary to protect undercover law enforcement operations from unauthorized disclosure. If the head of an agency issues an exemption under this section, the agency shall notify the Chair of CEQ [Council on Environmental Quality] in writing within 30 days of issuance of the exemption under this section. To the maximum extent practicable and without compromising national security, each agency shall strive to comply with the purposes, goals, and implementation steps in this order.
>
> (b) The head of an agency may exempt from the provisions of this order any vehicle, vessel, aircraft, or non-road equipment that is used in combat support, combat service support, military tactical or relief operations, or training for such operations or spaceflight vehicles, including associated ground-support equipment.
>
> (c) The head of an agency may submit to the President, through the Chair of CEQ, a request for an exemption of an agency activity and related personnel, resources, and facilities from this order for any reason not otherwise addressed by subsections (a) and (b) of this section.[69]

Clearly, the question then is what is the "maximum extent practicable" that the DOD can reduce emissions "without compromising national security." This is a balancing act weighing the certain consequences of climate change—which the DOD calls a threat multiplier—and those that might still be averted, against the *potential* risks of conflict that might be averted through other means. Which raises the larger question: how much the United States should rely on the military—as compared to the tools of diplomacy, economic incentives, and economic sanctions—to respond to and shape the international security environment.

If the U.S. military is reducing its emissions, does it matter whether or not they identify mitigation as a top priority? After all, since the DOD is the world's single largest energy consumer, any reductions in fossil fuel consumption are welcome. That climate change is often not the sole motive, or only one reason in a mix of motives for changing DOD policies, practices, and procurement, may not be important if the Pentagon is reducing emissions as much and as quickly as possible. But it has mattered. The emphasis has been on adaptation to the effects of climate change, and less on mitigation with the result that comparatively little effort has been taken to reduce Pentagon emissions.[70]

Finally, there is a bit of confusion about the difference between adaptation and prevention. Adaptation before a disaster is not true prevention. Prevention in the case of climate change would be emissions reduction and carbon sequestration. Only significant mitigation efforts so that the duration and frequency of extreme weather events would be reduced should be considered prevention. Installations constitute about 30 percent of DOD energy consumption; operational use is the other approximately 70 percent of DOD energy consumption. And thus, while green bases are important, reducing operational emissions is urgent. Operational emissions declined as the United States withdrew from Afghanistan and Iraq. Adaptation can reduce or even prevent damage from climate change, but it cannot prevent climate change itself. If the emphasis were more squarely and urgently on emissions reductions, then, ultimately, less adaptation would be required. Deeper operational emissions cuts, as I discuss in chapter 8, "The Path to Climate Security," are both possible and necessary.

The concept of "security" has flowed through this book like, well, oil—energy security, climate security, national security, and potentially human and ecological security. In the late 1990s, U.S. political leadership had a choice to make. The United States could emphasize national security, which has traditionally been understood as requiring military force to protect national borders and shape world events. This is security understood as the capacity to project power everywhere, essentially any time, to preserve U.S. global military dominance and promotfe its economic interests. This was the familiar path, rooted in deep cycles of consumption, fossil fuel demand, military forces to protect access to fossil fuels, increasing utilization of fossil fuels not only for power but also for plastics and industrial agriculture, fear of loss of access to oil, and back again, recursively, to ever higher levels of fossil fuel use and emissions.

Or the U.S. government and military leaders could have chosen an alternative path—to take advantage of the end of the Cold War to emphasize, human security, which depends on ecological security. Ecological security, the health and well-being of the biosphere, makes all life possible. Human security in the context of global warming is understood as securing people from the actual and potential effects of a dramatically warming planet not by military force, but by trade and diplomacy. It includes an investment in ways of growing food that do not depend on deforestation and nitrogen-based fertilizers and in modes of transportation that do not depend on petroleum. Climate change migrants from outside our borders are understood not as a threat, but as people who have been pushed out by the same droughts, fires, and floods that also affect many people in the United States and Europe. The workers in the fossil fuel industry, coal miners and oil industry workers who provide us with oil and natural gas, are not obstacles to change, but people who need assistance to find other, equally meaningful ways of making a living. The men and women who work in the military and military industry and in the towns that serve military bases are people who may need new jobs and a just transition.[1] The most vulnerable need help dealing with the consequence of choices that they themselves did not make, but which have put their lives and livelihood in jeopardy, nevertheless.

In a way, it is understandable that in the late 1990s the U.S. military and political elite chose a conception of security that stressed military dominance over human and ecological security. Recall that in 1997, the Department of Defense warned the White House of the dire consequences that could flow, not from global warming, but from the Kyoto Protocol. They said that "imposing greenhouse gas emissions limitations on tactical and strategic military systems would . . . adversely impact operations and readiness." The DOD said a 10 percent cut in fossil fuel use would reduce the U.S. advantage in tactics and training, which would then diminish U.S. combat effectiveness. If the U.S. Air Force cut 210,000 flying hours per year, fighter and bomber crews would "be unable to maintain full combat readiness," and a 10 percent cut would also reduce air lift and aerial tanker capacity.[2] DOD leaders concluded their argument this way: "While global climate change may be a serious threat to the nation's long-term interests, there are other threats we must not forget. We must not sacrifice our national security or our ability to offer humanitarian assistance to those in need to achieve reductions in greenhouse gases. We must not see this as an issue of being able to achieve either national security or protection of the global climate. The United States must pursue both objectives."[3] The DOD argued, "To accomplish this, the DOD strongly recommends that the United States insist on a national security provision in the climate change protocol now being negotiated."[4] The Pentagon essentially achieved their goal—ensuring that military emissions would be thought of separately from other emissions.

Congressman Henry Waxman was thus rowing against the tide in May 1998 when he articulated a new vision of security and urged Senate ratification of the Kyoto Protocol. Waxman said, "The Kyoto Protocol will improve the national security of the United States by reducing the risk of catastrophic climate change, which would create upheaval and unrest throughout the world, including the potential for millions of environmental refugees." Waxman also argued that "measures to implement the Kyoto Protocol can improve our security by reducing our dependence on imported oil through improved energy efficiency and increased reliance on domestic renewable energy resources."[5] Of course, the United States did not ratify the treaty even as it was able to have the rest of the world ratify a treaty that it in part wrote to protect military emissions from being counted and to preserve the United States' preeminent military power. It is not as if the Kyoto Protocol by itself would have saved the planet—the treaty was only a first step. Subsequent steps would have, among other measures, included capping the emissions of developing countries.

Just as Kyoto was a possible turning point, we are now at another moment when the United States has a choice between alternative approaches to security. The Biden administration's official position is that climate change poses a grave threat and "will increasingly exacerbate risks to U.S. national security interests as the physical impacts increase and geopolitical tensions mount about how to respond to the challenge."[6] In documents released by the Biden administration in anticipation of the United Nations' climate change Conference of the Parties (COP 26) meeting in Glasgow in 2021, the climate change–conflict nexus was asserted as a taken-for-granted and almost inevitable fate. Many people are already fleeing disasters caused by climate change—more intense hurricanes caused by higher ocean temperatures, giant and frequent wildfires that are fueled by extremely dry forests in drought-stricken regions, floods caused by torrential rainstorms and melting snow, rising sea levels that force people to flee island nations and coastlines, and a concern about diminished capacity of the earth to grow food. Although the streets of Glasgow were packed with people and non-governmental organizations demanding a different vision and a more urgent transition, including greater transparency around military emissions data and cuts in military emissions, there was little movement on these questions in the official meeting venues. This is the *now* familiar path. But does the U.S. military need to remain on its traditional path in order to provide for national security?

In this chapter I explore the consequences of business as usual by discussing the deep cycle of war, armaments production, industrialization, and rising emissions and whether and how the United States might take another path to security. The answer depends on thinking through the issues at several levels. I have already shown that U.S. military and military-industrial greenhouse gas emissions are significant. Are the military's current levels of fossil fuel consumption and their associated greenhouse gas emissions necessary to keep the United States the world's most militarily powerful nation? The military has already said that they can cut emissions. Assuming that military emissions will decline as the U.S. military gradually transitions off fossil fuels, but that the missions and force structure stay essentially the same, are those remaining emissions necessary because the United States needs a large military to defend against the major threats we face? Or can military emissions be further reduced?

I have shown that the military and war drive a much larger cycle of greenhouse gas emissions. I suggest that even as the United States reduces *some* of its military emissions and the defense industry becomes more efficient, the overall national security policy of the United States, and its associated military doctrines, *unless significantly restructured* will continue driving the military and military-industrial emissions of other countries—the United States' allies *and* its adversaries, most notably China. The fact is that the world must move much more quickly than current plans project to reduce greenhouse gas emissions in order to forestall and perhaps even prevent the worst outcomes that will occur as the atmosphere warms. Even in the context of greater fuel efficiency and continued emissions cuts in the Pentagon, continued military supremacy and a new cold war with China or Russia may come at the cost of a livable planet.

Understanding and breaking the deep cycle(s) of militarization and rising emissions is perhaps one of the most important things governments can do to make a more peaceful world more likely. The United States can be secure without having overwhelming military dominance. There are off ramps from the present path that offer the potential for transforming global security, moving from a confrontational and arguably quite risky trajectory, while much more dramatically reducing all nations' military and military-industrial greenhouse gas emissions. This is a path to climate security that is, fundamentally, how it will be possible to guarantee human and national security. I acknowledge that none of what I am proposing is easy or without risks. But in a time of crisis, it is worth thinking in new ways. Others have already begun to do so.

THE CURRENT PATH AND ITS CRITICS

There is no doubt that the United States is the most militarily powerful country in the world and that the Pentagon and U.S. taxpayers work very hard to keep it that way. With a budget authorized at over $730 billion in fiscal years 2020 and 2021 and $770 billion in fiscal year 2022, U.S. military spending was about half of the entire discretionary federal budget in each of those years. In fact, U.S. military spending is greater than the *combined* military budgets of the next fourteen highest countries.[7] Over the last decade, the United States has consistently spent more than *twice* the total *combined* military spending of its three most talked-about potential adversaries Russia, China, and Iran.[8] All that investment—including a substantial portion for fossil fuel—has assured that the U.S. military is dominant in quantity, quality, and mobility. While India and China each have more people in their active duty armed forces, and Russian active duty and reserve personnel outnumber U.S. active duty and reserves, it has long been U.S. policy to emphasize lethal firepower, superior technology, and outstanding training rather than numbers of troops.[9] The United States and its NATO allies are powerful adversaries. The U.S. alone has more capacity to project power and kill people—including more deployed nuclear warheads, military bases, long-range transport aircraft, amphibious assault ships and aircraft carriers—than any other country in the world. Indeed, since the Vietnam War, the United States often uses its airpower rather than, or before, sending in ground troops. The wealth of the U.S. economy, which has supported high levels of military spending, combined with the technological prowess of its military industries, has enabled the United States to sustain its dominant military position for decades. And, as the DOD avers, fuel has historically been and still remains essential to the U.S. military.

Many U.S. policy makers and military strategists assume that protecting access to Persian Gulf oil is a crucial, perhaps even vital interest of the United States. The United States has built its foreign policy and military forces on that assumption and on the view that U.S. dependency on oil imports is staying the same, or even increasing. But the policy consensus that developed in the wake of the 1973 oil embargo and the 1979 Iranian Revolution—that the United States must be ready, willing, and able to intervene to "protect access to Persian Gulf oil"—has been increasingly questioned.

In 2017, the American Security Project, composed of a number of former military and security experts, published a report *Powering the Department of Defense: Initiatives to Increase Resiliency and Energy*, suggesting that the United States should invest in alternative energy because "the military is vulnerable to sudden changes in the price of energy and disruption of supply." The argument was that "the military's overreliance on petroleum-based fuel causes operational, strategic, and financial risks and endangers critical missions. On installations in the homeland, the military relies on an electricity grid that is stressed and vulnerable to both extreme weather and enemy action. This reliance could negatively affect its projection of power and capability for defending the homeland, as well as its operations abroad." The group stressed an estimate that the United States had "spent $8 trillion protecting oil cargos in the Persian Gulf since 1976, with at least one aircraft carrier being stationed in the region during any given time."[10] In 1998, all of these assumptions— that the United States military had to provide access to oil, and that it was worth killing people and spending trillions of dollars to do so—were taken for granted, yet this group of former Pentagon and Washington insiders questioned them. They wanted, of course, to retain military supremacy, but they weren't sure about whether the current policies were correct.

In 2020 Ensign Benjamin Chiacchia probed a little deeper. In an article about the Great Green Fleet in the June 2020 issue of the U.S. Naval Institute's *Proceedings*, Chiacchia acknowledged that the aim of the Great Green Fleet was not to reduce emissions. The Great Green Fleet was, according to Chiacchia "too limited in scope" and it "ignored important areas such as environmental impact and force readiness at a time when climate change necessitates substantial changes to strategic thinking and planning."[11] Chiacchia noted that the

U.S. military is a large greenhouse gas emitter, and said, "The 'Green' aspect of the Great Green Fleet should be implemented through intentional policy changes. . . . Intentional changes to shore installations and assets also should have been included in the policy. Moving shore power to all renewable power would allow for direct, scalable provisions of electricity to bases irrespective of changes to the wider power grid, and would contribute to combating climate change's effects on environmental security and the viability of the Navy's numerous shore facilities (many of which remain vulnerable to the effects of climate change)." Chiacchia called for a radical rethinking of the U.S. Navy, one that takes a much larger view of security and the means to provide it, arguing that a smaller fleet would be nimbler, require fewer resources, and could also help lower the navy's carbon footprint. Chiacchia concluded that if the navy "does not change how it operates and climate change continues unabated, U.S. forces rightfully will be understood as a threat to vulnerable civilian populations the world over. The United States will lose credibility as a sincere and trustworthy ally, which would deal a blow to U.S. security strategy far greater than the marginal impact of drawing down its force presence abroad."[12]

Building on these and similar critiques some strategists argue that the United States should dramatically and safely reduce its military presence in the Persian Gulf and the greater Middle East.[13] These proposals to reduce U.S. presence or even leave the region generally rest on three propositions: (1) the United States is no longer dependent on Persian Gulf oil; (2) even if the United States were still dependent, the political and military costs of defending Persian Gulf oil are very high, and probably not worth it since the risks of a hostile takeover are low and the continued U.S. presence in the gulf creates problems; and (3) the United States could use much less military power and foreign aid to protect against any significant risks to the flow of Persian Gulf oil. Let us consider these arguments in turn.

The first proposition is that the United States is much less dependent on Persian Gulf oil than in the past. Still, we have to ask what would happen in the worst-case scenario—if oil flows from the Persian Gulf were curtailed or ceased for a week or several months. There would be some disruption to the U.S. economy and on the logic of uncertainty, there would be an increase in oil prices. But the economy would not collapse if oil flows from the Middle

East were disrupted for some time. Indeed, it is arguable that a total loss of Persian Gulf oil for up to three months would be, at worst, expensive. Higher costs may be painful for many who are dependent on fossil fuels, but those costs are generally not life-threatening. Further, in the short term, it is likely that the United States would adapt—as it did when the United States sanctioned Iraq and halted oil imports from Iraq and Kuwait after Iraq's invasion of Kuwait in 1990. The principal reaction of world markets would be an increase in the price of oil.

But the scenarios that propose an adversary would orchestrate a deliberate and total cutoff of Persian Gulf oil out of the blue are, in some senses, exaggerated. *The point of pumping oil is to use it or sell it.* A total cutoff of Persian Gulf oil by producers is thus unlikely. If one occurred, it would likely not be sustained for a long period of time. Even if a single country controlled supplies, that country would still almost certainly want to sell the oil. The exception to the rule that states want to sell their oil is when, during a war, a country hopes to deny its adversary either oil revenue or the oil itself. We saw this earlier, in the case of Iraq, which burned Kuwaiti oil fields when Iraqi forces retreated from Kuwait in 1991, and the Islamic State, which burned oil facilities as its soldiers retreated from 2015 to 2018. Similarly, the United States destroyed ISIS oil assets as a means to disrupt their source of revenue. While it has and might aggress elsewhere, Russia is unlikely to take the risk or bear the costs of invading the Persian Gulf since it has its own large reserves of petroleum. Further, Russia wants to retain the benefits of selling its oil and natural gas to Germany and other customers. Oil underwrites and bankrolls Russian military power.

The United States would be able to weather the unlikely event of a total cutoff of Persian Gulf oil by increasing its imports from other countries, by increasing domestic production, by drawing on the U.S. Strategic Petroleum Reserve (SPR) or by decreasing consumption. The SPR has over the last decade held between 726 and 634 million barrels of crude oil. This reserve could carry the United States through several months of shortages in supply if all OPEC countries were to cease imports into the United States, and much longer if *only* Persian Gulf supplies were cut off. For example, if the SPR were 630 million barrels at the start of a cutoff of Persian Gulf imports, and we assumed that the United States was depending on imports of one million barrels per day, the SPR could supply the U.S. market with the shortfall for more than a year and a half. If the United States had a shortfall of two million barrels per day, the SPR could make up the shortfall in imports for more than ten months. While it typically takes two weeks for the supply to begin moving out of the reserve storage and through the pipeline, the United States could draw oil at a much greater rate, up to 4.4 million barrels per day.[14] In addition, the U.S. military, which uses about 90–100 million barrels of oil each year in recent years, as I discussed earlier, has an average of a 50 million-barrel stockpile of fuel at any one time. Conservation and increased efficiencies could obviously increase the duration of the use of these supplies. Nevertheless, economists tend to associate oil price shocks due to restrictions on the flow of oil (such as the Arab Oil Embargo) with recessions. This is true. But the United States is much less dependent on Persian Gulf oil than in the past and has the capacity to increase its own production. Further, as the United States develops alternative energy, its vulnerability to oil supply disruption significantly declines.[15]

The second proposition is that the political and military costs of defending access to Persian Gulf oil are very high, and probably not worth it since the risks of a hostile takeover are low and the continued U.S. presence in the gulf generates political resistance among the populations of those states. In addition to the estimated $5–50 billion in annual costs of U.S. presence in the gulf described below, even when the United States is not at war in the region there are significant political costs for supporting some of the regimes there.

But those costs might be worth bearing if the risks of a hostile takeover of the Persian Gulf were high. The risks that are most often raised are these: (1) a hostile regional or local power will seize all or most of the oil by military means or a local power will establish hegemonic control over the Persian Gulf through ideological means; (2) a state will destroy the oil so that no one can use it; or (3) a state or pirates significantly raise the cost of oil by stealing oil or destroying oil production facilities and oil infrastructure.[16] The first scenario is the most likely of the three. The concern is that a hostile power would gain control of oil in the Persian Gulf—for instance, by occupying Saudi Arabia and Kuwait or by blocking the Strait of Hormuz—and be able to control world supply and increase the price of oil. It is in response to this fear that the United States created the Strategic Petroleum Reserve in 1975 and the Rapid Deployment Force (RDF) in 1979, whose specific mission was to defend U.S. interests in the Middle East, including oil. As Joshua Rovner, Eugene Gholz, and Barry Posen have argued in separate analyses there is little risk of a single entity, even Iran, having the political or even military power to take uncontested control of the Persian Gulf.[17] Keeping forces in the region there all the time, whether on ships or at bases, is expensive, not to mention fuel intensive given that the risks are actually not high. See figure 8.1.

The third proposition is that even if the United States were completely absent from the Persian Gulf, the United States could, in the unlikely event someone unfriendly to the United States took control of Persian Gulf oil, return and restore oil flow.[18] Indeed, the U.S. Navy with its global presence

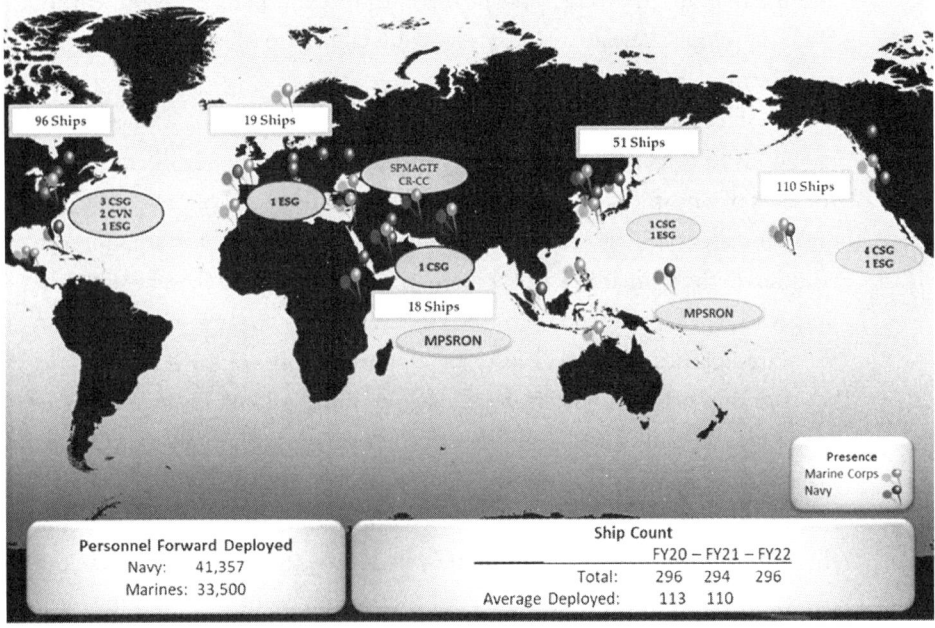

Figure 8.1
U.S. Navy presence in 2021, carrier strike groups and expeditionary strike groups.
Source: Department of Defense Comptroller, figure 10.2, https://comptroller.defense.gov/Portals/45/Documents/defbudget/FY2022/FY2022_Budget_Request_Overview_Book.pdf.

is quite capable of moving anywhere any time. Posen for example argues that if the United States reduced its presence in the region, it could use military forces—for instance, deployed on aircraft carriers—to return to the Gulf.

Because the United States would be able to return on relatively short notice, it could reap several benefits from reducing its presence in the region. Specifically, it is arguable that keeping U.S. forces in the region, to the extent that they appear threatening to local powers or even China, increases the level of militarization in the Persian Gulf. And, as I argued earlier, a constant presence further increases both military fuel use and military spending. An analysis by John Glaser for the CATO Institute, provocatively titled, "Withdrawing from Overseas Bases: Why a Forward-Deployed Military Posture Is Unnecessary, Outdated, and Dangerous," argued that the United States does not need a large forward presence any longer. "Even under a strategy of primacy—the view that a peaceful world order and our own national security depend on maintaining a preponderance of U.S. power—the extent of U.S. overseas basing creates needless cost and danger. A less aggressive strategy requiring fewer overseas bases would greatly reduce both military spending and security dangers to the United States."[19] Even some in the U.S. Navy itself are questioning the need for the United States to remain in the Persian Gulf. Commander Andrew Gustafson argued in the June 2020 issue of *Proceedings*, the journal of the United States Naval Institute, that the United States is no longer dependent on Persian Gulf oil and does not need a large presence there. The United States has friendly relations with most countries in the gulf, he argues, and Iranian oil is not worth fighting for. In fact, he concludes, "After 40 years of near continuous conflict in the region, it is time to completely reevaluate U.S. policy and, by extension, the U.S. Navy presence in the Arabian Gulf."[20]

Further, should the Persian Gulf region be threatened, it is unlikely that the United States would be alone in countering an extreme threat to the Persian Gulf. Recall that the United States was joined by a large coalition in 1990 to evict Iraq from Kuwait. Although it may have lost some credibility after the rationale for the invasion of Iraq in 2003 was exposed as false, the United States could still likely muster allies to deal with a major and obvious threat to Persian Gulf oil. As Herbert Feis argued during World War II, "In weighing the problem of military supply it should be borne in mind that, unless our diplomacy blundered tragically, the country would have allies or associates."[21] Feis concluded, "In the final analysis *the most satisfactory solution of the military problem is to minimize it*, through the combination of nations to maintain and enforce peace. Any nation that conducts it search for oil for its Armed Forces recklessly, heedless of antagonisms created, may ultimately prove to have increased its total oil requirements thereby."[22]

It is important to ask, however, whether a shortage of oil imports would hamper U.S. efforts to retake the Persian Gulf from a hostile power. In 2001, the Defense Science Board addressed this question: "DoD is the single largest fuel user in the US, and probably the world. However, it is unlikely that the DoD would ultimately be unable to obtain from the world market the fuel needed to conduct operations. The biggest impact to DoD resulting from any supply fluctuations is likely to be financial."[23] Again, the U.S. Defense Logistics Agency stockpile is adequate for quite a large war—enabling the United States to fight for about six months before running out of fuel; it could last much longer if the military economized and conserved its energy, since as we have seen, facilities fuel use is a significant consumer of DOD fuel. The SPR could also be tapped. In sum, as the Defense Science Board concluded, "The issue is not whether DoD will be able to obtain the oil it needs to provide for our national defense, because it will."[24]

The United States has an opportunity to get of this conflict–oil consumption cycle if it chooses to do so. As the United States has diversified its oil sources and overall petroleum demand has declined, the country has become less dependent on Persian Gulf oil. Notably, as figure 8.2 shows, total U.S. oil imports from the Persian Gulf peaked in 2001 and imports from OPEC peaked in 2005. Several trends converged after 2005 to bring down overall demand for oil imports: increased fuel efficiency that led to decreased consumption; increased domestic production of cheap natural gas and oil; and increased reliance on renewable energy resources. Specifically, in 2001, the United States imported 2,664 barrels per day from the Persian Gulf; in 2018, the United States imported 1,472 barrels per day from the Persian Gulf.[25] Overall, in 2016, Persian Gulf sources accounted for about 18 percent of all U.S. oil imports; by 2020, they accounted for less than 10 percent. In 2020,

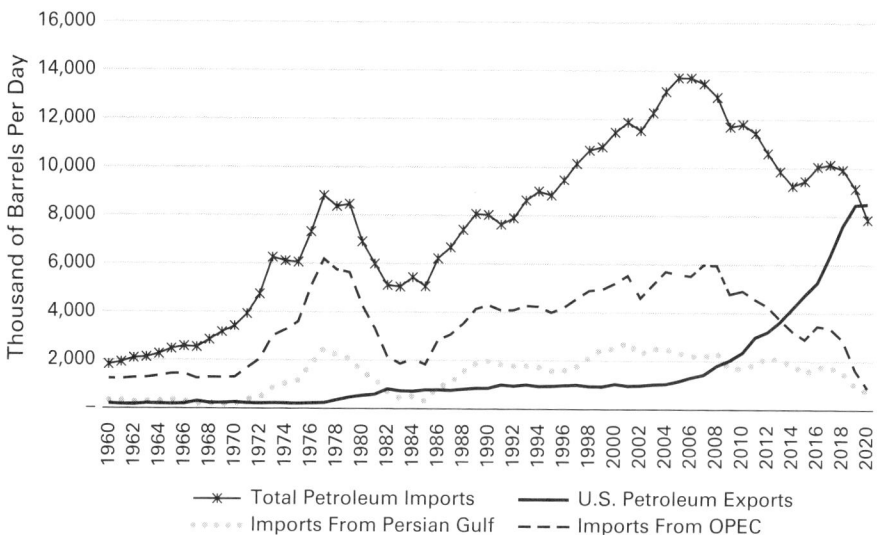

Figure 8.2

Average daily U.S. petroleum exports and imports, 1960–2020.

Source: U.S. Energy Information Administration data, Petroleum Trade Overview, https://www.eia.gov/totalenergy/data/monthly/#petroleum.

perhaps because of the decline in travel due to the pandemic, petroleum consumption declined dramatically, and natural gas and petroleum were consumed in nearly the same quantity.[26]

Overall, dependence on fossil fuels has declined in the United States. Total fossil fuel (coal, natural gas, and petroleum) energy consumption in the United States peaked in 2004 at 86.66 quadrillion BTUs. All non-nuclear renewable energy (hydropower, geothermal, wood, solar, and wind) energy sources have become increasingly important as a share of U.S. energy consumption since 1975, as figure 8.3 shows. In 2020 coal provided 10 percent of U.S. energy consumption, petroleum 35 percent, natural gas 34 percent, nuclear power 9 percent, and renewable energy 12 percent.

All forms of renewable energy have not grown at the same pace. Hydropower, which provides the largest share of non-nuclear renewable energy, has grown overall since 1949, although it fluctuates seasonally and has declined in recent years. See figure 8.4. By contrast, geothermal energy consumption has grown steadily. Wind and solar power consumption took off in the

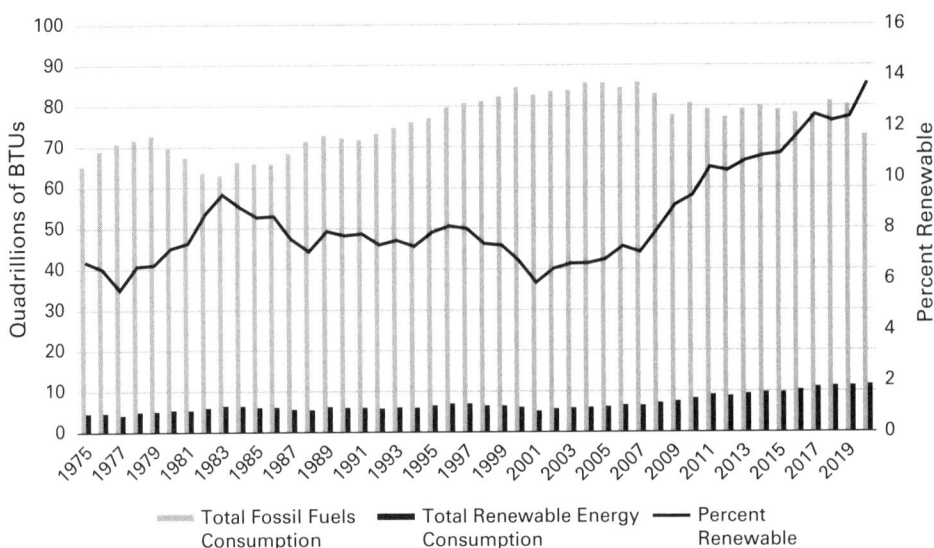

Figure 8.3

Fossil fuel and renewable energy consumption in the United States, 1975–2020.

Source: Data from U.S. Energy Information Administration, Primary Energy Consumption by Source.

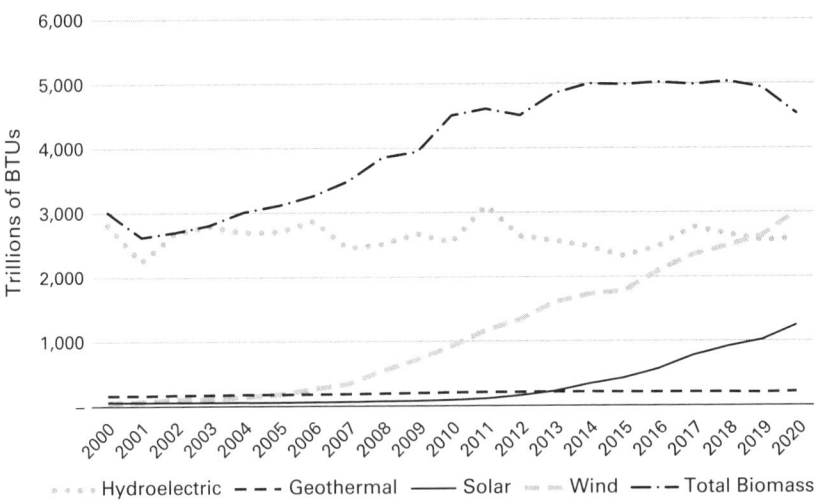

Figure 8.4

Biomass, hydropower, geothermal, solar, and wind energy consumption in the US, 2000–2020.

Source: Data from U.S. Energy Information Administration. Renewable Energy Consumption by Source, https://www.eia.gov/totalenergy/data/browser/?tbl=T10.01#/?f=A&start=1975 &end=2020&charted=6-7-8-9-14.

early twenty-first century. Biomass energy consumption has also risen fairly steadily. The takeoff point for wind energy consumption occured around 2007 and for solar in the period 2009–2012. These alternative forms of energy are poised to grow even more quickly in the next decade.

The United States has built its foreign policy and military forces on the assumption that U.S. dependency on oil imports is stable, or even increasing. This policy has been expensive and ultimately has arguably led to increased risks of conflict. I have also shown what has become obvious: U.S. dependency on Persian Gulf oil has declined even as the United States has not significantly changed its approach to the region. The United States does not need to keep such a significant military presence in the Persian Gulf to protect access to oil for which it has diminishing demand, and which is unlikely to be disrupted in any case. A smaller U.S. presence in the region would likely not increase the risk of war, and could simultaneously decrease U.S. dependency on Persian Gulf oil as U.S. military fossil fuel consumption due to the U.S. presence in the region declined.

THE POLITICAL ECONOMY OF THE DEEP CYCLE

Military industrialization and war have had a stimulus effect on greenhouse gas emissions that outlasts particular wars, in part because, although demobilization does occur, the U.S. military has tended not to demobilize after wars to its prewar size. Since the nineteenth century, wars have stimulated fossil fuel consumption, military emissions, and increases in civilian emissions. Major war and long-term mobilization stimulate enduring increases in overall greenhouse gas emissions because war not only causes temporary changes in the military-industrial economy, but because war tends to alter the wider civilian economy prompting enduring changes in the scale and pattern of fossil fuel consumption, tending to ratchet up fossil fuel demand. This is the deep cycle discussed earlier.

Specifically, as we saw in chapters 1 and 2, the military's reliance on fossil fuels that began with coal and continued with petroleum has shaped a military, foreign policy, and industrial system that used and depended on fossil fuels. Several iterations of this cycle deepened the dependence on fossil fuels as both the military and military industry required those fuels, and civilian industry and the civilian population became dependent on the goods and the lifestyle that fossil fuels made possible. Just as the industrial revolution transformed war, so warfare transformed industry. War was a stimulus to energy transitions—from wood and waterpower to coal and then oil. Concerns about energy security in war and during periods of mobilization and expansion were heightened as war became increasingly dependent on armed forces that used fossil fuel for vehicles and for the rubber, plastics, and weapons that made war possible. In other words, not only the military, but also military industry required fossil fuel. Fear phrased as a concern for energy security—which for some countries caused them to invade other countries to get oil, and which for the United States drove deeper engagements with oil-producing countries and the locations that could be used for coal and later petroleum refueling—is the engine of a deep, long-term, military-strategic cycle. The necessity for fuel to power armed conflict and military industry has caused states to extend their reach with military bases, and to make alliances with those who can provide access to oil or the bases needed to protect the flow of oil. Figure 8.5 illustrates this deep cycle in a diagram that starts with the availability of large quantities of cheap and versatile petroleum.

The increased dependence of the armed forces and military industry on coal and oil has shaped the larger economies of states. And in this deep oil-political economy cycle, the relationships were and are synergistic—the utility of coal and oil used in industry and war led to their increased value and perceived necessity; their increased value led to further investment in developing the industry and infrastructure necessary to secure those fuels. While not alone in promoting dependence on fossil fuels, war was an important factor—because wartime requires and promotes urgent and immediate mobilization and innovation—spurring the deep social, economic, technological transitions that resulted in a fossil fuel-dependent and industrial economy.

The synergies between industrialization and fossil fuels that began in the nineteenth century were thus bolstered and heightened by the demands of war. War thus became an important element in scholarship that explores long-term and deep political, economic, and technological changes that

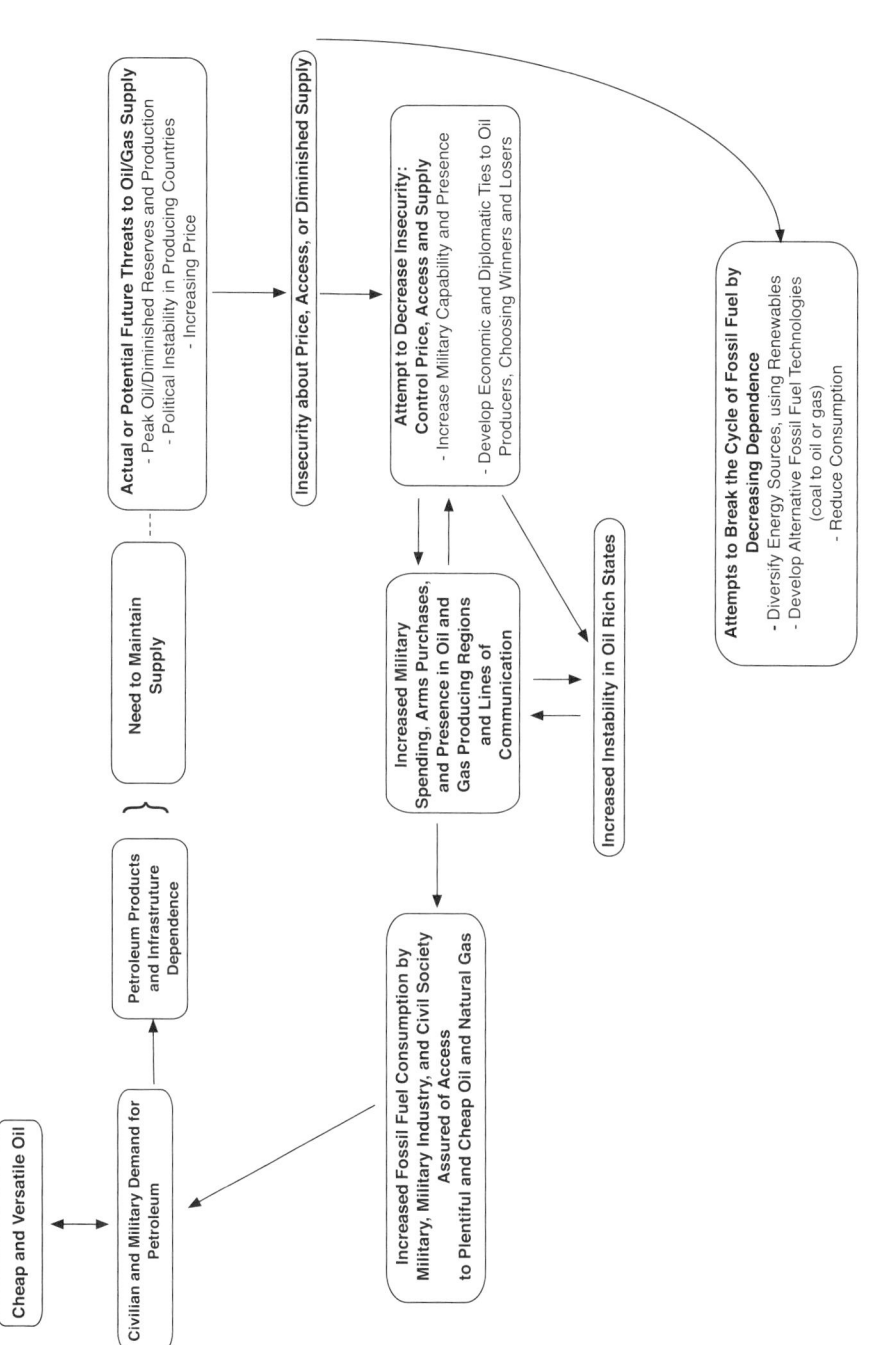

Figure 8.5
The deep cycle.

come to reinforce each other. The work on "deep transitions" by Johan Schot and Laur Kanger suggests that "changes across multiple systems can become connected and coordinated, developing a common directionality in the long run." Examples of deep transitions in this framework include reliance on fossil fuels, mechanization, resource intensity, and energy intensity. A deep transition is a "process of building connections between change processes in multiple systems" that Schot and Kanger say "takes on wave-type properties, unfolds through centuries, and is implicated in broader transformations of societies and economies."[27]

In the case of war, the mobilization of the economy and armed forces to meet a challenge can have cascading effects at home and in the international system. Gregory Hooks and Chad Smith, and other scholars, have called some of these effects the "treadmill of destruction."[28] In the treadmill of *production*, the benefits of economic growth and expansion are simply assumed as organizations use nature to accumulate wealth and power. In this process, the environmental consequences are minimized or discounted as "externalities." In the case of war and military mobilization, the highly destructive and specialized weapons of the twentieth century, including nuclear weapons, both demanded high-tech industrialization and were particularly toxic for the environment, becoming a treadmill of *environmental destruction*: "A treadmill of production is derived from and propels economic competition; a treadmill of destruction is fueled by geopolitical and military rivalries. . . . Each treadmill, operating independently, generates unsustainable withdrawals from and deposits to ecosystems. But these treadmills may operate simultaneously, often reinforcing and amplifying one another."[29]

High levels of military spending and the arms production that is required to equip a military can shape a domestic economy. Arms sales or gifts of military equipment, whether officially approved as foreign military sales to governments, or simple commercial transactions, bolster ties with allies and potential allies, thus shaping international politics, while also supporting the domestic arms industry. Processes in both a treadmill of production (driven by industrialization) and treadmill of destruction (driven by militarization and war) can hurt the environment. Treadmills of production and destruction are part of the deep cycle that I am describing.

War was thus both the trigger and the result of long-term processes in a complex system where part of what emerged was a deepening dependency on fossil fuels. War and consumer demand for materials and conditions that depended on fossil fuels for their production shaped the world we live in. That is not to say that these processes, economies, and institutions were inevitable. If humans had different beliefs and had sought to satisfy their desires through other means, another system could have emerged that would not have entailed massive climate changing emissions. But we are where we are because of the beliefs, micro-processes, and larger institutions and patterns of resource flows and consumption that have become dominant.[30]

One of the key points of the deep cycle is that it works as a positive feedback loop, amplifying the use of fossil fuels across systems. This means that the two causal pathways I mentioned earlier—one between war and energy transitions and the other between military spending, war, and industrialization, which drives fossil fuel use—are better pictured as cyclical and self-reinforcing pathways, rather than as linear sequences of cause and effect. The cycle is iterative and repeats, with each iteration deepening the perceived need for fuel, alliances and bases for refueling, military industry, and so on. An unintended emergent consequence of the actual, perceived, and constructed insecurity that flows through the cycle is, of course, climate-changing greenhouse gas emissions, which can then become another source of insecurity.

The two world wars are examples of the effects of the intersection of war, fossil fuel use, dependence on fossil fuels, and increased militarization. Specifically, as Phil Johnstone and Catriona McLeish argued, the world wars were shocks to the global economic system: "The heightened conditions of maintaining abundant and constant supply during the First and Second World Wars influenced the emergence of an 'age of oil.'"[31] In 1914, with oil providing less than 5 percent of world energy, coal was still dominant. During World War I, the consumption of petroleum increased 50 percent, with 90 percent of the oil provided to the British and French coming from the United States—the "wave of oil" that the British believed led to victory. World War I also accelerated industrialization, powered by coal and oil, and the use of fossil fuel-powered airplanes and automobiles. And as we have seen, the growing demand for oil accelerated the drive to control the Middle East—for bases, alliances, and commercial oil production arrangements. Oil was also a driver of strategy in World War II—for example, Germany's desire to get oil fields in Poland and Russia, Japan's invasions in Asia. The role of oil and other petroleum products in the conduct of World War I and World War II cannot be overstated.

World War II also accelerated the processes of industrial mechanization, promoted the development of fossil fuel industries, and developed the infrastructure—such as highways and airports used by gasoline- and jet fuel-powered vehicles—that remained after the war's end. In the reconstruction of Europe, the United States subsidized oil imports to Europe, with 10 percent of Marshall Plan loans related to the importation of oil to Europe.[32] Thus, as Johnstone and McLeish argue, "While the trend towards oil use accelerated faster in the mobility system after World War I and the growth of mechanised agriculture was modest, during World War II these systems become even more tightly integrated. Pipeline infrastructure, refinery capacity, petrochemicals, plastics, pesticides, increased mechanisation of agriculture, mass production of aeroplanes, construction of runways, and a host of other concurrent developments were accelerated by war."[33] They argue that wars led to the convergence of society's economic development around fossil fuels. "As a consequence of war time pressures, energy, food, and mobility were coordinated in a similar direction through the meta-rule of using oil, and became more closely integrated based around an increasing dependence on oil. Wars saw the rapid development of the technological, infrastructural, logistical, scientific, and institutional conditions underpinning oil-intense societies of the postwar economic 'golden age.'"[34]

The long Cold War and the regional rivalries and hot wars between countries that were a part of it also served to militarize states and their economies. To the extent that states had to produce their own weapons, their economies became more dependent on military industry.[35] If less powerful states could afford to purchase the most sophisticated weapons or receive them as part of a foreign aid package, they did not have to develop large arms industries. But those countries that were engaged in strategic rivalries, such as the United States, the UK, Russia, and increasingly China, developed military industries that were partially or almost entirely subsidized by their governments. All of this is the rather familiar dynamic of the arms race, but its impact on the environment and in particular on greenhouse gas emissions must be included in our understanding of why the post–World War II era saw a dramatic increase in emissions.

The deep cycle of fossil fuel demand and military industrialization can be reinforced when a state responds to anxiety about energy supplies by mobilizing its armed forces and foreign policy to protect petroleum. For example, when confronted in the oil-producing regions with unrest or the refusal of various regimes to go along with U.S. policy, the United States has doubled down on military solutions and, perhaps surprisingly, on reliance on fossil fuels. For example, in response to fears about the potential for restricted access to Persian Gulf oil, Presidents Nixon, Ford, and Carter each proposed substantial investments in energy independence. The Nixon administration's report on "Project Independence" published in November 1974 proposed several options for dealing with declining oil imports, assuming that "Foreign sources of oil have a significant probability of being insecure in the 1974–1985 time frame."[36] The report examined options that included reducing demand through conservation, but its scenarios assumed increasing demand. The goal was not independence from fossil fuel, however, and interest in developing alternative energy was minimal. Rather, since the formation of the Department of Energy, from 1978 to 2018, comparatively little funding had been allocated to renewables or energy efficiency. Of the $158.28 billion in funding for energy technology, from 1978 to 2018 renewable energy programs received $27.65 billion; energy efficiency programs received $24.73 billion and fossil fuel energy $37.8 billion.[37] Thus, as figure 8.6 illustrates, over the forty-year period, average spending on renewable and efficiency research and development were, respectively, about $700 million and $600 million per year. Government-sponsored research and development consistently reinforced the dominance of fossil fuel. A possible off ramp from the deep cycle was thus not developed with government support at the same time that the United States was heavily supporting fossil fuel-dependent sectors with funds for research and development.

Research on the relationship between military spending and greenhouse gas emissions shows that high military spending—owing to dependence on fossil fuels—is correlated with higher greenhouse gas emissions. Andrew

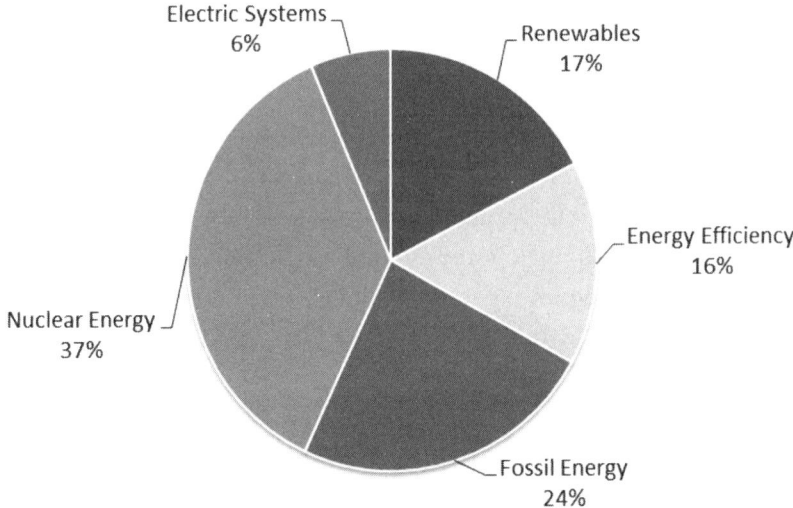

Figure 8.6
Department of Energy research and development funding, energy technology share of funding, 1978–2018.
Source: Corrie E. Clark, "Renewable Energy R&D Funding History: A Comparison with Funding for Nuclear Energy, Fossil Energy, Energy Efficiency, and Electric Systems R&D," Congressional Research Service, RS22858, June 18, 2018, 5, figure 3, https://fas.org/sgp/crs/misc/RS22858.pdf.

K. Jorgenson, Brett Clark, and Jeffrey Kentor's analysis of seventy-two countries' military spending per capita and emissions per capita between 1975 and 2000 showed that "it appears that all, else being equal, nations with more high-tech and labor intensive industries emit relatively higher overall levels and greater intensity of anthropogenic carbon dioxide gas."[38] Analysis by John Bradford and Alexander Stoner of countries between 1975 and 2014 showed similar results: "An enduring relationship between militarism and carbon emissions: countries that allocate relatively higher percentages of their total GDP to the military have higher average per capita CO_2 emissions, even after controlling for the size of the economy, urbanization, and adult population. The mutually reinforcing nature of military, political, and economic dominance could explain the cross-sectional associations between levels of military spending and carbon emissions."[39] While a more detailed analysis would be necessary to confirm preliminary findings, it seems that military-industrial processes tend to be higher tech and more greenhouse gas intensive than other industries.[40] In sum, cross-national comparisons find that "the amount of money devoted to the military and the scale of the military infrastructure varies between countries, but . . . militaries, once established, generate their own growth momentum, which may create a form of path dependency, as they incorporate more high-tech equipment or expand their ranks."[41] In sum, Clark, Jorgenson, and Kentor argue, "Militarization expands energy consumption, given the resources that are needed to sustain and support its infrastructure, equipment, soldiers, and support personnel."[42] Military procurement supports a greenhouse gas-intensive military industry.

If oil is vital for war and military industry, the source of the oil must be protected or controlled. Thus, as we saw earlier, after World War I the demand for oil for military purposes accelerated the drive to control the Middle East— for bases, alliances, commercial arrangements, and, increasingly, arms transfers to regimes that the United States hoped would be friendly to U.S. interests and fight regional threats. During the Cold War, the conventional military was increasingly supplemented by a nuclear arsenal, the building of which required even more industrialization and conventional energy consumption.[43]

The need to protect oil promoted higher military spending, and in turn, higher military spending promoted a fossil fuel-intensive economy. Military spending is usually understood in the U.S. to be good for economic growth, as it was during World War II when it created a large number of jobs. But that understanding depends on a narrow and short-term conception of the economy in which the immediate benefits to fossil fuel companies and military industries are measured without regard for public health, public safety, opportunity costs, or future costs. Just as the fossil fuel industry benefited from government-supported research on fossil fuels, military industries benefited from the research and development paid for or subsidized by the federal government, and cost-plus contracts that guaranteed a profit.[44]

Large, diverse economies have a great deal of complexity to them. Nevertheless, spikes in U.S. emissions appear to correlate with war. Figure 8.7 illustrates trends in greenhouse gas emissions of the United States for energy and cement production. Note that they are based on territorial fuel consumption and do not include emissions abroad, such as for the major wars.

Annual CO₂ emissions

Carbon dioxide (CO₂) emissions from the burning of fossil fuels for energy and cement production. Land use change is not included.

Figure 8.7

U.S. annual CO_2 emissions, 1845–2019.

Source: Oxford University, "United States: CO_2 Country Profile," Our World in Data, https://ourworldindata.org/co2/country/united-states?country=~USA#co2-emissions, accessed July 4, 2021.

Even still, it seems that changes in U.S. CO_2 emissions correspond to war and periods of territorial expansion, mobilization, and demobilization.

Indeed, it remains hard to reduce military industry and the armed forces because many people believe that high levels of both yield security, economic growth, and many well-paid jobs. Further, in the United States, Congress has tended, since World War II, to keep military spending up to "a persistent baseline of funding, regardless of whether or not the nation was engaged in a war."[45] This is in part because the DOD often makes multiyear contracts for procurement that extend beyond the authorization for a particular fiscal year and also in part because the infrastructure and personnel of the U.S. military vary little in size from year to year. The "enduring" or base budget grew incrementally, but as we see, it nearly doubled between 2001 and 2021, as figure 8.8 illustrates, even as war (categorized as "overseas contingency operations") spending has fluctuated and decreased.

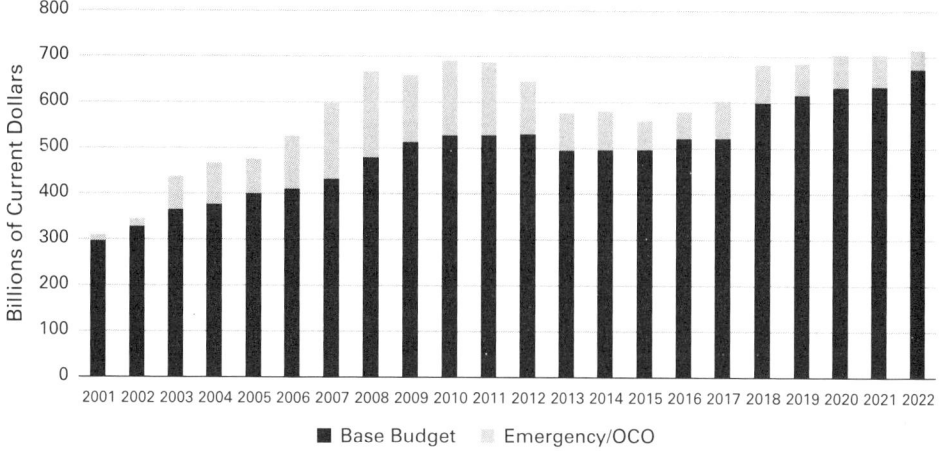

Figure 8.8

DOD annual "base" and "overseas contingency operations" war spending, FY2001–FY2022 in current dollars. FY2001–FY2021 is the authorized amount. FY2022 is the request. *Source*: Data from the Office of the Undersecretary of Defense, Comptroller, "Defense Budget Overview," various years.

Maintaining a global military presence is also expensive. Total military spending has averaged hundreds of billions each year for decades. Perhaps more relevant to understanding the budgetary effects of the dependence on fossil fuels are estimates of the portion of that military spending that can be attributed to defending oil supplies in the Middle East. Estimating those costs, as one might suspect, is not so easy. First, one has to count direct costs for war spending and the annual spending in nonwar years on the capability to intervene in the Middle East to protect oil. But the forces able to intervene in the Middle East may have other missions, such as providing capability to eliminate weapons of mass destruction or to retaliate against terrorists who might be based there. Indeed, Central Command forces were the ones used to prosecute the war in Afghanistan. And second, there are categories of spending that are indirectly related to supporting the capacity to intervene—such as military equipment and foreign aid given to U.S. allies in the Persian Gulf, the overall expense associated with pay and health care for a larger than might be required armed forces, training and exercises for Gulf War contingencies, and weapons acquisition. Further, the costs of the capacity to intervene in the Middle East can fluctuate. Sometimes these costs have even been heavily subsidized by allies, such as when the majority of the costs of the 1991 Gulf War were paid for by the United States' allies. The other major cost is the loss of life and injuries—for service members and contractors associated with military training and war, and for the civilians who are victims of it.

Several offices within the U.S. government have estimated the monetary costs of protecting access to Persian Gulf oil.[46] In 1991, for example, the Government Accounting Office reported that the Pentagon estimated the costs associated with "Southwest Asia." It noted, "For purposes of this report, only those countries in Southwest Asia that throughout the 1980s and early 1990s were considered of strategic importance to the United States are included. *These are oil producing countries in the Middle East, particularly those located in the Persian Gulf area, as well as non-oil producers bordering strategic transiting points and key regional allies.*"[47] The Pentagon told the GAO that between 1980 and 1990 the DOD spent $21.4 billion for military programs and activities "directly related to Southwest Asia-specific missions" and an additional $5.8 billion for programs and activities that were "primarily Southwest Asia-oriented." Furthermore, the DOD "invested about $272.6 billion in programs that, although motivated by requirements outside of Southwest Asia, have proven to be useful in meeting contingencies in this region. These costs are mostly related to forces available to the U.S. Central Command." Omitted were the costs incurred during specific military operations the United States held from 1980 through the 1991 Gulf War, "such as Operation Earnest Will—reflagging of the Kuwaiti oil tankers—and operations Desert Shield and Desert Storm."[48] All those operations totaled more than $61 billion, although the costs of the 1991 Iraq War were then being reimbursed by U.S. allies. To this was added the approximately $6 billion that the United States provided in foreign military assistance and economic assistance to countries "of strategic importance" in the region that were not Israel and Egypt.[49]

Scholars have also made varying estimates of the costs of protecting access to countries of "strategic importance" in the Middle East. In 2008, John Duffield estimated the additional annual costs of the capacity to intervene to be between $28–36 billion from 1980 to 1990, and he estimated those costs to be $30–51billion for the period 1991–2001 (in 2006 dollars) not including the major military operations the United States conducted there during those periods.[50] That same year, Mark Delucchi and James Murphy suggested that the cost for defending access to Persian Gulf oil in 2004 was $27–73 billion.[51] In 2016 Eugene Gholz estimated that the annual incremental cost of U.S. operations to protect against threats against Persian Gulf oil was about $5 billion.[52] By another estimate, at a minimum the United States spends about $81 billion annually defending the global oil supply.[53] Of course, adding to the complexity of estimating the cost of U.S. spending on military force to protect access to oil is the fact that the United States is increasingly concerned with Chinese military expansion not only in Asia and the Pacific, but in the Persian Gulf.

This process of ratcheting up military forces and military industry in wartime and increased demand is a self-reinforcing cycle that has rippled out into the larger economies of the United States' adversaries and allies. The deep cycle seems to be at work in Russia and China. Russia's abundant fossil fuel reserves can fuel its militarization, wars, and economic growth without concern for energy supply. On the other hand, while China has a great deal of coal, it lacks substantial reserves of petroleum. Both countries seem to be on the path of increased industrialization, fossil fuel dependency, and militarization.

China's military emissions are hard to estimate, but China also seems to be in the same deep cycle as it increases its military spending, armaments, and global presence. The Chinese military has many people under arms and nuclear weapons, but it is not yet a match for the United States. As China's energy consumption and dependence on crude oil increased, so has its military spending and engagement with oil-producing countries everywhere in the world where oil is to be found, and including passage through the Malacca Straits.[54] China's first overseas military base, established in Djibouti in August 2017, is positioned not only to enable it to protect China's access to the Persian Gulf, but also to protect its economic interests in Ethiopia, which include natural gas.[55] These activities make the U.S. military worry and are part of the rationale for arguing that China poses a growing threat. The Commander of CENTCOM argued in 2021 that Central Command should be preparing to respond to an increased Chinese presence: "China uses predominantly economic means to establish regional in-roads, with a long-term goal of expanding its military presence to secure vital routes of energy and trade."[56] As China's dependence on imported oil grows, so does its visible military presence near oil-producing countries, bolstering the sense that China is a rising military power, which it certainly is. The rise in China's economic power and military spending in turn is a rationale for increasing the overall U.S. defense budget and the U.S. presence in Asia and the Pacific.

Aaron Friedberg argues that China should be understood as comparatively weak and vulnerable to the superior power of the United States in the Pacific and Indian Oceans. Friedberg views the Chinese military as pursuing what is basically a defensive strategy against the United States—working to preserve and expand its economic growth through international trade. But because a large portion of its trade, and oil, must pass through the Malacca Straits in Southeast Asia, China may worry that it is vulnerable to a crippling blockade. In the Pacific, Friedberg suggests, the Chinese military's "theory of victory appears to envision a first strike that would effectively knock the United States out of the theater in the opening stages, accompanied by the seizure of key maritime terrain and establishment of a defensive perimeter along the first island chain, after which Beijing would presumably depend on economic suasion and threats of escalation to bring U.S. allies to terms and to discourage Washington from continuing the war." And absent "marked improvements in defensive capabilities," in the Indian Ocean region, the Chinese military's "surface ships would be vulnerable to attack by missiles and torpedoes, while its submarines would be susceptible to being tracked and targeted. Whatever bases or dual-use facilities support China's presence in peacetime would likely not survive the early stages of a war against a major power." Thus, he suggests, the Chinese military is preparing to fight and win through a preemptive attack if that is believed to be necessary. Any force capable of striking preemptively can appear to be a first-strike offensive force and observers may come to believe that this is exactly China's intention. Friedberg argues that China's economic investments are in some ways about ensuring its military capability in case of war, and that its military is about ensuring its economy has resources.

> The investments the regime has made over the past two decades in trans-Eurasian pipelines, roads, and railways sometimes appear irrational when assessed in purely economic terms. Under normal peacetime conditions, ships can carry far greater volumes of cargo much more cheaply. In the event of conflict, however, overland

links could provide a vital lifeline, potentially enabling the delivery of enough energy and raw materials to keep a wartime economy running.

Similarly, many of the ports and other facilities China has bought or built along the maritime axis of Xi Jinping's much-touted Belt and Road Initiative appear to be operating at a loss and may never turn a profit. Control of these logistical hubs could nevertheless ensure that, in a crisis, China-bound cargo (much of it carried on vessels built, owned, and operated by Chinese state-owned enterprises) would receive top priority. Some of the facilities springing up around the rim of the Indian Ocean also could eventually handle visiting naval vessels and military aircraft.[57]

China's deep cycle is bolstered by the U.S. reaction to its economic and military expansion. If the United States continues to make the Pacific and the "threat" from China a focus, it is likely that China will continue to increase its military spending and armament. And thus China's military and military-industrial greenhouse gas emissions will increase. China's total national emissions exceeded the total national emissions of the United States for the first time in 2006. To the extent that nations decide to participate in military mobilization, U.S. and Chinese military spending will drive their own and other countries' military-industrial emissions. China's military industry is currently expanding as China builds its armed forces. While the U.S. military remains the world's largest purchaser of military goods and services, China's arms imports and exports increased between 2009 and 2019 and the sales of its top four weapons manufacturers grew from $53.7 million to $56.7 million from 2015 to 2019.[58]

U.S. military industry may go through cycles of boom and bust as the United States is or is not at war. Specifically, U.S. military industry did contract after the Cold War ended, but it rebounded in the post-9/11 war era. While military industry might be expected to be more sensitive to whether the United States is at war, the military-industrial sector seemed to do quite well even as the U.S. military presence in Afghanistan and Iraq declined. The United States has, since 1980, experienced a steady decline in total manufacturing jobs, while military-industrial employment is larger than it was at the end of the Cold War, and from 2000 to 2015, the U.S. manufacturing sector lost about five million jobs. By contrast, the military-industrial sector has

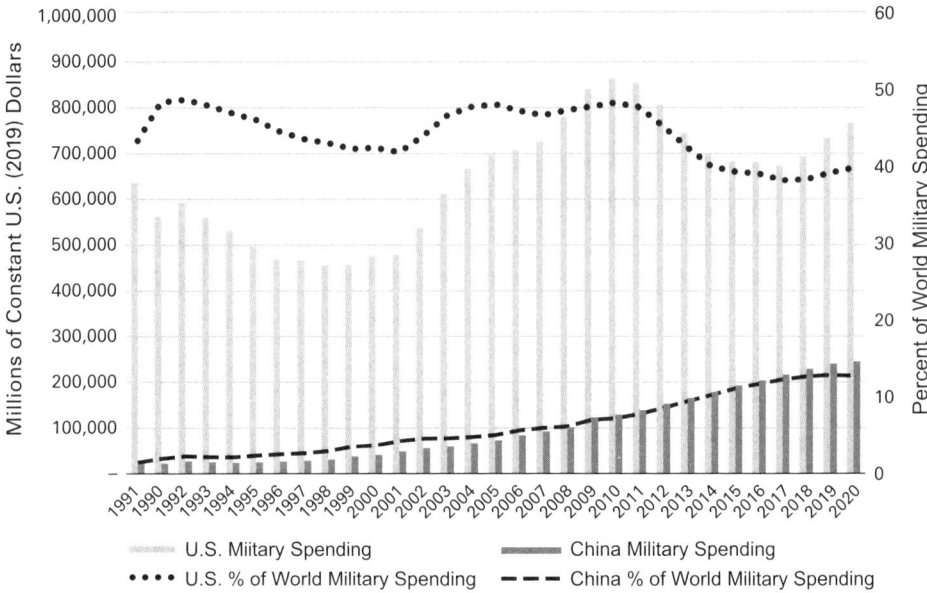

Figure 8.9

U.S. and Chinese military spending, 1990–2020 in constant (2019) U.S. dollars.

Source: Stockholm International Peace Research Institute, World Military Spending Database, https://www.sipri.org/databases/milex.

been holding its own; in 1997, U.S. military industry employed about 2.1 million people; in 2019, aerospace and defense industries employed about 2.2 million people.[59] As figure 8.10 illustrates, the earnings of the "big six" U.S. military industries and the next twenty-five mid-tier defense contractors were also strong from fiscal years 2014–2019, the period when the United States was drawing down its military forces from Iraq and Afghanistan.

Further, U.S. military spending and modernization tends to increase its allies' spending. Specifically, NATO's guideline for members to spend a minimum of 2 percent of GDP on defense is intended to increase alliance burden sharing. But, although not always followed, the 2 percent guideline tends to keep military and military-industrial emissions higher than they perhaps need to be. This may make NATO members feel secure, but it can also serve as a rationale or pretext for Russian military spending, mobilization, and military industrialization.

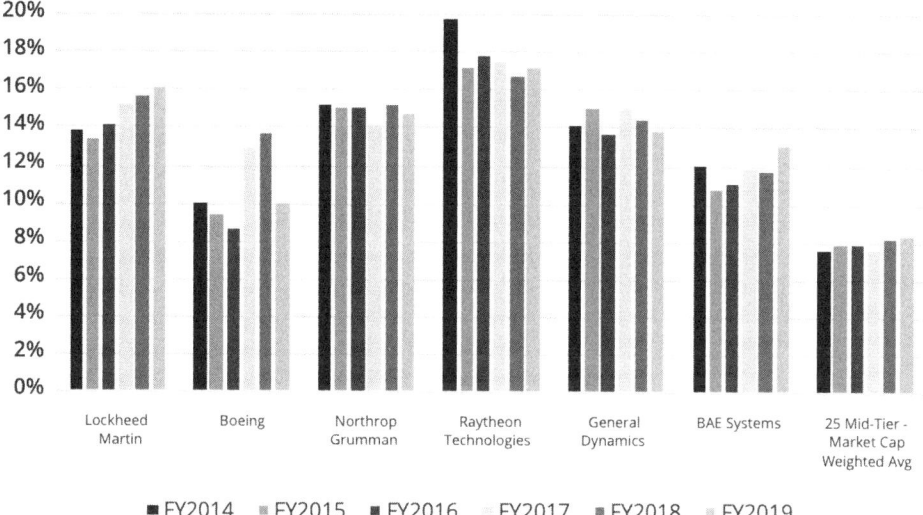

Figure 8.10

U.S. defense suppliers earnings before interest, tax, depreciation, and amortization, FY2014–FY2019.

Source: Department of Defense, "Figure 6.3 Big 6 DoD Prime & 25 Mid-Tier Market Cap Weighted Average EBITDA Margin [FY2014-FY2019]" from Department of Defense, Fiscal Year 2020 Industrial Capabilities Report to Congress," January 2021, https://media.defense .gov/2021/Jan/14/2002565311/-1/-1/0/FY20-INDUSTRIAL-CAPABILITIES-REPORT.PDF.

THE TRAGEDY OF THE DEEP CYCLE

There is something tragic here. Two tragedies, really. For most of the last 250 years humans have carried on with making "progress"—industrialization in the service of the good life—assuming that fossil fuels were an essential ingredient in shaping the world we were certain would be better than the one we were leaving behind. Our grand strategy for security took for granted that we would need fossil fuel for industry and fossil fuel for war. Only in the last fifty years or so has it become clear that burning all that fuel, and at the same time destroying the forests and the wetlands that take up the carbon released by the fire, is not just leaving that past behind, but destroying the possibility of preserving what is increasingly—in essential, life-giving respects—understood as a better world. As Hannah Arendt once said, "The road to hell may just as well be paved with no intentions as well as the proverbial good ones."[60]

Many of our leaders—scientists, policy makers, and the captains of the fossil fuel industry—have understood, however, that the fires of the past, and the fire right now, would bring both the fire and the floods of tomorrow. And we have reached tomorrow. It is now. Our greenhouse gas emissions—made to render us safe from extremes of heat, cold, hunger, and violence, which gave generations both prosperity and a measure of independence from the elements—are guaranteeing that future generations will suffer greater insecurity as the elements become less hospitable for humans and all other life.

The military has understood the science and the consequences of global warming quite well for decades. They paid for much of that research. National security strategists have sounded muted alarms, the Pentagon has adapted some of their equipment and operations, and experts have imagined scenarios of increasingly dire complex emergencies and catastrophes and climate change-caused wars. The Pentagon's leaders have helped us through the first iterations of these complex catastrophes, the wildfires, the floods, the hurricanes, and they have prepared early warning systems for the next iteration, and the next, and the ones after that. Faced with the incontrovertible reality of global warming, Pentagon leaders have been farsighted and tactically flexible. Some of the smartest, best-trained, and most determined people on this planet, given the resources of the richest nation on earth, the people at the Pentagon are trying to make things better. They are on the job. They have developed better batteries, put up solar arrays at bases, and even thought about moving some bases. Like the rest of the government, the armed forces have promoted "green procurement," and they include sustainability in the evaluation of equipment and supplies. So, for example, the army has announced it "will adopt a Buy Clean policy for procurement and construction materials with lower embodied carbon emissions from manufacturing, transportation, installation, maintenance, and disposal sub-processes."[61] In sum, the DOD has responded to climate change with adaptation and preparedness for all sorts of scenarios, including climate change-caused war, and set goals for reducing emissions.

And yet at the same time, the Pentagon has been strategically inflexible and blind. This is the second tragedy. For the most part, the armed forces, and our political leaders, have not put away the tools and the habits of mind that got us here in the first place. Our grand strategy for national security has not fundamentally changed. In some ways it can't, because it is premised on the anticipation and fear of war—the idea that for us to be safe, we have to be prepared to meet every threat anywhere at any time with overwhelming force. The national security strategy is also premised on the idea that force works, that the threat of coercion and the actuality of destruction can get us what we want. So, once we believe those things—that war is possible and may be imminent, that we must have a capacity to make war that far exceeds our enemies' abilities, and that coercion and destruction are effective—it seems that the only way to deal with the threat of climate change-caused war is to prepare for more war. Of course, in preparing for more war, governments give the armed forces everything they need: money, weapons, people, bases, and fossil fuels. We protect and defend the oil we think we need to defend ourselves. We treat arms makers well because they supply the weapons that will keep us safe. We put bases all over and everywhere we can think of, either at the destination itself that we want to defend, or as steppingstones and refueling stations to those places. At the same time that we are making our weapons more energy efficient and the installations more resilient, we scarcely question whether war is inevitable or in fact made more likely by our bases and our burning fuel to be the most powerful nation on the earth.

The DOD leadership assumes that climate change will be a disaster for the institution and the planet no matter what they do, even as they believe that they must continue to protect access to Persian Gulf oil, and counter China's efforts to have access to that oil in a war, so that the United States and the rest of the world can burn as much oil as it wants at as low a price per barrel as possible. The Pentagon focuses its efforts on adapting to climate change, on "consequences management," by preparing for climate caused-insecurity, even as it continues to ensure that Americans continue to have relatively inexpensive access to imported oil. The military is inadvertently or perhaps deliberately *militarizing* climate change. There is some evidence for that interpretation in the Biden administration's approach. For example, while the DOD has long argued that the Arctic will be an area of greater competition for economic and potentially military advantage, the DOD's 2021 *Department of Defense Climate Risk Analysis* warns that "competitors such as China may try to take advantage of climate change impacts to gain influence."[62]

THE WAY FORWARD

As I argued at the beginning of this chapter, the United States' foreign policy decision makers and the public have important policy decisions to make. Does the United States continue to orient its foreign policy and military forces toward ensuring access to fossil fuels? Does the United States dramatically reduce the use of fossil fuels, including the military's own dependency, and thus reduce the perceived need to preserve access to oil resources? Does the United States continue to militarize its relationship with China, which is itself seeking to ensure access to petroleum? Does the U.S. military put most of its effort into adaptation, or does it redouble its efforts to reduce emissions and thereby diminish the likelihood of the worst and longest-lasting outcomes associated with global warming?

The DOD can take steps that include increasing efficiency and resilience as well as reducing overall energy consumption. These can have enormous effects. The U.S. military is already achieving some reductions in operational fuel and emissions use by increasing fuel economy and switching fuels. It has been demonstrated that biofuels could power most U.S. military aircraft. It should continue to replace its fuel-guzzling equipment with vehicles that use either renewable energy or biofuels. And in many cases, equipment will not have to be replaced. Mixes of biofuels and conventional fossil fuel, or even 100 percent biofuel, can already power many U.S. vehicles from ships to aircraft. The army has plans to purchase efficient tactical and non-tactical vehicles—including Tactical Vehicle Electrification Kits (TVEK)—which can reduce average fuel consumption by 25 percent.[63] But the U.S. military should not go on a buying spree of alternative fuel-powered vehicles, for two reasons: any increased demand for hardware will potentially ramp up the military-industrial sector, and, absent a larger game plan, it is not clear that a one-for-one replacement of military equipment is wise.

I have shown that the militaries of the United States (and the UK) often balked in the nineteenth and twentieth century when they were asked to transition to new technologies—in the case of the U.S. Navy, from sail to the coal-powered steamer and then from coal to oil, and eventually from oil to nuclear power. And in each instance, the military adapted. Indeed, the U.S. military is apparently eager to adopt technologies that are both efficient and effective.

Further, by stressing energy efficiency and the transition to renewable energy, the DOD can make market-shaping investments through the different purchases and choices of the smaller—but likely still enormous—armed forces. The DOD is already doing some of this work by promoting changes in the supply chain that are lower in greenhouse gas emissions. If the military converted more of its energy consumption to renewable energy, this would stimulate the renewable energy industry with important benefits for the entire U.S. economy.[64]

However, while the emissions reductions that have already occurred and which are on the table seem, and are, significant, there is room for even larger cuts in U.S. military emissions if the United States makes large-scale reductions in the size and operations of the U.S. military. Specifially, the military, supported by Congressional action, should take three additional steps.

First, on the side of installation energy use, the United States should not only continue to find reduce energy use at individual installations; it could relatively easily close about 20 percent of its installations and bases.[65] The Department of Defense has already made the case for closing about 20 percent of U.S. military bases. As Christopher Mann of the Congressional Research Service found, "Between FY2012 and FY2018, the Department consistently argued for a new BRAC [base realignment and closure], asserting that 'absent another BRAC round, the Department will continue to operate some of its installations sub-optimally as other efficiency measures, changing force structure, and technology reduce the number of missions and personnel.'"[66] In March 2016 the DOD said that it had an excess capacity of 22 percent in its military bases. In October 2017 the DOD modified its estimate to an excess capacity of 19 percent and argued that it was time to engage in a base realignment and closure process.[67]

Bases that were converted to civilian use, and remaining military installations, could perhaps contribute to mitigation by switching land use so that bases become sites of carbon sequestration (taking carbon out of the atmosphere and fixing it in the soil and trees). Again, even before the end of the U.S. War in Afghanistan, the military acknowledged that it had about 20 percent excess installation capacity. A plan to close and then convert closed military bases into civilian uses would smooth local transitions. Fortunately, the United States has been through several iterations of base closures in the recent past, and there is ample evidence that this process does not have to result in economic pain for local communities. In fact, communities can be safer and more prosperous after bases close when some of the resources used to keep them open are used in redevelopment efforts.

Despite the fact that the U.S. military acknowledges that it has excess capacity at its installations, there has not been a strong push for base realignment and closure.[68] Some base closures will be necessitated by climate change itself. More significant reductions in greenhouse gas emissions will be gained by restructuring the U.S. military posture, including reducing U.S. military operations and installations worldwide, and closing bases in the United States abroad that are no longer necessary. Which bases are no longer necessary? The United States does not, as I have argued, need to direct so much effort to patrolling the Persian Gulf. Given that the United States and the world have reduced dependency on oil, the U.S. military does not need to be there all the time.[69] Any base closures should be accompanied by a plan to both clean up the toxic wastes that have been left there and convert those bases into something else.

Second, the United States should rethink its global force posture, starting with the Persian Gulf. If the United States were to immediately and dramatically decrease its dependence on oil by reducing both military and civilian fossil fuel use, it could reduce the political and fuel resources it uses to defend access to oil. If the United States further reduced its imports of oil from the Persian Gulf, including fuel used by the military to protect those imports, it could then reevaluate the size of the U.S. military presence in the region and reevaluate its relationship with Saudi Arabia and other allies in the region. The United States would reap political and security benefits, including reducing the dependence of troops in the field on oil and decreasing dependence on oil and those who provide it. By spending less money on fuel and operations to provide secure access to petroleum, the United States could, in the long run, decrease its military spending and reorient its economy to more economically productive activities.

Of course if the U.S. military dramatically reduced its size, and other countries followed suit, some sectors, such as the extremely profitable arms industry, would not be building and selling as many weapons to the United States, and arms exports would also decline. But these industries can and already do make other valuable commodities. A plan for economic conversion of military industry would have to be part of an overall climate change and national security strategy.

This brings me to the third and in some ways the most important step. A policy of armed competition that is premised on fear, and great-power rivalry ratchets up tensions and therefore military forces, military industry, and greenhouse gas emissions. And it risks war. The U.S. military should use the opportunity of the end of the war in Afghanistan and declining dependence on oil to rethink its entire force posture and the exercises and operations that support it. U.S. allies would also necessarily rethink their force posture. If climate change is as serious a threat as the DOD, White House, and National Intelligence Council have said in the climate change assessments they released in September and October 2021, then the United States should fundamentally rethink the meaning of national security and the means it uses to provide that security.

If climate change is a more certain national security threat than many other potential threats, the military might move beyond adaptation and "consequence management" to true climate change-related conflict prevention by further reducing fuel use and greenhouse gas emissions. This is a win-win-win strategy. By reducing the use of greenhouse gas-emitting fuels (coupled with emission reductions in other sectors) the Pentagon would decrease its contribution to the associated potential climate change threats to national security. Indeed, the Pentagon could play a major role in preventing or reducing the worst effects of climate change, and any potential security consequences of global warming, rather than reacting to climate change-caused emergencies or cleaning up after those effects have occurred.

These three steps are related. A significant reduction in the size of the military and its infrastructure is possible if the United States rethinks its overall approach to potential conflict and changes its military posture.[70] A smaller military would be a less expensive military. A reduction in the size and operations of the U.S. military would allow the resources freed by reduced military spending to be used to support a major transition in the energy infrastructure of the Unites States.

Of course it matters what adversaries do. It is essential that the United States use its other, nonmilitary tools—including economic sanctions—to deal with aggressive regimes, including Putin's Russia. A transition away from fossil fuels is part of the U.S. and NATO toolbox. If exports of Russian oil and natural gas to decline, Russian revenue and influence will also wane.

In the long run, climate change is the chaos coming to a neighborhood near you, near all of us. Thus, it is imperative that Russia's wars of aggression, Chinese economic expansion, and other crises and challenges not imperil deep cuts in military industrial and military emissions. Again, military industrial and military emissions are not the only emissions that need to be reduced. The point, however, is that these emissions are large, and they are often taken for granted as sacrosanct.

Deep reductions in Pentagon fossil fuel use could have enormous positive implications for the global climate and the U.S. economy, creating a positive feedback loop as powerful as the deep cycle that has amplified global military and military-industrial emissions. Reductions in fuel use save money and make the U.S. military less vulnerable to fuel shortages; in the long run, reductions in fuel use and conversion of bases by reforestation decrease climate change-caused impacts including insecurity; and conversion to renewable energy sources and alternative fuels could significantly boost the renewable energy and electric car industries in the United States. The path to climate security does not have to lie in preparing for climate war. If we want peace, we must prepare for peace.

Acknowledgments

The joy of teaching comes in part from regular contact with people who look at the world with fresh eyes, and also in good measure from the necessity of having to clearly explain complex ideas. And interdisciplinary teaching requires fresh eyes in spades. Without Anna Henchman, Les Kaufman, Nathan Phillips, and Adam Sweeting, my colleagues over several years in developing and then for four years teaching our course Interdisciplinary Perspectives on Global Challenges: Climate Change at Boston University, I would not have started the research that became this book. I am also grateful to the extremely bright and curious students in the Kilachand Honors College who took that course. Thank you, Carrie Preston, for being such a generous and inspired leader of Kilachand—we couldn't have done the course without you. The ideas explored here grew out of a simple question that I wanted to answer for one lecture—one slide, one number on that slide.

I wrote this book when I was still at Boston University, but I have moved to Oxford University in the UK and I no longer have the pleasure of teaching with Anna, Les, Nathan, and Adam and of working with Carrie. I have learned from all of you. I will miss running into you on campus and, most of all, just talking with you.

I am more than grateful to the Carnegie Corporation of New York for funding the larger work of which this book project is a part. I am also grateful to my colleagues at Ohio State University, St. Andrews University, London School of Economics and Politics, Stanford University, MIT, Columbia University, Cornell University, Brown University, and Boston University—who participated in seminars where I made presentations about this research and expanded on the findings of a paper on this topic that I wrote for the Costs of War Project. I learned a great deal from your questions and comments. You pushed me to go deeper. And so did the many journalists and activists who were curious about Pentagon fuel use and military emissions. I hope this work helps answer your questions.

I thank Cathy Lutz, Stephanie Savelle, Heidi Peltier, Miriam Pemberton, Ann Markuson, and Judith Reppy. I thank Deborah Burton and Ho-Chin Lin of Tipping Point whom I first met in a London café in January 2020 so we could talk about our mutual interests, and who generously shared their work with me. I am also grateful that Ben Neimark of Lancaster University met us in the café and because we were so excited, joined us for dinner so we could continue comparing different methods of counting and estimating military greenhouse gas emissions. Ben and the rest of the "carbon boot-print" team, Oliver Belcher and Patrick Bigger, were also generous interlocutors and shared some of their data. I thank Carol Cohn for reading an early draft and writing both perceptive and hilarious comments in the margins, and Jill Breitbarth for talking about the book as we walked around the Mystic River in search of birds, turtles, and butterflies. I am *deeply* grateful (pun intended) to Anna Henchman for reading portions of a later draft of the book and helping me see how to be a better writer. I thank Matt Evangelista, Joan Filler, and Henry Shue for ideas and meals along the way. Jeffrey and Patti Buck were excellent companions during the final stretch—generous with food, fun and purpose. What a joy to play with you both. Thank you Patti for that and for so many other things.

I—sort of—thank my editor at the MIT Press, Beth Clevenger, who wrote to me out of the blue and convinced me to drop everything—including a book I was nearly done writing—to write this book. I say "sort of" because Beth put me on a strict timetable that would only let me leave from my desk for more than a few hours a day until the first draft and then the second draft were finished. I'm still not sure why I agreed to do it. I am also appreciative of Anthony Zannino, Judith Feldmann, and the production staff at the MIT Press. Thanks also to the four anonymous reviewers; you can't know this, but you all had different, and in some cases completely contradictory, suggestions. You will be the judge of whether I have threaded the needle.

More than anything, I am grateful to my daughter Rose Jordan Crawford, who said basically the same thing nearly every time she inquired about my progress and about the chances of me meeting the MIT Press deadline. After I said, "Well, I'm still on chapter 2," or was stuck somewhere else, Rose would always reply with some version of "You're doing great mom. You've got this." I appreciate your love Rose, and your faith in me. But you must know that your curiosity, your passion to make the world better, your incredible work ethic, and the creativity that you manifest on canvas are a wonder to behold. I write for you and your generation.

Appendix: Estimating U.S. Military Greenhouse Gas Emissions

The military greenhouse gas emissions for the United States and other countries are not so easy to discern using either the United Nations IPCC guidelines or national inventories. As Michael Durant Thomas suggests, this is not an easy exercise. "Some of the challenges in obtaining clear information on military GHG data include: (1) a lack of a clear guiding international or national regulatory or legal requirement for nations to account for and report military emissions; (2) difficulties in data collection relating to offshore/non-domestic consumption and usage of fossil fuels by military forces involved in permanent off-shore basing arrangements, exercises, routine patrols and operations, or war; (3) an unwillingness by nations to report military emissions for operational or security-related reasons; and (4) challenges in developing an appropriate set of emissions boundaries that attribute responsibility in an equitable manner."[1]

While it is possible to use some existing data to understand U.S. military emissions in some years (for fiscal years 2010 through 2019, see chapter 4), there is currently no public data set available that would give a sense of the long-term trends and total greenhouse gas emissions over a longer time period. Academics have used several methods for estimating the greenhouse gas emissions of the Pentagon and other countries.[2]

This appendix discusses alternative ways to estimate military emissions and summarizes the sources, methods, assumptions, and calculations I used to estimate total military greenhouse gas emissions from fiscal years 1975–2019 and the ways I checked the likelihood that the estimates were reasonable.[3]

ALTERNATIVE APPROACHES TO ESTIMATING MILITARY EMISSIONS

One method uses a combination of military spending and "emission factors" associated with petroleum use to estimate total emissions. As Ho-Chih Lin and Deborah Burton argue in their report *Indefensible: The True Cost of the Global Military to Our Climate and Human Security*, "any given nation's annual military spending is a good predictor of annual energy/fuel usage of that nation's military." They suggest, "The higher the military spending, the greater the number of big-ticket items procured, and, therefore, the greater the amount of fossil fuel consumed, which in turn requires even higher military spending to cover the increased maintenance and operation costs. In this way, it is reasonable to suppose annual fuel usage of a military is proportional to its government's annual military spend."[4] Lin and Burton estimate that in 2017, the total CO_2e emissions of world's militaries and associated defense industries was 445 MMT and that global militaries account for "at least 1% of the total global greenhouse gas emissions" but perhaps as much as 5 percent of world emissions.[5]

Another method calculates emissions based on energy use—the total amount of energy consumed multiplied by coefficient for CO_2e emissions per BTU, or a finer-grained analysis, starting with the category of fuel used (or a combination of both energy and fuel). See table A.1. Oliver Belcher, Patrick Bigger, Ben Neimark, and Cara Kennelly use this method to calculate emissions based on declassified Department of Defense comprehensive data for fuel use that they obtained through a Freedom of Information Act

Table A.1
Emissions for different types of fuel

Fuel	Pounds of CO_2 emissions per million BTUs
Anthracite coal	228.6
Bituminous coal	205.7
Diesel	161.3
Gasoline	157.2
Natural gas	117.0

Source: EPA emissions data, https://www.eia.gov/tools/faqs/faq.php?id=73&t=11.

Request.[6] Because it uses fuel consumption data provided by the Pentagon, this method is close to the EPA method. But because it is comprehensive, starting with the raw data, their estimate includes both domestic and international fuel use. They are thus able to estimate fuel used abroad in operations, at military bases, and by service.

Adam Liska and Richard Perrin use a method that combined actual fuel use and imputed emissions based on military spending. For the military itself, they calculated emissions based on fuel consumed in British Thermal Units and CO_2e per trillion BTU for each type of fuel. They further assumed that military industry was a sizable portion of the civilian economy and that military-industrial emissions are the same greenhouse gas emission intensity factor as all manufacturing sector emissions, 0.300 MMT of CO_2e per billion dollars of goods produced. Using their calculation of fuel use emissions, they calculate a total implied intensity factor of .0289 MMT of CO_2e per billion dollars of conventional DOD expenditures in 2008.[7]

Because the major portion of military fuel consumption is operational, and thus is driven by diesel and jet fuel usage, fuel use closely tracks the size and scope of military operations; when operations increase, fuel use rises and so does military spending. In my own work I have not used military spending as a proxy for U.S. greenhouse gas emissions for two reasons. First, the annual data for military consumption in major categories of fuel use are available and overall military emissions can be calculated, as I will describe, using that data. Second, although total military spending *should* track closely with operations—and often does—it sometimes does not. For instance, in the U.S. post-9/11 wars, total military spending has often increased or remained steady even as the portion of that spending that was dedicated to actual warfighting, and the military activity in the war zone itself, declined. The level of military spending is durable for several reasons, not least of which is the fact that a good portion of military spending in any year is for personnel pay and health care which also increases during wars but does not go down as quickly as operational spending unless the total size of the force significantly decreases. Nevertheless, as figure A.1 indicates, there is a strong correlation between military spending and total DOD emissions for the period 2008–2019, but the durability of military spending in the later years of this period despite reductions in operations is also illustrated here.

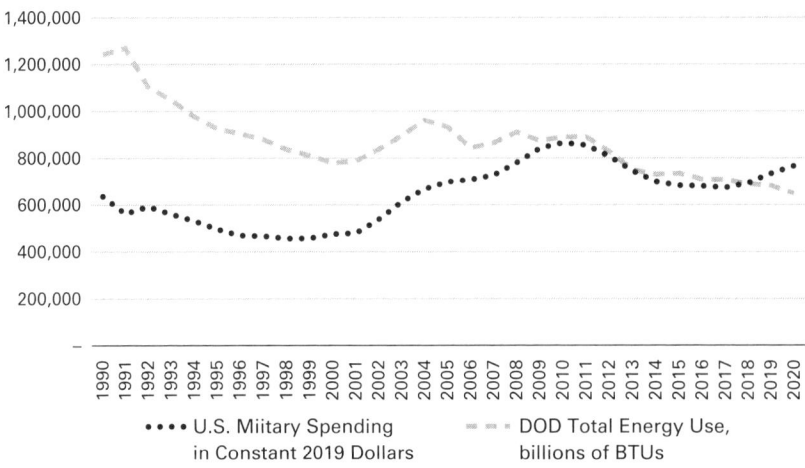

Figure A.1

U.S. military spending, in constant 2019 dollars and BTUs per fiscal year.

Sources: Stockholm International Peace Research Institute, SIPRI, "Military Expenditure Database," https://www.sipri.org/databases/milex; Department of Energy, Comprehensive Annual Energy Data and Sustainability Performance, https://ctsedwweb.ee.doe.gov/Annual /Report/HistoricalFederalEnergyConsumptionDataByAgencyAndEnergyTypeFY1975To Present.aspx.

The third method calculates emissions using publicly available Department of Energy data on energy consumption for the Department of Defense and standard methods for calculating CO_2 equivalent greenhouse emissions. It then estimates total emissions based on the known data. The explanation for the method I used follows.

SOURCES

The Department of Energy database supplies greenhouse gas emissions in terms of metric tons of CO_2 equivalent (CO_2e) for the U.S. government and all its departments, including the Department of Defense, for FY2008 and FY2010–2020. Table A.2 is a reproduction of the DOE database information about the DOD for FY2020.[8] Scope 1 emissions are those produced on site and scope 2 emissions are from purchases of energy.

The DOE also provide data for buildings, vehicles, and equipment for those years in the form shown in table A.3, which is a reproduction of the

Table A.2

DOE report of Department of Defense greenhouse gas inventory, CO_2e, FY2020

Scope and category	GHG emissions from standard operations (MT CO_2e)	GHG emissions from nonstandard operations* (military operations, law enforcement, and other) (MT CO_2e)	Total quantity emitted biogenic (MT CO_2e)
Scope 1: On-site fuel consumption at federal facilities	5,371,797.2	0.0	66,893.1
Scope 1: mobile emissions—vehicles, aircraft, ships, and equipment	961,687.8	31,369,399.8	7,002.4
Scope 1: mobile emissions—passenger fleet vehicles	390,149.6	75,729.0	19,806.6
Scope 1: fugitive emissions—fugitive fluorinated gases and other fugitive emissions	309,213.2	504,653.1	
Scope 1: fugitive emissions—on-site wastewater treatment	7,848.6		1,669.8
Scope 1: fugitive emissions—on-site landfills and municipal solid waste facilities	278,599.5		73,846.1
Scope 1: manufacturing and industrial process emissions	0.0	0.0	
Subtotal Scope 1	**7,319,295.9**	**31,949,781.9**	**169,218.0**
Scope 2: purchased electricity	11,975,932.3	634.2	0.0
Scope 2: purchased biomass energy	3,595.9		311,117.4
Scope 2: purchased steam and hot water	444,162.7	0.0	0.0
Scope 2: purchased chilled water	1,283.8	0.0	0.0
Scope 2: purchased combined heat and power electricity, steam, and hot water	0.0	0.0	0.0
Subtotal Scope 2	**12,424,974.7**	**634.2**	**311,117.4**
Scope 2: reductions from renewable energy use	−225,221.8		0.0
Subtotal Scope 1 and 2	**19,519,048.9**	**31,950,416.1**	**480,335.4**

Source: Department of Energy, "Greenhouse Gas Inventory (Metric Tons of Carbon Dioxide Equivalent," Comprehensive Annual Energy Data and Sustainability Performance, https://ctsedwweb.ee.doe.gov/Annual/Report/ComprehensiveGreen houseGasGHGInventoriesByAgencyAndFiscalYear.aspx.

*Non-standard operations are vehicles, vessels, aircraft, and other equipment used by federal government agencies in combat support, combat service support, tactical or relief operations, training for such operations, law enforcement, emergency response, or spaceflight (including associated ground-support equipment). Non-standard operations also include generation of electric power produced and sold commercially to other parties.

Table A.3

Department of Defense site-delivered energy use and costs, FY2019

Goal-subject buildings	Native units	Quantity units	Billion BTU	Cost (in 2019 $)
Electricity	26,743,441.8	MWh	91,248.6	2,322,252,970.2
Fuel oil	88,189.8	Thou. gallons	12,170.2	248,040,261.1
Natural gas	71,486,043.3	Thou. cu. ft.	73,344.7	468,597,132.4
Lpg propane	11,323.6	Thou. gallons	1,041.8	16,347,333.7
Coal	234,750.8	Short tons	4,988.9	25,411,871.1
Purchased steam	3,148.5	BBTU	3,148.5	74,681,035.8
Purchased renewable energy	1,374.9	BBTU	1,374.9	24,618,341.0
On-site renewables and adjustments	3,921.3	BBTU	3,921.3	32,209,133.0
Other	292.3	BBTU	292.3	4,398,719.6
End-use sector total			191,531.1	3,216,556,797.9
Gross square feet	1,893,896.0	Thou. GSF		
Goal-excluded facilities	**Native units**	**Quantity units**	**Billion BTU**	**Cost (in 2019 $)**
Electricity	2,483,961.1	MWh	8,475.3	250,877,358.2
Fuel oil	756.1	Thou. gallons	104.3	1,779,629.0
Natural gas	551,912.1	Thou. cu. ft.	566.3	3,315,190.2
Lpg propane	0.0	Thou. gallons	0.0	0.0
Coal	0.0	Short tons	0.0	0.0
Purchased steam	70.8	BBTU	70.8	681,320.0
Purchased renewable energy	0.0	BBTU	0.0	0.0
On-site renewables and adjustments	489.6	BBTU	489.6	0.0
Other	0.0	BBTU	0.0	101.0
End-use sector total			9,706.3	256,653,598.4
Gross square feet	18,113.9	Thou. GSF		
Vehicles and equipment	**Native units**	**Quantity units**	**Billion BTU**	**Cost (in 2019 $)**
Auto gas	92,896.8	Thou. gallons	11,612.1	293,306,905.1
Dist-diesel	701,387.0	Thou. gallons	96,854.1	2,077,192,714.3
Lpg propane	9.4	Thou. gallons	0.9	22,043.0
Aviation gas	340.5	Thou. gallons	40.9	2,141,326.9
Jet fuel	2,749,960.8	Thou. gallons	371,244.7	8,479,528,246.1
Navy Special Fuel Oil (NSFO)	0.0	Thou. gallons	0.0	0.0

Table A.3 (continued)				
Goal-subject buildings	Native units	Quantity units	Billion BTU	Cost (in 2019 $)
Other	1,157.5	BBTU	1,157.5	31,403,311.5
End-use sector total			480,910.1	10,883,594,546.8
Department of Defense Total (FY2019)			682,147.5	14,356,804,943.1

Source: Department of Energy, "Site-Delivered Energy Use and Costs by End-Use Sector and Energy Type," Government Total (FY2020), Comprehensive Annual Energy Data and Sustainability Performance, https://ctsedwweb.ee.doe.gov/Annual/Report/SiteDeliveredEnergyUseandCostsbyEndUseSectorAndEnergyTypeByFederalAgencyNativeUnitsAndBillionBtu.aspx; https://ctsedwweb.ee.doe.gov/Annual/Report/BEAReport.aspx?ef=Excel&fy=1&yo=&ag=6&au=false.

DOE's "Site-Delivered Energy Use and Costs by End Use Sector and Energy Type" for fuel and electricity consumption for the DOD in fiscal year 2020. Within the DOD, the Defense Logistics Agency has reports on its website that detail the purchases of fossil fuel products from commercial refineries and other sources for fiscal years 1999–2020, and the sales of that fuel to its customers.[9] But the DOD does not make that information available for prior years, nor does it report detailed fuel *consumption* information by type of fuel either on its website or to Congress in its annual budget requests.[10] Further, as noted in chapter 3, not all the fuel purchased by the DOD in a year is necessarily consumed in that fiscal year. An accurate estimate of emissions should be based on the amount of fuel consumed in that year. The DOE also has fuel consumption data for vehicles and equipment and energy consumed for buildings for other years, going back to 1975.

METHODS FOR ESTIMATING DOD GREENHOUSE GAS EMISSIONS

While data for standard and nonstandard greenhouse gas emissions by U.S. government departments are published for fiscal year 2008, and fiscal years 2010–2020 on the Department of Energy website, there is no accounting of standard and nonstandard emissions for other years.[11] The total emissions for each year were estimated by using vehicle and equipment emissions as the starting point.

Vehicle and equipment emissions for each fuel type used by the DOD in these years were estimated by calculating emissions for Vehicles and Equipment found in the Department of Energy's Comprehensive Annual Energy Data and Sustainability annual reports, which reported DOD nonstandard fuel consumption for the fiscal years 1975–2007 and 2009.[12]

Most operational energy consumed is in the form of "bulk fuel" purchases of jet (JP-8 and JP-5) and diesel fuel. A calculation of CO_2 equivalent (CO_2e) emissions for U.S. DOD jet fuel consumption in 2017 is illustrated in table A.4. The calculation of emissions of jet fuel makes the following assumptions: each gallon of jet fuel produces 0.135 HHV MMBtu per gallon. Using the standard emission factors for jet fuel—CO_2 of 72.22 kg/MMBtu; for CH_4 (methane) of .003 kg/MMBtu; and for N_2O (nitrous oxide) of .0006 kg/MMBtu—one can calculate the greenhouse gas, CO_2 equivalent, emissions for a given quantity of jet fuel.[13]

Table A.4

Calculating greenhouse gas emissions for U.S. military jet fuel consumption, FY2017

	Jet fuel	Unit of measure
Annual consumption GHG nonstandard operations	2,915,738.50	thousand gallons
Total energy consumed	393,624,693.30	MMBTU
Cost	$6,681,061.20	
Unit cost	$2.29	
Anthropogenic CO_2 emission factor	72.2	kg CO_2e/MMBTU
CH_4 emission factor	0.003	kg CH_4/MMBTU
N_2O emission factor	0.0006	kg N_2O /MMBTU
Total quantity emitted anthropogenic CO_2	28,427,575,352.60	kg
Total quantity emitted CH_4	1,180,874.10	kg
Total quantity emitted N_2O	236,174.80	kg
GWP factor for CO_2	1	CO_2e
GWP factor for CH_4	25	CO_2e
GWP factor for N_2O	298	CO_2e
Total quantity emitted (CO_2e)	28,527,477,299.80	kg CO_2e
Total quantity emitted (MT CO_2e)	28,527,477.30	MT CO_2e

The impact of methane and nitrous oxide gases on atmospheric warming, their global warming potentials (GWPs), are significantly higher than CO_2; these greenhouse gases are thus scaled into an equivalent relation to carbon dioxide, which has, by definition, a GWP of 1. For these calculations I use the EPA and Department of Energy GWP 100-year values of 25 for methane and 298 for nitrous oxide.[14]

In this way, DOE fuel-consumption data for the DOD were used to calculate annual emissions for DOD vehicle- and equipment-caused greenhouse gas emissions for each fuel type—gasoline, diesel, LPG/propane, aviation gas, jet fuel, and Navy Special Fuel Oil—based on fuel energy consumption.[15]

Although nonstandard fuel consumption is *not* the same as all vehicle and equipment emissions, since the vast majority of vehicle emissions were nonstandard emissions, I used vehicle emissions as a proxy for nonstandard emissions. In those years for which I have Department of Energy and DOD data for all DOD emissions (fiscal years 2008 and 2010–2019), for nonstandard emissions and standard emissions, nonstandard emissions were on average 63 percent of total emissions. Using the vehicle emissions as a proxy for nonstandard emissions, the total emissions and standard emissions were estimated assuming that nonstandard emissions were 63 percent of total emissions, also assuming that the ratios were the same throughout the entire period. The results are shown in table A.5 and graphed in chapter 4, figure 4.1.

CONSERVATIVE ESTIMATES

While the ratio between standard and nonstandard emissions varies, these estimates of total U.S. military greenhouse gas emissions are still likely conservative for four reasons.

First, the focus is on the major emissions of the DOD. I have not included all emissions; for example, biogenic emissions—usually comparatively small—are not included in these estimates. In fiscal years 2017 and 2018, for example, U.S. DOD biogenic emissions were .57 million MT CO_2e and .49 million MT of CO_2e. Also not included are Scope 3 emissions from, for example, employee air and ground business travel, wastewater treatment, and solid waste disposal. For example, U.S. DOD scope

Table A.5

Estimated annual DOD greenhouse gas emissions, FY1975–2018

Fiscal year	DOD standard emissions	DOD nonstandard emissions	Total DOD CO$_2$e emissions
1975	40.4	68.7	109
1976	34.6	58.9	93
1977	35.1	59.7	95
1978	33.9	57.7	92
1979	35.0	59.6	95
1980	35.9	61.1	97
1981	38.6	65.7	104
1982	39.5	67.2	107
1983	39.2	66.7	106
1984	40.2	68.4	109
1985	39.1	66.6	106
1986	38.8	66.1	105
1987	40.3	68.6	109
1988	35.3	60.1	95
1989	40.2	68.5	109
1990	38.7	65.9	105
1991	40.8	69.4	110
1992	32.4	55.2	88
1993	32.0	54.5	87
1994	29.7	50.5	80
1995	28.1	47.9	76
1996	27.3	46.4	74
1997	26.6	45.3	72
1998	25.5	43.4	69
1999	24.6	41.9	66
2000	23.1	39.3	62
2001	23.5	40.0	63
2002	26.0	44.2	70
2003	29.0	49.3	78
2004	31.6	53.9	85

Fiscal year	DOD standard emissions	DOD nonstandard emissions	Total DOD CO_2e emissions
2005	30.6	52.0	83
2006	27.2	46.3	73
2007	28.3	48.2	76
2008	26.9	50.4	*77*
2009	28.4	48.3	77
2010	27.0	49.5	*77*
2011	25.7	48.8	*74*
2012	24.4	44.9	*69*
2013	24.1	39.5	*64*
2014	23.8	38.1	*62*
2015	23.9	38.7	*63*
2016	22.1	37.2	*59*
2017	21.3	37.1	*58*
2018	20.9	34.5	*55*
2019	20.7	34.1	*55*
2020	19.5	32.0	*51*
Total	1,399	2,390	3,789

Table A.5
(continued)

Source: Calculated from Department of Energy fuel consumption data rounded to the nearest metric ton. Figures in italics, for FY2008 and FY2010–2020, are the Department of Energy reported emissions figures. The italicized data for the corresponds to the DOD report of emissions.

3 emissions in 2008 were 7.6 million MT CO_2e, and in 2016, 7 million MT CO_2e. Further, although contractors are an increasingly important part of the mix of operations in war zones, performing functions that militaries used to do—such as providing security for military convoys, refueling, preparing meals, and maintaining equipment, telecommunications, and other infrastructure—their vehicle emissions are difficult to track and are not counted in Pentagon emissions.

Second, while I use EPA and Department of Energy global warming potentials (GWPs), the U.S. EPA uses lower GWPs than the international standard. Specifically, as noted earlier, the DOE and the EPA use the U.S.

EPA 100-year GWPs that scale the GWP of methane, CH_4, at 25 and nitrous oxide, N_2O, at 298 over 100 years.[16] The Intergovernmental Panel on Climate Change Fifth Assessment Report uses a GWP of 34 for methane's CO_2 equivalent.[17] If the IPCC global warming potentials were used, estimates of U.S. DOD greenhouse gas emissions for vehicles would be higher. Further, I have also not calculated the emissions from fluorinated gases from U.S. vehicles and equipment.

Third, there are some sources of DOD facilities energy for which the source fuel is unclear. Specifically, table A.3 shows Department of Energy data for DOD site-delivered energy use in fiscal year 2019 including electricity and purchased steam for facilities from external power suppliers, but the DOE does not provide detail about the fuel used to produce that electricity. The source of purchased electricity or steam in those categories could be nuclear power, which would have no greenhouse gas emissions, or coal, which could have a significant greenhouse gas footprint, or natural gas, which would have the lowest emissions of the fossil fuels. And further, the emissions factors for purchases of electricity and steam vary from location to location. More significantly, the mix of fuels powering U.S. power plants has gradually shifted away from coal since 1975. Thus, for the earlier years, say 1975 to 2000, I am probably underestimating the standard emissions at facilities.

Fourth, it may be that we are underestimating the impact of jet fuel emissions, which are the major source of vehicle nonstandard military greenhouse gas emissions. While CO_2 is the major product of jet fuel consumption, jet fuel combustion emissions at high altitude also contain water vapor, a global warming gas, which itself causes the formation of cirrus clouds. The DOD also puts additives in its jet fuels to ensure that they perform according to military requirements. For instance, because military jets fly at much higher altitudes than commercial jets, they use additives to ensure that the fuel lines do not freeze. Any emissions from those additives and warming from water vapor are not counted in the standard estimates of jet fuel emissions. Yet, climate scientists suggest that even though CO_2 is the major product of jet fuel consumption, the impact of these other greenhouse gases should not be discounted. While the Department of Energy figures and the calculations here include nitrous oxide and methane, it is possible that the additional effects of additives for jet fuel combustion and high altitude water vapor, which are not included in these calculations, are significant. As the European Environmental Agency notes, "Non-CO_2 impacts cannot be ignored as they potentially represent approximately 60% of total climate impacts that are important in the shorter term (excluding cloudiness impacts)."[18] In sum, this means that the impact of military aviation emissions when all greenhouse gases are included may be higher than those estimated here. Military jet contrails, leading to the creation of cirrus clouds, is a significant contributor to global warming.[19]

CALCULATING FUEL USE IN A WAR ZONE

Absent a full Pentagon accounting of their fuel consumption and emissions by operation, there are various ways to estimate DOD greenhouse gas emissions in the post-9/11 wars. To estimate the war-related emissions for a war, one needs data for the entire period and then. based on some estimate of the portion of total military activity dedicated to war, one could estimate the amount of total greenhouse gas emissions that are due to any war.

One could base an estimate of total greenhouse gas emissions that should be attributed to the war on the proportion of the total military budget that were spent on the war. In the case of the Overseas Contingency Operations of the post-9/11 wars, these were funded by special appropriations. It is possible to determine the amount of money authorized for those wars—although there will be war-related spending in the base budget. In other words, one can use the average portion of the DOD budget spent on Overseas Contingency Operations as an approximate measure of energy use related to the war effort and assume that some portion of the base budget, and therefore base/non-war operations and installation energy use, is correlated to war-related spending. The Overseas Contingency Operations budget for the major war zones accounted for an average of 17 percent of the entire DOD (top line) budget from fiscal year 2001 to 2018. But this rule of thumb would give an estimate of war-related emissions that would be too low, since nonstandard emissions account for such a high proportion of all DOD fuel use.

A better way to estimate total greenhouse gas emissions for Overseas Contingency Operations would be to focus on operational fuel consumption, defined in DOE parlance as nonstandard fuel consumption. Between fiscal years 2010 and 2019, the Department of Energy attributed an average of 63 percent of all DOD greenhouse gas emissions to nonstandard operations. However, to assume that *all* nonstandard fuel use was for the major wars would yield an estimate that is probably too high, since the DOD performs other, non-post-9/11 war-related missions—such as exercises with its allies or operations on the southern border of the United States.

Another, and arguably better, method would be to base estimates of greenhouse gas emissions during the major post-9/11 wars on the proportion of fuel use by Central Command and other war zones. In fiscal year 2014 (see figure 4.10), Central Command used about 24 percent of the total operational fuel consumption by the DOD. But because U.S. post-9/11 counterterror operations were underway all over the world (in about 80–90 countries) the Central Command was not the only war zone in the war on terrorism. The portion of all greenhouse gas emissions related to Central Command including overseas contingency operations and the Global War on Terrorism is estimated to be about 35 percent of total greenhouse gas emissions for nonstandard and standard operations. Because the war in Afghanistan started so late in fiscal year 2001, I have not included emissions for that year in my estimate. Thus (see table A.6), my rough estimate of the emissions that may be attributable to the post-9/11 wars is 458 million metric tons of CO_2e.

Table A.6

Rough estimate of DOD and war-related overseas contingency operation greenhouse gas emissions, millions of metric tons CO_2e, FY2002–2020

	Total DOD CO_2e emissions in millions of metric tons FY2002–2020	OCO-related CO_2e emissions in millions of metric tons FY2002–2020
Standard	481	*168*
Nonstandard (directly support combat)	829	*289*
Total FY2002–2020	1,308	*458*

Source: Rounded to the nearest million metric tons. Based on Department of Energy data.

EPA EMISSIONS REPORTING

More subtly, although some of the basic information is there, the EPA does not make it easy to determine total military emissions for domestic and overseas fuel use. For example, consider the emissions from military aviation, the largest portion of DOD emissions where one has to look at several different places in the *Inventory of U.S. Greenhouse Gases and Sinks: 1990–2019* to put together total military aviation emissions. In one table on transportation emissions, the EPA reports emissions for all domestic military aviation in the category "other aircraft," which includes *both* domestic military and general aviation emissions. The EPA explanatory note for the category says, "Consists of emissions from jet fuel and aviation gasoline consumption by general aviation and military aircraft."[20] These emissions are distinct from those of commercial aircraft, as table A.7 shows. The information the EPA

Table A.7

Inventory of transportation-related greenhouse gas emissions (MMTCO$_2$ eq).*

Gas/vehicle	1990	2005	2015	2016	2017	2018	2019
Commercial aircraft	110.9	134.0	120.1	121.5	129.2	130.8	135.4
Other aircraft	78.3	59.7	40.4	47.5	45.6	44.7	45.7

Source: Environmental Protection Agency, *Inventory of U.S. Greenhouse Gases and Sinks: 1990–2019*, 2-38–2-39, table 2-13, Transportation Related Greenhouse Gas Emissions.

Table A.8

CO$_2$ Emissions from fossil fuel combustion in transportation end-use sector (MMTCO$_2$ eq.)

Fuel/vehicle type	1990	2005	2015	2016	2017	2018	2019
Jet fuel	184.2	189.3	157.6	166.0	171.8	172.3	177.8
Commercial aircraft	109.9	132.7	119.0	120.4	128.0	129.6	134.2
Military aircraft	35.0	19.4	13.5	12.3	12.2	11.8	11.9
General aviation aircraft	39.4	37.3	25.1	33.4	31.5	30.9	31.7
International bunker fuels	38.0	60.1	71.9	74.1	77.7	80.8	80.7
International bunker fuels from commercial aviation	30.0	55.6	68.6	70.8	74.5	77.7	77.6

Source: Environmental Protection Agency, *Inventory of U.S. Greenhouse Gases and Sinks: 1990–2019*, 3-27–3-28, table 3-13. The table is labeled CO$_2$e, but the figures appear to be only CO$_2$.

provides for "other aircraft" follows. While the aggregation in the category "other aircraft" does preserve "confidentiality," as suggested by the IPCC, it is not meant to be too precise. We can see, however that the figure suggests that emissions from military aircraft used in domestic settings is declining. Incidentally, we can see from this that U.S. commercial aircraft emissions have increased dramatically since 1990.[21]

Elsewhere in the EPA report, it is possible to get a more precise sense of U.S. domestic military aircraft emissions in the data on jet fuel emissions by category. This allows us to distinguish domestic military greenhouse gas

Table A.9
EPA report of international bunker fuel emissions (MMTCO$_2$).

International bunker fuel	1990	2005	2015	2016	2017	2018	2019
Total CO$_2$e for international bunker fuels	104.5	114.3	112.0	117.7	121.3	123.3	117.2
Military aviation bunker fuels	8.1	4.5	3.3	3.3	3.2	3.1	3.1

Source: Table 3-104, *Environmental Protection Agency, Inventory of U.S. Greenhouse Gases and Sinks: 1990–2019*, 3–118, https://www.epa.gov/sites/production/files/2021-04/documents/us-ghg-inventory -2021-main-text.pdf.

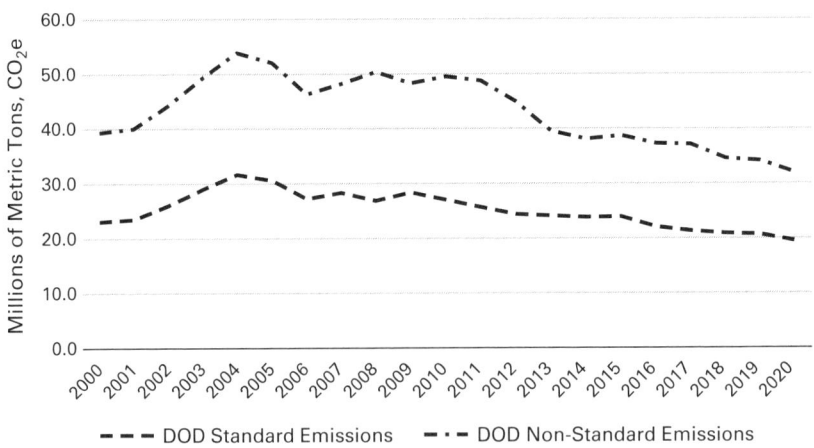

Figure A.2
Estimate of DOD standard and nonstandard greenhouse gas emissions, millions of metric tons CO$_2$e, FY2000–2020.
Source: Based on Department of Energy data, https://ctsedwweb.ee.doe.gov/Annual/Report /ComprehensiveGreenhouseGasGHGInventoriesByAgencyAndFiscalYear.aspx.

emissions from general aviation emissions in the EPA report. Domestic military aviation emissions declined 66 percent (23.1 MMTCO$_2$) between 1990 and 2019, while general aviation emissions declined 17 percent.[22]

Table A.8 shows how the EPA has distinguished jet fuel emissions from international bunker fuels from commercial aviation from a larger category of international bunker fuels from commercial aviation. The EPA says that "Emissions from International Bunker Fuels include emissions from both civilian and military activities; these emissions are not included in the transportation totals."[23] Later in the report, the EPA distinguishes military aviation bunker from total international bunker fuels, which includes U.S. Navy

Table A.10

Direct private sector defense-related employment in the U.S. and portion of total U.S. GDP

	1977*	1987*	1996	2011	2016
Total direct and indirect defense-related employment in civilian sector	1,652,000	3,544,000	2,123,000	4,457,213**	4,112,603**
Total direct defense-related employment in civilian sector	930,000	1,997,000	1,180,000	1,004,215	839,171
Employment in defense-related construction	100,100	164,000	96,800		
Employment in defense-related manufacturing	965,500	1,801,800	879,200		
Employment in defense-related transportation	96,600	154,400	122,900		
Employment in defense-related services	228,200	895,700	716,500		
Percent of total employment	.97	1.71	.88		
Military purchases as % of GDP total and noncompensation	6.2 3.	7.2 4.5	4.6 2.9		

Source: For 1977, 1987, and 1996, see data from tables 1 and 2 in Allison Thomson, "Defense-related Employment and Spending, 1996–2006," *Monthly Labor Review* (July 2008): 14–33; 15. For 2011 and 2016 data from figure 13, see Deloitte, "2017 Aerospace and Defense Sector Export and Labor Market Study," 13, https://www2.deloitte.com/content/dam/Deloitte/us/Documents/manufacturing/us-2017-us-A&D-exports-and-labor-market-study.pdf. For estimating indirect employment, Deloitte used an employment multiplier of 2.36.

*Employment numbers, rounded to the nearest 1,000.

**Total direct employment in aerospace and defense; see "Defense-related Employment and Spending, 1996–2006," *Monthly Labor Review* (July 2008): 14–33.

and other international bunker fuels emissions. See table A.9. However, the bunker emissions from Naval ships are even more difficult to identify in the EPA report and separate from commercial marine emissions.[24] The EPA also notes, "These data may not include fuel used in aircraft and ships as a result of a Service procuring fuel from, selling fuel to, trading fuel with, or giving fuel to other ships, aircraft, governments, or other entities."[25] Further, the EPA acknowledges that "Uncertainties exist with regard to the total fuel used by military aircraft and ships."[26]

As noted in chapter 4, military aviation fuel consumption at U.S. territories is not part of the data collected for domestic use and there is uncertainty about bunker fuel consumption in U.S. territories. As I said previously, the EPA notes, "The United States does not collect energy statistics for its territories at the same level of detail as for the fifty states and the District of Columbia."[27] The EPA acknowledged several other uncertainties in calculating historical data as well for both aircraft and naval vessels.[28]

ESTIMATING MILITARY-INDUSTRIAL EMISSIONS USING EMPLOYMENT DATA

As I discussed in chapter 4, it is possible to use employment to approximate military-industrial emissions. Table A.10 provides some employment data from several different sources.

Abbreviations

AEMR	Annual Energy Management and Resilience Report, Department of Defense
BRAC	Base Realignment and Closure
CENTCOM	Central Command
CH_4	Methane
CO_2	Carbon Dioxide
DLA	Defense Logistics Agency
DOD	Department of Defense
DOE	Department of Energy
EPA	Environmental Protection Agency
FY	Fiscal Year
GHG	Greenhouse Gases
GWP	Global Warming Potential
HFCS	Hydrofluorocarbons
IMCCS	International Military Council on Climate and Security
IPCC	Intergovernmental Panel on Climate Change
$MMTCO_2e$	Million Metric Tons of Carbon Dioxide equivalent
NATO	North Atlantic Treaty Organization
N_2O	Nitrous Oxide
RDT&E/RDTE	Research, Development, Test, and Evaluation
SIPRI	Stockholm International Peace Research Institute

Notes

INTRODUCTION

1. Intergovernmental Panel on Climate Change, *Climate Change 2021*, Sixth Assessment Report (United Nations Environment Program and World Meteorological Association, 2021), SPM-37.

2. International Military Council on Climate and Security (IMCCS), https://imccs.org.

3. See Michael Klare, *All Hell Breaking Loose: The Pentagon's Perspective on Climate Change* (New York: Metropolitan Books, 2019).

4. Bryan Bender, "Chief of US Pacific Forces Calls Climate Biggest Worry," *Boston Globe*, March 9, 2013, https://www.bostonglobe.com/news/nation/2013/03/09/admiral -samuel-locklear-commander-pacific-forces-warns-that-climate-change-top-threat /BHdPVCLrWEMxRe9IXJZcHL/story.html.

5. Gwynne Dyer, *Climate Wars: The Fight for Survival as the World Overheats* (Oxford: One World Publications, 2010); and Harald Welzer, *Climate Wars: What People Will Be Killed for in the 21st Century* (Cambridge, UK: Polity, 2017).

6. President Joseph R. Biden Jr., "Interim National Security Strategic Guidance," March 2021, The White House, Washington, DC, https://www.whitehouse.gov/wp-content/uploads /2021/03/NSC-1v2.pdf, 14.

7. The FY2021 National Defense Authorization Act (NDAA) mandates that the DOD "report on the total level of greenhouse gas emissions for each of the last 10 fiscal years. Such emissions levels shall include the agency wide total, breakdowns by military department, and delineations between installation and operational emissions" (Section 328, https://www.congress .gov/bill/116th-congress/house-bill/6395/text). That report is Office of the Assistant Secretary of Defense for Sustainment, "Report on Greenhouse Gas Emission Levels" (Department of Defense, August 2021). The FY2022 National Defense Authorization Act required that the "Secretary of Defense submit to Congress a plan to reduce the greenhouse gas emissions of the Department of Defense" no later than September 30, 2022 (Section 323, https://www .congress.gov/bill/117th-congress/senate-bill/1605/text).

8. Department of Defense total scope 1 and 2 emissions as reported by the Department of Energy, "Comprehensive Annual Data and Sustainability Performance," https://ctsedwweb .ee.doe.gov/Annual/Report/ComprehensiveGreenhouseGasGHGInventoriesByAgencyAnd FiscalYear.aspx. CO_2 equivalent is discussed further in the appendix.

9. Department of Energy, "Comprehensive Annual Data and Sustainability Performance."

10. The label "bootprint" for military emissions is from Oliver Belcher, Patrick Bigger, Ben Neimark, and Cara Kennelly, "Hidden Carbon Cost of the 'Everywhere War': Logistics, Geopolitical Ecology, and the Carbon Bootprint of the U.S. Military," *Transactions of the Institute of British Geographers* 45, no. 1 (March 2020): 65–80.

11. In 2021, eight aerospace and defense companies were in the Fortune 500, and an additional twelve companies were ranked in the top 1,000 by Fortune. See Fortune, "Fortune 500," https://fortune.com/fortune500/search/.

12. Environmental Protection Agency, "Data Highlights: Inventory of U.S. Greenhouse Gas Emissions and Sinks, 1990–2019," https://www.epa.gov/sites/production/files/2021-04 /documents/us-ghg-inventory-1990-2019-data-highlights.pdf.

13. See Environmental Protection Agency, "Greenhouse Gas Inventory Explorer," https://cfpub .epa.gov/ghgdata/inventoryexplorer/#allsectors/allgas/econsect/current.

14. Office of the Assistant Secretary of Defense for Sustainment, *Department of Defense Annual Energy Management and Resilience Report* (AEMRR) FY2019, Department of Defense, June 2020, 43, https://www.acq.osd.mil/eie/ie/FEP_Energy_Reports.html.

15. Calculated from U.S. Energy Information Administration data, table 2.7, https://www.eia .gov/totalenergy/data/monthly/pdf/sec2_14.pdf, accessed September 24, 2021.

16. M. Crippa et al., *Fossil CO_2 Emissions of All World Countries: 2020 Report* (Joint Research Centre, European Commission, 2020), https://publications.jrc.ec.europa.eu/repository/handle /JRC121460. Also see country emissions data from Climate Watch, World Resources Institute, Washington, DC, https://www.climatewatchdata.org/.

17. Stuart Parkinson, "The Environmental Impacts of the UK Military Sector," Scientists for Global Responsibility, May 2020, https://www.sgr.org.uk/sites/default/files/2020-05/SGR -DUK_UK_Military_Env_Impacts.pdf; Stuart Parkinson and Linsey Cottrell, *Under the Radar: The Carbon Footprint of Europe's Military Sectors: A Scoping Study* (Conflict and Environment Observatory and Scientists for Global Responsibility, February 2021), https://www .sgr.org.uk/sites/default/files/2021-02/EU-MCE-report-by-SGR-CEOBS-GUE.pdf.

18. Kurt M. Campbell, Jay Gulledge, J. R. McNeill, John Podesta, Peter Ogden, Leon Fuerth, R. James Woolsey, Alexander T. J. Lennon, Julianne Smith, Richard Weitz, and Derek Mix, *The Age of Consequences: The Foreign Policy and National Security Implications of Global Climate Change* (Center for Strategic & International Studies and Center for a New American Security, November 2007), 9, https://csis-website-prod.s3.amazonaws.com/s3fs-public/legacy_files /files/media/csis/pubs/071105_ageofconsequences.pdf.

19. National Oceanic and Atmospheric Administration, "Carbon Dioxide Levels Race Past Troubling Milestone," September 30, 2016, https://www.noaa.gov/stories/carbon-dioxide-levels-race-past-troubling-milestone.

20. See Peter Merriman and Kimberley Peters, "Military Mobilities in an Age of Global War, 1870–1945," *Journal of Historical Geography* 58 (October 2017): 53–60; Andrew K. Jorgenson, Brett Clark, and Jeffrey Kentor, "Militarization and the Environment: A Panel Study of Carbon Dioxide Emissions and the Ecological Footprints of Nations, 1970–2000," *Global Environmental Politics* 10, no. 1 (February 2010): 7–29; 22. Also see Melike Bildirici, "The Causal Link among Militarization, Economic Growth, CO2 Emission, and Energy Consumption," *Environmental Science and Pollution Research International* 24, no. 5 (February 2017): 4625–4636; Brett Clark, Andrew K. Jorgenson, and Jeffrey Kentor, "Militarization and Energy Consumption: A Test of Treadmill of Destruction Theory in Comparative Perspective," *International Journal of Sociology* 40, no. 2 (Summer 2010): 23–43; Phil Johnstone and Catriona McLeish, "World Wars and the Age of Oil: Exploring Directionality in Deep Energy Transitions," *Energy Research and Social Science* 69 (November 2020): 101732–101744.

21. On human security and climate change, see W. N. Adger et al., "Human Security," in *Climate Change 2014: Impacts, Adaptation, and Vulnerability. Part A: Global and Sectoral Aspects. Contribution of Working Group II to the Fifth Assessment Report of the Intergovernmental Panel on Climate Change* (Cambridge: Cambridge University Press, 2014), 755–791.

22. For a short discussion, see Parkinson and Cottrell, *Under the Radar*, 9.

23. Conger quoted in Anna Mulrine Grobe, "Why the Pentagon Is Serious about Reducing Its Carbon Footprint," *Christian Science Monitor*, March 16, 2021, https://www.csmonitor.com/Environment/2021/0316/Why-the-Pentagon-is-serious-about-reducing-its-carbon-footprint.

24. Laura Potter, quoted in Karlie Goldenberg, "Army Grapples with Very Serious Climate Change Threat," *Association of the United States Army*, October 13, 2021, https://www.ausa.org/news/army-grapples-very-serious-climate-change-threat.

25. President Biden, The White House, "Executive Order on Catalyzing Clean Energy Industries and Jobs Through Federal Sustainability," December 8, 2021, https://www.whitehouse.gov/briefing-room/presidential-actions/2021/12/08/executive-order-on-catalyzing-clean-energy-industries-and-jobs-through-federal-sustainability/.

26. The White House, "Fact Sheet: President Biden Sets 2030 Greenhouse Gas Pollution Reduction Target Aimed at Creating Good-Paying Union Jobs and Securing U.S. Leadership on Clean Energy Technologies," April 22, 2021, https://www.whitehouse.gov/briefing-room/statements-releases/2021/04/22/fact-sheet-president-biden-sets-2030-greenhouse-gas-pollution-reduction-target-aimed-at-creating-good-paying-union-jobs-and-securing-u-s-leadership-on-clean-energy-technologies/; United States, "The United States of America Nationally Determined Contribution: Reducing Greenhouse Gasses in the United States: A 2030 Emissions Target," April 21, 2021, https://www4.unfccc.int/sites/ndcstaging

/PublishedDocuments/United%20States%20of%20America%20First/United%20
States%20NDC%20April%202021%202021%20Final.pdf, accessed June 1, 2021.

CHAPTER 1

1. Ralph Waldo Emerson, "Wealth," in R. W. Emerson, *The Conduct of Life* (Boston: Ticknor and Fields, 1860), 73–110, 74.

2. Alfred T. Mahan, quoted in John H. Maurer, "Fuel and the Battle Fleet: Coal, Oil and American Naval Strategy, 1898–1925," *Naval War College Review* 34, no. 6 (November–December 1981), 60–77; 63.

3. The strategy of destroying the other's fuel is a continuation of the older strategy of plunder and pillage or laying waste to a land—the "feed fight"—so that the adversary would have little food or fodder with which to fuel their armies.

4. William H. McNeill, *The Pursuit of Power: Technology, Armed Force and Society since A.D. 1000* (Chicago: University of Chicago Press, 1982), 242.

5. Lindsay Schakenbach Regele, *Manufacturing Advantage: War, the State and the Origins of American Industry, 1776–1848* (Baltimore: Johns Hopkins University Press, 2019), 62, 49.

6. Regele, *Manufacturing Advantage*, 64–65.

7. Bernard Brodie, *Sea Power in the Machine Age* (Princeton, NJ: Princeton University Press, 1941), 18–21.

8. Sam Schurr and Bruce Netschert, with Vera Eliasberg, Joseph Lerner, and Hans Landsberg, *Energy in the American Economy, 1850–1975: An Economic Study of Its History and Prospects* (Baltimore: Johns Hopkins Press, 1960), 47.

9. McNeill, *Pursuit of Power*, 225.

10. Peter A. Shulman, *Coal and Empire: The Birth of Energy Security in Industrial America* (Baltimore: Johns Hopkins Press, 2015), 37.

11. See David Vine, *The United States of War: A Global History of America's Endless Conflicts, from Columbus to the Islamic State* (Oakland: University of California Press, 2020), 102–103.

12. Whales were then still a major supplier of fuel for lamps, only gradually replaced by kerosene starting in the 1850s.

13. James C. Dobbin, "Steam Navy of the United States: Letter from the Secretary of the Navy Transmitting Papers Giving Information in Reference to the Steam Navy of the United States," 33d Congress, 1st Session, 1854, 18; K. Jack Bauer, *The Mexican War, 1846–1848* (Lincoln: University of Nebraska Press, 1992), 111. Also see James Fenimore Cooper, *History of the Navy of the United States of America* (New York: Stringer and Townsend, 1856).

14. Dobbin, "Steam Navy of the United States," 13.

15. Secretary of the Navy, *Annual Report of the Secretary of the Navy*, 1855 (Washington, DC: United States Government Printing House, 1855), 131.

16. Daniel Webster to John H. Aulick, June 10, 1851, National Archives, Record Group 59, Special Missions, quoted in Kenneth E. Shumaker, "Forging the 'Great Chain': Daniel Webster and the Origins of American Foreign Policy Toward East Asia and the Pacific, 1841–1852," *Proceedings of the American Philosophical Society* 129, no. 3 (September 1985): 225–259; 248.

17. Webster to Aulick, June 10, 1851, quoted in Robert Hopkins Miller, *The United States in Vietnam, 1787–1941* (Washington, DC: National Defense University Press, 1990), 59.

18. Seward W. Livermore, "American Naval-Base Policy in the Far East, 1850–1914," *Pacific Historical Review* 13, no. 2 (June 1944): 113–135; 113.

19. Shulman, *Coal and Empire*, 79–91.

20. Matthew C. Perry, *A Paper by Commodore M. C. Perry, U.S.N. Read before the American Geographical and Statistical Society, at a Meeting Held March 6th, 1856* (New York: D. Appleton and Company, 1856), 8.

21. Perry, *A Paper by Commodore M. C. Perry*, 28–29.

22. Perry, 29.

23. See Lance E. Davis and Stanley Engerman, *Naval Blockades in Peace and War: An Economic History since 1750* (Cambridge: Cambridge University Press, 2006), 139–140.

24. Schurr et al., *Energy in the American Economy*, 52.

25. Matthew Carr, *Sherman's Ghosts: Soldiers, Civilians and the American Way of War* (New York: The New Press, 2015), 19.

26. Lisa M. Brady, *War upon the Land: Military Strategy and the Transformation of Southern Landscapes during the American Civil War* (Athens: University of Georgia Press, 2012), 43–44.

27. Quoted in Brady, *War upon the Land*, 122.

28. Megan Kate Nelson, *Ruin Nation: Destruction and the American Civil War* (Athens: University Georgia Press, 2012), 135, 152, and 267.

29. Livermore, "American Naval-Base Policy in the Far East, 1850–1914," 114–115.

30. U.S. Navy Department, *Annual Report of the Secretary of the Navy*, 1899 (Washington, DC: United States Navy, 1899), 304.

31. Rear Admiral Robley Evans, "Reserve Anthracite for Our Navy," *North American Review* 607 (February 1, 1907): 246–253; 250.

32. Walter Millis, *Arms and Men: A Study in Military History* (New Brunswick: Rutgers University Press, 1981), 123.

33. E. G. Campbell, "The United States Military Railroads, 1862–1865: Wartime Operation and Maintenance," *Journal of the American Military Foundation* 2, no. 2 (Summer 1938): 70–89.

34. See Allan R. Millett, Peter Maslowski, and William B. Feis, *For the Common Defense: A Military History of the United States from 1607 to 2012* (New York: Free Press, 2012).

35. William G. Thomas, *The Iron Way: Railroads, the Civil War, and the Making of Modern America* (New Haven, CT: Yale University Press, 2011).

36. Millis, *Arms and Men*, 125.

37. The future president had done his own naval history. See Theodore Roosevelt, *The Naval War of 1812, or the History of the United States Navy during the Last War with Great Britain, to Which Is Appended an Account of The Battle of New Orleans* (New York: G. P. Putnam's Son's 1882).

38. Alfred T. Mahan, *The Influence of Sea Power upon History, 1660–1783* (Boston: Little, Brown, 1890), 31.

39. Emphasis in the original. Quoted in Andrew Yeo and Stacie Pettyjohn, "Bases of Empire? The Logic of Overseas U.S. Military Base Expansion, 1870–2016," *Comparative Strategy* 40, no. 1 (2020): 18–35; 23.

40. Yeo and Pettyjohn, "Bases of Empire?," 24.

41. Benjamin Harrison, "The Inaugural Address of Benjamin Harrison," Yale Law School, The Avalon Project, https://avalon.law.yale.edu/19th_century/harris.asp.

42. Shulman, *Coal and Empire*, 7.

43. Mahan, *The Influence of Sea Power upon History*, 83.

44. Mahan, *Naval Strategy*, quoted in Philip A. Crowl, "Alfred Thayer Mahan: The Naval Historian," in *Makers of Modern Strategy: From Machiavelli to the Nuclear Age*, ed. Peter Paret (Princeton, NJ: Princeton University Press, 1986), 444–477; 460.

45. Mahan, *The Influence of Sea Power upon History*, 329n.

46. U.S. Navy Department, *Annual Report of the Secretary of the Navy*, 1899, 22.

47. U.S. Navy Department, *Annual Report of the Secretary of the Navy*, 1861, 1863, and 1864 (Washington, DC: United States Navy, 1861, 1863, and 1864), and Evans, "Reserve Anthracite for Our Navy."

48. Naval History and Heritage Command, "Spanish–American War," https://www.history.navy.mil/research/publications/documentary-histories/united-states-navy-s/coal.html, accessed 28 May 2021. Also see Shulman, *Coal and Empire*, 42–53.

49. U.S. Navy Department, *Annual Report of the Secretary of the Navy*, 1899, 27.

50. Coal for the war was also stored at Frenchman's Bay, ME; Portsmouth, NH; Boston; New London, CT; New York; League Island, PA; Washington, DC; Norfolk, VA; and Port Royal, SC. U.S. Navy Department, *Annual Report of the Secretary of the Navy*, 1899, 25.

51. Maurer, "Fuel and the Battle Fleet," 62.

52. Naval History and Heritage Command, "Spanish–American War."

53. U.S. Navy Department, *Annual Report of the Secretary of the Navy*, 1899, 22.

54. U.S. Navy Department, 1899, 27–28 and 304–306.

55. U.S. Navy Department, 1899, 303.

56. U.S. Navy Department, 1899, 305.

57. Asa Walker, "The Battle of Manila Bay," Unpublished Manuscript, Record Group 14, Naval War College Archives, Newport, RI, quoted in Maurer, "Fuel and the Battle Fleet," 60–77; 60–61.

58. Evans, "Reserve Anthracite for Our Navy," 253.

59. See Millis, *Arms and Men*, 192–197; Mike McKinley, "Cruise of the Great White Fleet," Naval History and Heritage Command, https://www.history.navy.mil/research/library/online-reading-room/title-list-alphabetically/c/cruise-great-white-fleet-mckinley.html, accessed May 29, 2021.

60. Christopher McMahon, "The Great White Fleet Sails Today?," *Naval War College Review* 71, no. 4 (Autumn 2018): 67–90; 74.

61. Naval Institute Archives, "December 16, 1907: The Great White Fleet Departs Hampton Roads for Circumnavigation," *Naval History Blog*, December 16, 2012, https://www.navalhistory.org/2012/12/16/december-16-1907-the-great-white-fleet-departs-hampton-roads-for-circumnavigation.

62. On Barak, *Powering Empire: How Coal Made the Middle East and Sparked Global Carbonization* (Oakland: University of California Press, 2020), 4.

63. Maurer, "Fuel and the Battle Fleet," 68.

64. "Lessons and Results of the Battleship Cruise," *Scientific American* 100, no. 8 (February 20, 1909): 146.

65. "Lessons and Results of the Battleship Cruise," 146.

66. Steven Gray, "Fueling Mobility: Coal and Britain's Naval Power, c. 1870–1914," *Journal of Historical Geography* 58 (October 2017): 92–103; 93.

67. Maurer, "Fuel and the Battle Fleet," 69; John A. DeNovo, "Petroleum and the United States Navy before World War I," *Mississippi Valley Historical Review* 41, no. 4 (March 1955): 641–656; 645.

68. Brodie, *Sea Power in the Machine Age*, 117.

69. Gray, "Fueling Mobility," 101. Also see Erik J. Dahl, "Naval Innovation: from Coal to Oil," *Joint Forces Quarterly* 27 (Winter 2000–2001): 50–56.

70. Shulman, *Coal and Empire*, 197; DeNovo, "Petroleum and the United States Navy," 644; Maurer, "Fuel and the Battle Fleet," 70; Naval History and Heritage Command, "U.S. Ship Force Levels, 1866–Present," https://www.history.navy.mil/research/histories/ship-histories/us-ship-force-levels.html#1910, accessed May 31, 2021.

71. DeNovo, "Petroleum and the United States Navy," 656.

72. Quoted in DeNovo, 650.

73. U.S. Navy Department, *Annual Report of the Navy Department and the Secretary of the Navy*, 1914, 17. Washington, DC: United States Navy, 1914).

74. Shulman, *Coal and Empire*, 176–177.

75. Harold F. Williamson, Ralph L. Andreano, Arnold R. Daum, and Gilbert C. Klose, *The American Petroleum Industry, Volume II: The Age of Energy 1899–1959* (Evanston, IL: Northwestern University Press, 1969), 262–263.

76. Isabel V. Hull, *A Scrap of Paper: Breaking and Making International Law during the Great War* (Ithaca, NY: Cornell University Press, 2014), 168.

77. Williamson et. al, *American Petroleum Industry*, 273.

78. Brodie, *Sea Power in the Machine Age*, 317. Also see Williamson et al., *American Petroleum Industry*, 272; Daniel Yergin, *The Prize: The Epic Quest for Oil, Money & Power* (New York: Free Press, 1991), 160–161.

79. Yergin, *The Prize*, 164–165.

80. Quoted in Williamson et al., *American Petroleum Industry*, 261.

81. Fiscal year ending on June 30. Williamson et al., *American Petroleum Industry*, 266.

82. Shulman, *Coal and Empire*, 179.

83. Shulman, 177.

84. Shulman, 178–179.

85. Williamson et al., *American Petroleum Industry*, 285.

86. John Williamson, "Federal Aid to Roads and Highways since the 18th Century: A Legislative History," Congressional Research Service, R42140, January 6, 2012, 5, https://fas.org/sgp/crs/misc/R42140.pdf.

87. See the National Museum of the United States Army, "National Highway System," https://www.thenmusa.org/armyinnovations/nationalhighwaysystem/.

88. Williamson et al., *American Petroleum Industry*, 294–295.

89. Herbert Feis, "Petroleum and American Foreign Policy," *Commodity Policy Studies*, no. 3, Food Research Institute, Stanford University, March 1944, 3; Shulman, *Coal and Empire*, 199.

90. These reserves were transferred to the Department of the Interior in 1922.

91. DeNovo, "Petroleum and the United States Navy," 651.

92. Shulman, *Coal and Empire*, 195–196; Williamson et al., *American Petroleum Industry*, 308–310.

93. Coolidge quoted in Williamson et al., 311.

94. Williamson et al., 303.

95. Williamson et al., 729. Also see Rosemary A. Kelanic, "The Petroleum Paradox: Oil, Coercive Vulnerability, and Great Power Behavior," *Security Studies* 25, no. 2 (May 2016): 181–213.

96. Robert Goralski and Russell W. Freeburg, *Oil & War: How the Deadly Struggle for Fuel in WWII Meant Victory or Defeat* (New York: William Morrow, 1987), 25.

97. Williamson et al., *American Petroleum Industry*, 729.

98. Goralski and Freeburg, *Oil & War*, 31.

99. Goralski and Freeburg, 84.

100. Hitler quoted in Matthieu Auzanneau, *Oil, Power and War: A Dark History* (White River Junction, VT: Chelsea Green Publishing, 2018), 160.

101. Goralski and Freeburg, *Oil & War*, 53–56. Peter Liberman, *Does Conquest Pay? The Exploitation of Occupied Industrial Societies* (Princeton, NJ: Princeton University Press, 1996), 51–52.

102. Williamson et al., *American Petroleum Industry*, 44.

103. Goralski and Freeburg, *Oil & War*, 45. For an overview, see Davis and Engerman, *Naval Blockades in Peace and War*, 239–320.

104. Vine, *The United States of War*, 138–148.

105. Williamson et al., *American Petroleum Industry*, 753.

106. Auzanneau, *Oil, Power and War*, 157.

107. Yergin, *The Prize*, 310.

108. Auzanneau, *Oil, Power and War*, 160.

109. Williamson et al., *American Petroleum Industry*, 772.

110. John W. Frey and H. Chandler Ide, *A History of the Petroleum Administration for War, 1941–1945* (Washington, DC: U.S. Government Printing Office, 1946), 4.

111. Frey and Ide, *A History of the Petroleum Administration for War, 1941–1945*, 7.

112. Frey and Ide, 7.

113. Williamson et al., *American Petroleum Industry*, 170.

114. Johnstone and McLeish, "World Wars and the Age of Oil."

115. Daniel Immerwahr, *How to Hide an Empire: A History of the Greater United States* (New York: Picador, 2019) 269.

116. Goralski and Freeburg, *Oil & War*, 167; Williamson et al., *American Petroleum Industry*, 791.

117. John D. Millet, *The Organization and Role of the Army Services Forces* (Washington, DC: Center of Military History, 1987), 241–242.

118. Auzanneau, *Oil, Power and War*, 169; Davis and Engerman, *Naval Blockades in Peace and War*, 334–335.

119. United States Strategic Bombing Surveys, *Summary Report* (Washington, DC: U.S. Government Printing Office, 1 July 1946), 85.

120. Frey and Ide, *A History of the Petroleum Administration for War, 1941–1945*, 7.

121. Goralski and Freeburg, *Oil & War*, 310.

122. Yergin, *The Prize*, 346; Davis and Engerman, *Naval Blockades in Peace and War*, 374.

123. Salim Yaqub, "The United States and the Persian Gulf: 1941 to the Present," in *Crude Strategy: Rethinking the US Military Commitment to Defend Persian Gulf Oil*, ed. Charles L. Glaser and Rosemary A. Kelanic (Washington, DC: Georgetown University Press, 2016), 21–47; 23–24.

124. Haywood S. Hansell, *The Air Plan That Defeated Hitler* (New York: Arno Press, 1972), 202, 216–228.

125. United States Strategic Bombing Surveys, *Summary Report*, 21.

126. Speer, quoted in United Kingdom Chiefs of Staff Committee, *Oil as a Factor in the German War Effort, 1933–1945* (London: Offices of the Cabinet and Minister of Defense, March 1946), 66.

127. United States Strategic Bombing Surveys, *Summary Report*, 23.

CHAPTER 2

1. James A. Baker, quoted in Thomas Friedman, "Mideast Tensions; U.S. Jobs at Stake in Gulf, Baker Says," *New York Times*, November 14, 1990, https://www.nytimes.com/1990/11/14/world/mideast-tensions-us-jobs-at-stake-in-gulf-baker-says.html.

2. General David Petraeus, quoted in Department of Energy, "Energy for the Warfighter: The Department of Defense Operational Energy Strategy," June 14, 2011, https://www.energy.gov/articles/energy-war-fighter-department-defense-operational-energy-strategy.

3. John Glaser, "Withdrawing from Overseas Bases: Why a Forward-Deployed Military Posture Is Unnecessary, Outdated, and Dangerous," *CATO Institute Policy Analysis*, no. 816 (July 18, 2017): 4, https://www.cato.org/policy-analysis/withdrawing-overseas-bases-why-forward-deployed-military-posture-unnecessary.

4. Yergin, *The Prize*, 392–404.

5. Bruce A. Beaubouef, *The Strategic Petroleum Reserve: U.S. Energy Security and Oil Politics, 1975–2005* (College Station: Texas A&M University Press, 2007): 3.

6. Edward Walsh, "Carter Finds 'Crisis of Confidence,'" *Washington Post*, July 16, 1979, https://www.washingtonpost.com/archive/politics/1979/07/16/carter-finds-crisis-of-confidence/9c2c9663-49b6-44cd-b922-879534ad95bf/.

7. Daniel Moran and James A. Russell, "Introduction: The Militarization of Energy Security," in *Energy Security and Global Politics: The Militarization of Resource Management*, ed. Daniel Moran and James A. Russell (New York: Routledge, 2009): 1–18, 4 and 7; and Michael Klare, "Petroleum Anxiety and the Militarization of Energy Security," in Moran and Russell, *Energy Security and Global Politics*, 39–61.

8. Quoted in Yergin, *The Prize*, 377.

9. Feis, "Petroleum and American Foreign Policy," 29.

10. Quoted in Irvine H. Anderson, *Aramco, the United States, and Saudi Arabia: A Study of the Dynamics of Foreign Oil Policy, 1933–1950* (Princeton, NJ: Princeton University Press, 1981), 135.

11. Quoted in Robert Vitalis, *America's Kingdom: Mythmaking on the Saudi Oil Frontier* (Stanford: Stanford University Press, 2007), 81.

12. Melvin P. Leffler, *A Preponderance of Power: National Security, the Truman Administration, and the Cold War* (Stanford: Stanford University Press, 1992), 79–80.

13. Vine, *The United States of War*, 158–165.

14. Vine, 167.

15. Vine, 180.

16. Vine, 185, 187.

17. Peter W. DeForthe, "U.S. Naval Presence in the Persian Gulf: The Mideast Force since World War II," *Naval War College Review* 28, no. 1 (Summer 1975): 28–38.

18. Yeo and Pettyjohn, "Bases of Empire?," 27.

19. David A. Rosenberg, "The U.S. Navy and the Problem of Oil in a Future War," *Naval War College Review* 29, no. 3 (Summer 1976): 53–64; 55, https://digital-commons.usnwc.edu/cgi/viewcontent.cgi?article=5839&context=nwc-review.

20. Yaqub, "The United States and the Persian Gulf," 24.

21. Rosenberg, "The U.S. Navy and the Problem of Oil in a Future War," 57–58.

22. Immerwahr, *How to Hide an Empire*, 273.

23. Michael Pollen, "What's Eating America," *Smithsonian*, June 15, 2006, https://michaelpollan.com/articles-archive/whats-eating-america/.

24. Environmental Protection Agency, https://www.epa.gov/ghgemissions.

25. Richard F. Weingroff, "The Man Who Changed America, Part I," *Public Roads* 66, no. 5 (March/April 2003), https://www.fhwa.dot.gov/publications/publicroads/03mar/05.cfm.

26. Yergin, *The Prize*, 532–536.

27. See Peter L. Hahn, "Securing the Middle East: The Eisenhower Doctrine of 1957," *Presidential Studies Quarterly* 36, no. 1 (March 2006): 38–47.

28. See Yaqub, "The United States and the Persian Gulf," 21–47.

29. Philip Shabecoff, "Nixon Offers Broad Plan for More 'Clean Energy,'" *New York Times*, June 5, 1971.

30. Beaubouef, *The Strategic Petroleum Reserve*, 11.

31. U.S. House Committee on Foreign Affairs, Subcommittee on the Near East, *The United States and the Persian Gulf*, Report of the Subcommittee on the Near East of the Committee on Foreign Affairs (Washington, DC: U.S. Government Printing Office, 1972).

32. House Committee on Foreign Affairs, *The United States and the Persian Gulf*, 1–2.

33. House Committee on Foreign Affairs, 13.

34. House Committee on Foreign Affairs, 13.

35. Heather L. Greenley, "The Strategic Petroleum Reserve: Background, Authorities, and Considerations," Congressional Research Service, R46355, May 13, 2020, 2, https://www.everycrsreport.com/files/20200513_R46355_a7b0f9897caa032cac61287f9bfc4d010f0b8dc0.pdf.

36. Yergin, *The Prize*, 576.

37. Auzanneau, *Oil, Power and War*, 328; Rachel Bronson, *Thicker Than Oil: America's Uneasy Partnership with Saudi Arabia* (Oxford: Oxford University Press, 2008): 118.

38. See Yergin, *The Prize*, 584–591.

39. Yergin, 596.

40. Richard Nixon, "The Energy Emergency: The President's Address to the Nation Outlining Steps to Deal with the Emergency," November 7, 1973, https://www.cvce.eu/content/publication/2003/7/3/1158015d-8cf9-4fae-8128-0f1ee8a8d292/publishable_en.pdf.

41. John Finney, "Schlesinger Testifies Fuel Allocations to Military Must Be Raised This Spring," *New York Times*, February 1, 1974, https://www.nytimes.com/1974/02/01/archives/schlesinger-testifies-fuel-allocations-to-military-must-be-raised.html.

42. Quotes from Office of Historian, Foreign Service Institute, U.S. Department of State, "Foreign Relations of the United States, 1969–1976, volume XXXVI, Energy Crisis, 1969–1974," https://history.state.gov/historicaldocuments/frus1969-76v36/d244.

43. Office of Historian, "Foreign Relations of the United States, 1969–1976."

44. Glenn Frankel, "U.S. Mulled Seizing Oil Fields in 1973," *Washington Post*, January 1, 1974, https://www.washingtonpost.com/archive/politics/2004/01/01/us-mulled-seizing-oil-fields-in-73/0661ef3e-027e-4758-9c41-90a40bbcfc4d/.

45. Quoted in Bronson, *Thicker Than Oil*, 119.

46. Quoted in Bronson, 119.

47. Robert McFadden, "Aramco Concedes Denying Oil to U.S. Military since October," *New York Times*, January 26, 1974, https://www.nytimes.com/1974/01/26/archives/aramco-concedes-denying-oil-to-us-military-since-october-could-lose.html?searchResultPosition=2.

48. Special Subcommittee on Investigations of the Committee on International Relations, *Oil Fields as Military Objectives: A Feasibility Study*, Prepared for the Special Subcommittee on Investigations, 94th Congress, 1st Session, August 21, 1975 (Washington, DC: U.S. Government Printing Office, 1975), 4; John M. Collins, "Defense Trends in the United States 1952–1973," Congressional Research Service, May 14, 1974, 19.

49. Bronson, *Thicker Than Oil*, 120.

50. Schlesinger quoted in Special Subcommittee on Investigations, *Oil Fields as Military Objectives*, 81.

51. Kissinger quoted in Special Subcommittee on Investigations, 79.

52. Ford quoted in Special Subcommittee on Investigations, 78.

53. Special Subcommittee on Investigations.

54. Special Subcommittee on Investigations, 5 and 11.

55. Beaubouef, *The Strategic Petroleum Reserve*, 13.

56. Richard Scott, *The History of the International Energy Agency: The First 20 Years, Volume 1: Origins and Structures of the IEA* (Paris: OECD, 1994): 56–57. Also see Beaubouef, *The Strategic Petroleum Reserve*, 27.

57. Greenley, "The Strategic Petroleum Reserve," 3.

58. Jared G. Carter quoted in Beaubouef, *The Strategic Petroleum Reserve*, 25.

59. Greenley, "The Strategic Petroleum Reserve," 4.

60. Beaubouef, *The Strategic Petroleum Reserve*, 28.

61. Quoted in Beaubouef, *The Strategic Petroleum Reserve,* 48.

62. See Andrew J. Bacevich, *America's War for the Greater Middle East: A Military History* (New York: Random House, 2016), 22–23.

63. President Jimmy Carter, State of the Union Address, January 23, 1980.

64. President Jimmy Carter, State of the Union Address, January 23, 1980.

65. Bacevich, *America's War*, 25.

66. General Paul Kelley, "One Telephone Call Gets It All: Maritime Prepositioning for Crisis Response," in *The Legacy of American Naval Power: Reinvigorating Maritime Strategic Thought*, ed. Paul Westermeyer (Quantico, VA: United States Marine Corps History Division, 2019), 209–215; 212.

67. Reagan quoted in Steven R. Weisman, "Reagan Says U.S. Would Bar a Takeover in Saudi Arabia That Imperiled Flow of Oil," *New York Times*, October 2, 1981, https://www.nytimes.com/1981/10/02/world/reagan-says-us-would-bar-a-takeover-in-saudi-arabia-that-imperiled-flow-of-oil.html.

68. John D. Mayer, *Rapid Deployment Forces: Policy and Budgetary Implications*, a CBO Study (Washington, DC: Congress of the United States, Congressional Budget Office, February 1983), 2, https://apps.dtic.mil/sti/pdfs/ADA474780.pdf.

69. Quoted in Bacevich, *America's War*, 35.

70. Michael Armacost, "U.S. Policy in the Persian Gulf and Kuwaiti Reflagging," statement presented by Undersecretary Armacost before the Senate Foreign Relations Committee, June 16, 1987, https://apps.dtic.mil/sti/pdfs/ADA496911.pdf.

71. Joint Chiefs of Staff, *United States Military Posture, FY1986*, Department of Defense, Washington, DC, 1986, 3–4.

72. President George H. W. Bush, *National Security Strategy of the United States*, March 1990, The White House, Washington, DC, 1990, 22, https://history.defense.gov/Portals/70/Documents/nss/nss1990.pdf?ver=x5cwOOez0oak2BjhXekM-Q%3d%3d.

73. President George H. W. Bush, *National Security Strategy of the United States*, March 1990, 22.

74. "World Oil Reserves 1948–2001: Annual Statistics and Analysis," *Energy Exploration & Exploitation* 19, no. 2–3 (2001): 261–265; 263, https://journals.sagepub.com/doi/pdf/10 .1260/0144598011492561.

75. UN Security Council, Resolution 661 (1990), https://digitallibrary.un.org/record/94221 ?ln=en.

76. President George H W. Bush, National Security Directive 45, August 20, 1990, The White House, Washington, DC, https://fas.org/irp/offdocs/nsd/nsd_45.htm.

77. Yergin, *The Prize*, 755.

78. Yergin, 759.

79. Elliot Cohen, ed., *Gulf War Air Power Survey Volume III, Logistics and Support* (Washington, DC: U.S. Government Printing Office, 1993), 180, https://media.defense.gov/2010/Sep/27 /2001329815/-1/-1/0/AFD-100927-063.pdf.

80. Cohen, *Gulf War Air Power Survey Volume III*, 123.

81. The offloading of fuel averaged 2.1 million gallons per day during Desert Shield and 15.5 million gallons per day during Desert Storm. Cohen, 181 and 199.

82. The equivalent of 1,882,670,174 gallons. Cohen, 14.

83. The equivalent of 1,757,075,564 gallons. Cohen, 15. The exceptions were Navy JP-5 fuel and Jet Propellant Thermally Stable (JPTS) fuel.

84. President George H. W. Bush, *National Security Strategy of the United States*, 1991, The White House, Washington, DC, August 1991, 1. https://history.defense.gov/Portals/70 /Documents/nss/nss1991.pdf?ver=3sIpLiQwmknO-RplyPeAHw%3d%3d.

85. President George H. W. Bush, *National Security Strategy of the United States*, August 1991, 3.

86. President George H. W. Bush, *National Security Strategy of the United States*, August 1991, 10.

87. President George H. W. Bush, *National Security Strategy of the United States*, August 1991, 21.

88. President George H. W. Bush, *National Security Strategy of the United States*, August 1991, 21.

89. President George H. W. Bush, *National Security Strategy of the United States*, August 1991, 21–22.

90. President George H. W. Bush, *National Security Strategy of the United States*, August 1991, 22.

91. Les Aspin, *Report on the Bottom Up Review*, Department of Defense, Washington, DC, October 1993, 8, https://apps.dtic.mil/sti/pdfs/ADA359953.pdf,

92. Aspin, *Report on the Bottom Up Review*, 8.

93. Aspin, 7.

94. Quoted in Elaine Sciolino, "To U.S., Afghanistan Seems to Move Farther Away," *New York Times*, February 12, 1989, https://www.nytimes.com/1989/02/12/weekinreview/the-world -to-us-afghanistan-seems-to-move-farther-away.html; and Bacevich, *America's War*, 58.

95. Rumsfeld quoted in John Esterbrook, "Rumsfeld: It Would Be a Short War," *CBS News*, November 15, 2002, https://www.cbsnews.com/news/rumsfeld-it-would-be-a-short-war/.

96. Alan Greenspan, *The Age of Turbulence: Adventures in A New World* (New York: Penguin, 2007), 463, quoted in Adam J. Liska and Richard K. Perrin, "Securing Foreign Oil: A Case for Including Military Operations in the Climate Change Impact of Fuels," *Environment* 52, no. 4 (July/August 2010): 9–22; 15.

97. Council on Foreign Relations, *National Security Consequences of Oil Dependency: Report of an Independent Task Force*, Council on Foreign Relations, Washington, DC, 2006, 29, https://cfrd8-files.cfr.org/sites/default/files/report_pdf/0876093659.pdf.

98. Peter Baker, "Bush Says U.S. Pullout Would Let Radicals Use Oil as a Weapon," *Washington Post*, November 5, 2006, http://www.washingtonpost.com/wp-dyn/content/article/2006/11/04/AR2006110401025.html.

99. Michael A. Vane and Paul E. Roege, "The Army's Operational Energy Challenge," *Army* (May 2011): 36–43; 37, https://apps.dtic.mil/dtic/tr/fulltext/u2/a543153.pdf.

100. On troop numbers, see Amy Belasco, "The Cost of Iraq, Afghanistan, and Other Global War on Terror Operations Since 9/11," Congressional Research Service, RL33110, December 8, 2014, 9, https://fas.org/sgp/crs/natsec/RL33110.pdf.

101. Vane and Roege, "The Army's Operational Energy Challenge," 3.

102. Deloitte, *Energy Security: America's Best Defense*, 2009, 15, https://www.offiziere.ch/wp-content/uploads/us_ad_EnergySecurity052010.pdf.

103. See Robert S. Dudney, "The Gulf War II Air Campaign, by the Numbers," *Air Force Magazine* (July 2003): 36–42, https://www.airforcemag.com/PDF/MagazineArchive/Magazine%20Documents/2003/July%202003/0703Numbers.pdf.

104. United States, Central Command Combined Forces Air Component Commander Airpower Statistics, Operation Inherent Resolve Strike Releases, Sorties with at least one weapon release, https://www.afcent.af.mil/About/Airpower-Summaries/.

105. Michael Birnbaum and Missy Ryan, "U.S. Defense Secretary Mark Esper Says U.S. Will Leave Forces in Syria to Defend Oil Fields from Islamic State," *Washington Post*, October 25, 2019, https://www.washingtonpost.com/world/us-defense-secretary-mark-esper-says-us-will-leave-forces-in-syria-to-defend-oil-fields-from-islamic-state/2019/10/25/fd131f1a-f723-11e9-829d-87b12c2f85dd_story.html.

106. Dana Farrington, "Read: Trump Announcement on Baghdadi's Death," October 27, 2019, https://www.npr.org/2019/10/27/773842999/read-trump-statement-on-baghdadis-death.

107. General Kenneth F. McKenzie, Jr., "Posture Statement of General Kenneth F. McKenzie, Jr., Commander, United States Central Command Before the Senate Armed Services Committee," April 22, 2021, https://www.centcom.mil/ABOUT-US/POSTURE-STATEMENT/.

108. Stockholm International Peace Research Institute Data, https://twitter.com/SIPRIorg/status/1371463863355473929/photo/1.

109. Duane Chapman, "Gulf Oil and International Security: Can the World's Only Superpower Keep the Oil Flowing?," in Moran and Russell, *Energy Security and Global Politics*, 75–94, 82.

110. Chapman, "Gulf Oil and International Security," 82 and 84.

111. Also see Klare, "Petroleum Anxiety," in Moran and Russell, *Energy Security and Global Politics*; and Michael T. Klare, *Blood and Oil: The Consequences and Dangers of America's Growing Dependence on Imported Petroleum* (New York: Metropolitan Books, 2004).

CHAPTER 3

1. Groves quoted in Arthur Neslen, "Pentagon to Lose Emission Exemption Under Paris Climate Deal," *Guardian*, December 14, 2015, https://www.theguardian.com/environment/2015 /dec/14/pentagon-to-lose-emissions-exemption-under-paris-climate-deal.

2. See Neslen, "Pentagon to Lose Emission Exemption"; Parkinson and Cottrell, *Under the Radar*, 9.

3. The White House, "Executive Order on Catalyzing Clean Energy Industries and Jobs through Federal Sustainability," December 8, 2021, https://www.whitehouse.gov/briefing-room /presidential-actions/2021/12/08/executive-order-on-catalyzing-clean-energy-industries -and-jobs-through-federal-sustainability/.

4. Schurr et al., *Energy in the American Economy*, 485.

5. Eunice Foote, "Circumstances Affecting the Heat of the Sun's Rays," *American Journal of Science and Arts* 22 (November 1856): 382–383, paper read before the American Association for the Advancement of Science, August 23, 1856, https://archive.org/details /mobot31753002152491/page/v/mode/2up?view=theater; John Schwartz, "Overlooked No More: Eunice Foote, Climate Scientist Lost to History," *New York Times*, April 21, 2020.

6. CO_2 emissions from fossil fuel combustion have caused the largest share of global warming, about 60 percent. The other gases are nevertheless significant causes of increasing temperature. Methane or CH_4 is emitted from the extraction of coal and oil, distribution (leaks and deliberate flaring to reduce pressure) and combustion of fossil fuel, industrial processes, the enteric fermentation of ruminant animals such as cows and goats, rice cultivation, manure management, and the waste sector, including landfills and food waste. Nitrous oxide or N_2O is mostly emitted from industrial processes, agricultural soils, manure management, and wastewater; fluorinated gases or F-gases such as chlorofluorocarbons (CFCs) are mostly emitted from industrial processes. Methane and nitrous oxide are both shorter-lived in the atmosphere and much more heat trapping than CO_2, and thus reducing emission of those gases will have a significant impact on overall global warming. See Shilpa Rao and Keywan Riahi, "The Role of Non-CO_2 Greenhouse Gases in Climate Change Mitigation: Long-term Scenarios for the 21st Century," *Energy Journal* (2006): 177–200; and S. A. Montzka, E. J. Dlugokencky, and J. H. Butler, "Non-CO_2 Greenhouse Gases and Climate Change," *Nature* 476, no. 7358 (August 4, 2011): 43–50.

7. Letter from Assistant Secretary of Defense Donald A Quarles to Dr. Joseph Dodge regarding the DOD's interest in the IGY Program, March 19, 1954, Eisenhower Library, https://www

.eisenhowerlibrary.gov/sites/default/files/research/online-documents/igy/1954-3-19.pdf; *National Science Foundation, National Academy of Sciences; report on the International Geophysical Year. Hearings before the subcommittee of the Committee on Appropriations, House of Representatives, Eighty-sixth Congress, first session, February 1959* (Washington, DC: U.S. Government Printing Office, 1959), https://catalog.hathitrust.org/Record/007157947.

8. United States, Joint Task Force One, *Operation Crossroads: The Official Pictorial Record* (New York: W. H. Wise & Co., 1946), 30, https://archive.org/details/operationcrossro00unit/mode/2up.

9. Roger Revelle testimony before the House Committee on Appropriations, February 8, 1956, in *Making Climate Change History: Documents from Global Warming's Past*, ed. Joshua P. Howe (Seattle: University of Washington Press, 2017), 60–63; 62–63.

10. Roger Revelle and Hans E. Seuss, "Carbon Dioxide Exchange between Atmosphere and Ocean and the Question of an Increase in Atmospheric CO_2 During the Past Decades," *Tellus* 9, no. 1 (1957): 18–27.

11. Gilbert N. Plass, "The Carbon Dioxide Theory of Climatic Change," *Tellus* 8, no. 2 (1956): 140–154; 152.

12. Charles D. Keeling, "The Concentration and Isotopic Abundances of Carbon Dioxide in the Atmosphere," *Tellus* 12, no. 2 (1960): 200–203.

13. Quoted in Naomi Oreskes, *Science on a Mission: How Military Funding Shaped What We Do and Don't Know about the Ocean* (Chicago: University of Chicago, 2021), 397–398.

14. See James Gustave Speth, *They Knew: The Federal Government's Fifty-Year Role in Causing the Climate Crisis* (Cambridge, MA: MIT Press, 2021).

15. Roger Revelle, Wallace Broeker, Harmon Craig, C. D. Keeling, and J. Smagorinski, "Appendix Y4: Atmospheric Carbon Dioxide," in *President's Science Advisory Committee, Restoring the Quality of Our Environment: Report of the Environmental Pollution Panel*, The White House, Washington, DC, November 1965, 126–127, https://assets.documentcloud.org/documents/3227654/PSAC-1965-Restoring-the-Quality-of-Our-Environment.pdf.

16. Daniel P. Moynihan Memorandum for John Ehrlichman, Associated Press, "Declassified Documents Show Nixon Warned of Global Warming 30 Years Ago," *New York Daily News*, July 3, 2010, https://www.nydailynews.com/news/national/declassified-documents-show-nixon-warned-global-warning-30-years-article-1.463318, accessed 30 June 2021.

17. Speth, *They Knew*, 17.

18. World Meteorological Organization, "Declaration of the World Climate Conference," in *World Climate Conference: A Conference of Experts on Climate and Mankind. Declaration and Supporting Documents*, Geneva, February 1979, 714, https://library.wmo.int/doc_num.php?explnum_id=3778.

19. Philip Shabecoff, "Global Warming Has Begun, Expert Tells Senate," *New York Times*, June 24, 1988, https://www.nytimes.com/1988/06/24/us/global-warming-has-begun-expert-tells-senate.html.

20. President George H. W. Bush, *National Security Strategy of the United States*, The White House, Washington, DC, March 1990, 22.

21. J. T. Houghton, G. J. Jenkins, and J. J. Ephraums, eds., *Climate Change: The IPCC Scientific Assessment* (Cambridge: Cambridge University Press, 1990), https://archive.ipcc.ch/publications_and_data/publications_ipcc_first_assessment_1990_wg1.shtml.

22. Thatcher quoted in Jeremy Leggett, "The Cut Thatcher Does Not Want to Make," *Independent*, May 28, 1990.

23. Terry P. Kelley, *Global Climate Change Implications for the United States Navy*, United States Naval War College, Advanced Research Program, Newport, RI, May 1990, 27 and 31–32, https://documents.theblackvault.com/documents/weather/climatechange/globalclimatechange-navy.pdf.

24. Kelley, *Global Climate Change Implications for the United States Navy*, 1–2.

25. United Nations General Assembly, Rio Declaration on Environment and Development in *Report of the United Nations Conference on Environment and Development* (A/Conf.151/26) August 12, 1992, https://www.un.org/en/development/desa/population/migration/generalassembly/docs/globalcompact/A_CONF.151_26_Vol.I_Declaration.pdf.

26. United Nations, United Nations Framework Convention on Climate Change, 1992, FCCC/Informal/GE.05–62220 (E) 200705, https://unfccc.int/resource/docs/convkp/conveng.pdf.

27. Department of Defense, "DOD Background Paper on a National Security Provision for the Proposed Climate Change Protocol," September 5, 1997, reprinted in *Inside the Pentagon* 13, no. 41 (October 9, 1997): 1, 10–13; 12.

28. Department of Defense, "DOD Background Paper," 12.

29. Department of Defense, "DOD Background Paper," 13.

30. Department of Defense, "DOD Background Paper," 13.

31. Department of Defense, "DOD Background Paper," 10.

32. Cable, State Department, State 202013, to U.S. Del Mark Hambley and All NATO Post Collective, Subject: National Security Exemption on Climate Change, October 26, 1997 [Confidential], National Security Archive, https://nsarchive.gwu.edu/document/27354-document-1-cable-state-department-state-202013-us-del-mark-hambley-and-all-nato-post.

33. Undersecretary of State Stuart Eizenstat testimony, *Implications of the Kyoto Protocol on Climate Change*, Hearing before the Committee on Foreign Relations, United States Senate, February 11, 1998, https://www.govinfo.gov/content/pkg/CHRG-105shrg46812/pdf/CHRG-105shrg46812.pdf.

34. Statement, United States Delegation to the Conference of Parties to the United Nations Framework Convention on Climate Change Subsidiary Body Meetings, as delivered to Russian Delegation, October 31, 1997 [Classification Unknown], https://nsarchive.gwu.edu/document/27357-document-4-statement-united-states-delegation-conference-parties-united-nations.

35. Background Paper, Council on Environmental Quality, Subject: Climate Treaty—National Security Exemption, November 23, 1997 [Classification Unknown], https://nsarchive.gwu.edu/document/27358-document-5-background-paper-council-environmental-quality-subject-climate-treaty.

36. Cable, State Department, State 237825, to All Diplomatic Posts, Subject: Climate Change: Outcomes from the Kyoto Conference, December 19, 1997 [Confidential], https://nsarchive.gwu.edu/document/27369-document-16-cable-state-department-state-237825-all-diplomatic-posts-subject-climate.

37. Eizenstat testimony, *Implications of the Kyoto Protocol on Climate Change*, 9.

38. Eizenstat testimony, 46.

39. Eizenstat testimony, 8–9.

40. Senator Joe Biden, *Implications of the Kyoto Protocol on Climate Change*, Hearing before the Committee on Foreign Relations, United States Senate, February 11, 1998, 15. Emphasis added. Also see the Biden Statement on C-span at minute 49, https://www.c-span.org/video/?100413-1/kyoto-global-climate-change-agreement&start=60&stop=780.

41. Eizenstat testimony, *Implications of the Kyoto Protocol on Climate Change*, 16.

42. Senator John Kerry, *Implications of the Kyoto Protocol on Climate* Change, Hearing before the Committee on Foreign Relations, United States Senate, February 11, 1998, 31.

43. Senator Kerry, *Implications of the Kyoto Protocol on Climate Change*, 31.

44. COMPASS Letter quoted in Danielle Knight, "Climate: U.S. Exempts Military from Kyoto Treaty," *Inter Press Service News Agency*, May 20, 1998, http://www.ipsnews.net/1998/05/climate-us-exempts-military-from-kyoto-treaty/. Also see Greenpeace, "Documents Related to COMPASS," http://research.greenpeaceusa.org/?a=download&d=4196.

45. Eizenstat testimony, *Implications of the Kyoto Protocol on Climate Change*, 31.

46. Eizenstat testimony, 31–32.

47. Eizenstat testimony, 31–32.

48. Senator Chuck Hagel, prepared statement, *Implications of the Kyoto Protocol on Climate* Change, Hearing before the Committee on Foreign Relations, United States Senate, February 11, 1998, 40.

49. Eizenstat testimony, *Implications of the Kyoto Protocol on Climate Change*, 33.

50. Memorandum, The White House, James. B. Steinberg, et al., to President Clinton, Subject: Climate Change/Military, March 3, 1998 [Classification Unknown], https://nsarchive.gwu.edu/document/27376-document-23-memorandum-white-house-james-b-steinberg-et-al-president-clinton-subject.

51. Frank Carlucci, "Making Military Sense Out of Kyoto," *Washington Times*, May 18, 1998, quoted in Knight, "Climate: U.S. Exempts Military from Kyoto Treaty."

52. Section 1232 (a) of the United States 1999 National Defense Authorization Act, Public Law 105–261, October 17, 1998, https://www.govinfo.gov/content/pkg/PLAW-105publ261/pdf /PLAW-105publ261.pdf.

    Also see Roy K. Salomon, "Global Climate Change and U.S. Military Readiness," *Federal Facilities Environmental Journal* 10, no. 2 (Summer 1999): 133–142.

53. Section 1232 (c) of the United States 1999 National Defense Authorization Act.

54. For a general discussion, see Tamara Lorincz, "Demilitarization for Deep Decarbonization: Reducing Militarism and Military Expenditures to Invest in the UN Green Climate Fund and to Create Low-Carbon Economies and Resilient Communities," International Peace Bureau, September 2014, 22–25.

55. Decision 2/CP.3, "Methodological Issues," United Nations Framework Convention on Climate Change, Report of the Conference of the Parties on Its Third Session, Held at Kyoto from 1 to 11 December 1997, Addendum Part Two: Action Taken by the Conference of the Parties at Its Third Session, March 25, 2998, FCCC/CP/1997/7/Add.1, 31, https://unfccc.int/resource /docs/cop3/07a01.pdf.

56. See IPCC, *2006 IPCC Guidelines for National Greenhouse Gas Inventories*, vol. 2: "Military activity is defined here as those activities using fuel purchased by or supplied to military authorities in the country. . . . Data on military fuel use should be obtained from government military institutions or fuel suppliers. If data on fuel split are unavailable, all the fuel sold for military activities should be treated as domestic," 3.53, https://www.ipcc-nggip.iges.or .jp/public/2006gl/pdf/2_Volume2/V2_3_Ch3_Mobile_Combustion.pdf. (IPCCC, "Good Practice Guidance and Uncertainty Management in National Greenhouse Gas Inventories," https://www.ipcc-nggip.iges.or.jp/public/gp/bgp/2_5_Aircraft.pdf)

57. IPCC, *2006 IPCC Guidelines for National Greenhouse Gas Inventories*, vol. 2, 3.53.

58. IPCC, 3.53 and 3.67; also 3.32. Further, "Confidentiality may also be a problem for reporting military aviation in a transparent manner," 3.73, https://www.ipcc-nggip.iges.or.jp/public /2006gl/pdf/2_Volume2/V2_3_Ch3_Mobile_Combustion.pdf. Military aircraft emissions from military transport within countries are reported in the category "other," which includes "all remaining emissions from non-specified fuel combustion." IPCC, "Revised 1996 IPCC Guidelines for National Greenhouse Gas Inventories: Reporting Instructions," Common Reporting Framework, I.5 and 1.6, https://www.ipcc-nggip.iges.or.jp/public/gl/guidelin/ch1ri .pdf.

59. UNFCC "Reporting Requirements," https://unfccc.int/process-and-meetings/transparency -and-reporting/reporting-and-review-under-the-convention/greenhouse-gas-inventories -annex-i-parties/reporting-requirements. Further, "36. Emissions and removals should be reported at the most disaggregated level of each source/sink category, taking into account that a minimum level of aggregation may be required to protect confidential business and military information." UNFCC, Report on the Conference of the Parties on its 19th Session, Held in Warsaw from 11 to 23 November 2013," UN FCCC /CP/2013/10/Add.3, https://unfccc .int/resource/docs/2013/cop19/eng/10a03.pdf#page=2.

60. IPCC, *Revised 1996 IPCC Guidelines for National Greenhouse Gas Inventories*, vol. 3. Energy Section, I.4.

61. The revised 1996 guidelines say, "International bunker fuels are combusted in ships at sea and by airplanes (both undertaking international movements) and therefore should be included in global greenhouse gas estimations. Following guidance from the INC, the IPCC recommends that every country estimate emissions from international bunker fuels sold within national boundaries, but that these emissions would be reported separately and, as far as possible, excluded from national totals." IPCC, Revised 1996 IPCC Guidelines for National Greenhouse Gas Inventories, Reference Manual, vol. 3. Energy Section, I.6n, https://www.ipcc-nggip.iges .or.jp/public/gl/guidelin/ch1ref1.pdf.

62. "Emissions resulting from multilateral operations pursuant to the Charter of the United Nations should not be included in national totals, but reported separately." IPCC, *2006 IPCC Guidelines for National Greenhouse Gas Inventories*, vol. 2, 3.53.

63. IPCC, *2006 IPCC Guidelines for National Greenhouse Gas Inventories*, vol. 2, 3.53.

64. Alex Michaelowa and Tobias Koch, "Military Emissions, Armed Conflicts, Border Changes and the Kyoto Protocol," *Climatic Change* 50, no. 4 (September 2001): 384–394; 387, https:// www.researchgate.net/publication/226447633_Military_Emissions_Armed_Conflicts_Border _Changes_and_the_Kyoto_Protocol.

65. Excerpt of the "Department of Defense Inventory of Greenhouse Gas Emissions & Sinks, 1990 and 1996, July 2000" in Richard Truly and Alvin Alm, *Report of the Defense Science Board on More Capable Warfighting Through Reduced Fuel Burden*, Office of the Under Secretary of Defense for Acquisition and Technology, Washington, DC, May 2001, D-3n, https://apps.dtic .mil/sti/pdfs/ADA392666.pdf.

66. Excerpt of the "Department of Defense Inventory of Greenhouse Gas Emissions & Sinks, 1990 and 1996, July 2000," in Truly and Alm, *Report of the Defense Science Board*.

67. Environmental Protection Agency, "Inventory of U.S. Greenhouse Gas Emissions and Sinks: 1990–2000," Washington, DC, April 2002, annex I, 305–306. Also see excerpt of the "Department of Defense Inventory of Greenhouse Gas Emissions & Sinks, 1990 and 1996, July 2000," in Truly and Alm, *Report of the Defense Science Board*, D-7.

68. "Emissions from International Bunker Fuels are not included in totals." "These values are presented for informational purposes only, in line with the 2006 IPCC Guidelines and UNFCCC reporting obligations." Note from table 3–1 in Environmental Protection Agency, *Inventory of U.S. Greenhouse Gases and Sinks: 1990–2019*, 3–4, https://www.epa.gov/sites /production/files/2021-04/documents/us-ghg-inventory-2021-main-text.pdf.

69. Environmental Protection Agency, *Inventory of U.S. Greenhouse Gases and Sinks: 1990–2019*, 3–28.

70. Environmental Protection Agency, 3–37.

71. Environmental Protection Agency, 3-120–121, https://www.epa.gov/sites/production/files /2021-04/documents/us-ghg-inventory-2021-main-text.pdf.

72. See National Inventory Submissions to the UNFCCC at the United Nations Climate Change, https://unfccc.int/ghg-inventories-annex-i-parties/2021.

73. Department of Defense, *Report on Greenhouse Gas Emission Levels* (Office of The Assistant Secretary of Defense, August 2021).

74. United States Arctic Research Commission, "The Arctic Ocean and Climate Change: A Scenario for the U.S. Navy," United States Arctic Research Commission, 2000, 7, https://wayback.archive-it.org/3286/20180205132306/https://storage.googleapis.com/arcticgov-static/publications/other/arctic_and_climate_change.pdf.

75. William S. Cohen, *Annual Report of the Secretary of Defense to the President and Congress* (Washington, DC: U.S. Government Printing Office, 1998), 173.

CHAPTER 4

1. Office of the Assistant Secretary of Defense for Sustainment, *Annual Energy Management and Resilience Report (AEMRR) Fiscal Year 2020* (September 2021), 13, https://www.acq.osd.mil/eie/ie/FEP_Energy_Reports.html.

2. Department of Defense, *Report on Greenhouse Gas Emission Levels* (Office of The Assistant Secretary of Defense, August 2021).

3. The DOE relies on data provided by federal agencies and departments. These are Department of Defense total scope 1 and 2 emissions as reported by the Department of Energy, https://ctsedwweb.ee.doe.gov/Annual/Report/ComprehensiveGreenhouseGasGHGInventoriesByAgencyAndFiscalYear.aspx, accessed on September 24, 2021. CO_2 equivalent, or CO2e, is a measure of all greenhouse gases that takes into account the different warming potentials of different gases. CO_2 equivalent is discussed further in the appendix.

4. There is a lack of comprehensive federal fuel-use data prior to 1975 and therefore I am unable to estimate the military's greenhouse gas emissions before 1975.

5. U.S. Air Force, "Air Force Global Strike Flyover to Support Super Bowl 55 in Tampa," Secretary of the Air Force Public Affairs, January 25, 2021, https://www.af.mil/News/Article-Display/Article/2481317/air-force-global-strike-flyover-to-support-super-bowl-55-in-tampa/.

6. U.S. Navy, "Blue Angels," https://www.blueangels.navy.mil/default.htm.

7. See Tom Demerly, "All We Know About the U.S. B-2 Bombers 30-hour Round Trip Mission to Pound Daesh in Libya," *The Aviationist*, January 29, 2017, https://theaviationist.com/2017/01/20/all-we-know-about-the-u-s-b-2-bombers-30-hour-round-trip-mission-to-pound-daesh-in-libya/.

8. Richard Sisk, "B-52 Bombers Carry Out First Airstrikes Against ISIS in Iraq," *Military.com*, April 20, 2016, https://www.military.com/daily-news/2016/04/20/b52-bombers-carry-out-first-airstrikes-against-isis-in-iraq.html.

9. Brian Everstine, "More B-52s Land in the Middle East as Afghanistan Withdrawal Continues," *Air Force Magazine*, May 5, 2021, https://www.airforcemag.com/more-b-52s-land-in-middle-east-as-afghanistan-withdrawal-continues/.

10. Environmental Protection Agency, *Inventory of U.S. Greenhouse Gas Emissions and Sinks: 1990–2000* (Washington, DC: U.S. EPA, April 2002), annex I, 305–306.

11. Arthur Neslen and Climate Desk, "Why the U.S. Military Is Losing Its Carbon-Emissions Exemption," *The Atlantic*, December 15, 2015, https://www.theatlantic.com/science/archive/2015/12/paris-climate-deal-military-carbon-emissions-exemption/420399/.

12. Because I focus on greenhouse gas emissions resulting from fossil fuel use, biogenic emissions are not included in the estimates of military emissions provided here.

13. McKinley, "Cruise of the Great White Fleet."

14. The services may purchase fuel locally and be reimbursed by the Defense Logistics Agency. United States General Accountability Office, *Bulk Fuel: Actions Needed to Improve DOD's Fuel Consumption Budget Data* (GAO-16–664) (September 2016), 6, https://www.gao.gov/assets/680/679682.pdf.

15. Defense Logistics Agency Energy, *Fiscal Year 2020 Fact Book*, 3, https://www.dla.mil/Portals/104/Documents/Energy/Publications/DLAEnergyFactBookFY20_lowres2.pdf?ver=VE-mCUImzFiKKnG1uajkxg%3d%3d, accessed June 23, 2021.

16. Defense Energy Support Center, *Fact Book FY08*, 45, https://www.dla.mil/Portals/104/Documents/Energy/Publications/DESC%20Fact%20Book%202008.pdf?ver=2019-05-24-140841-620; and Defense Energy Support Center, *Fact Book FY2009*, 47, https://www.dla.mil/Portals/104/Documents/Energy/Publications/DESC%20Fact%20Book%202009.pdf?ver=2019-05-24-140749-637, accessed June 23, 2021.

17. The DLA describes the price system this way. "The standard price of fuel is a tool that was created by Department of Defense fiscal managers to insulate the military services from the normal ups and downs of the fuel marketplace. It provides the military services and the Office of the Secretary of Defense with budget stability despite the commodity market swings, with gains or losses being absorbed by a revolving fund known as the Defense Working Capital Fund. In years that the market price of fuel is higher than the standard price, the DWCF loses money. In years that the market price is lower than the standard price, it makes money. This gain or loss can be made up by adjusting future standard prices or by providing our DoD customers with a refund. This decision is typically made by the Office of the Secretary of Defense Comptroller. However, the DWCF must remain cash solvent. As a result, in rare instances the standard price is changed during the fiscal year so the fund remains solvent." Defense Logistics Agency, "Standard Prices," https://www.dla.mil/Energy/Business/StandardPrices/, accessed June 24, 2021.

18. See DLA fact books, "Worldwide Bulk Fuel Ending Inventory," recent years.

19. McKinley, "Cruise of the Great White Fleet."

20. The lower number in this range assumes that the U.S. Navy was burning 430,000 short tons of bituminous coal. If the navy had been burning anthracite coal, emissions would been about 887,000 metric tons of CO_2e. Emissions for subbituminous coal would be closer to emissions for anthracite coal.

21. Schurr et al., *Energy in the American Economy*, 223.

22. Schurr et al., 224.

23. Schurr et al., 224.

24. Truly and Alm, *Report of the Defense Science Board on More Capable Warfighting*, 22. https://apps.dtic.mil/sti/pdfs/ADA392666.pdf.

25. Truly and Alm, *Report of the Defense Science Board on More Capable Warfighting*, 21–23 and D-7–8.

26. Truly and Alm, 21 and 22.

27. For a discussion of alternative methods of estimating U.S. military emissions, see the appendix in this volume.

28. Liska and Perrin, "Securing Foreign Oil," 11.

29. Belcher et al., "Hidden Carbon Cost of the 'Everywhere War.'"

30. On the role of military contractors in war zones, see Heidi Peltier, "The Growth of the 'Camo Economy' and the Commercialization of the Post-9/11 Wars," Costs of War Project, June 30, 2020, https://watson.brown.edu/costsofwar/files/cow/imce/papers/2020/Peltier%202020%20-%20Growth%20of%20Camo%20Economy%20-%20June%2030%202020%20-%20FINAL.pdf; P. W. Singer, *Corporate Warriors: The Rise of the Privatized Military Industry, Updated Edition* (Ithaca, NY: Cornell University Press, 2008).

31. Heidi M. Peters, "Department of Defense Contractor and Troop Levels in Afghanistan and Iraq: 2007–2020," Congressional Research Service, February 22, 2021, https://fas.org/sgp/crs/natsec/R44116.pdf.

32. Department of Energy, "All Agency Energy Cost Data by End Use Sector in FY 2019 (Billion Btu)," https://ctsedwweb.ee.doe.gov/Annual/Report/AgencyMasterDataByYear.aspx; Statement of Honorable Lucian Niemeyer, Assistant Secretary of Defense, Energy, Installations and Environment, before the Senate Committee on Appropriations, Subcommittee on Military Construction, Veterans Affairs, and Related Agencies," April 26, 2018, 13, https://www.acq.osd.mil/eie/Downloads/Testimony/FY19%20EI&E%20Posture%20Statement%20-%20SAC-M.pdf; see Office of the Assistant Secretary of Defense for Energy, Installations, and Environment, "Department of Defense Annual Energy Management and Resilience (AEMR) Report, Fiscal Year 2016 (July 2017), 15, https://www.acq.osd.mil/eie/Downloads/IE/FY%202016%20AEMR.pdf.

33. Department of Defense, "2016 Operational Energy Strategy," 2016, 3–4, https://www.acq.osd.mil/eie/Downloads/OE/2016%20DoD%20Operational%20Energy%20Strategy%20WEBc.pdf.

34. Department of Energy, "Greenhouse Gas Inventory (Metric Tons of Carbon Dioxide Equivalent)," All Agencies, 2020, Comprehensive Annual Energy Data and Sustainability Performance, http://ctsedwweb.ee.doe.gov/Annual/Report/ComprehensiveGreenhouseGasGHGInventoriesByAgencyAndFiscalYear.aspx. These categories do not correspond to the EPA Greenhouse Gas Inventory reports, nor to the DOD categories. And then there are the scope 1, 2, and 3

emissions, where scope 1 emissions are produced by the department itself; scope 2 are indirect, from purchased energy, and scope 3 are emissions from other sources such as employee travel and waste disposal.

35. Besides the DOD, only the Department of Homeland Security performs a significant amount of "non-standard operations." Department of Energy, Comprehensive Annual Energy Data and Sustainability Performance, https://ctsedwweb.ee.doe.gov/Annual/Report/Comprehen siveGreenhouseGasGHGInventoriesByAgencyAndFiscalYear.aspx.

36. Department of Energy, Greenhouse Gas Inventory, https://ctsedwweb.ee.doe.gov/Annual /Report/ComprehensiveGreenhouseGasGHGInventoriesByAgencyAndFiscalYear.aspx.

37. The Department of Energy was established on October 1, 1977, by amalgamating the Federal Energy Administration and the Energy Research and Development Administration. The Energy Information Administration was established in 1977 out of the Office of Energy Information and Analysis.

38. On U.S. bases, see David Vine, "Lists of U.S. Military Bases Abroad, 1776–2021," American University Digital Research Archive, 2021, https://doi.org/10.17606/7em4-hb13.

39. John T. Correll, *The Air Force in the Vietnam War*, Air Force Association, December 2004, https://secure.afa.org/Mitchell/Reports/1204vietnam.pdf.

40. Office of the Assistant Secretary of Defense for Sustainment, *Department of Defense Annual Energy Management and Resilience Report*, 2020, 6.

41. Department of Energy, *All Agency Energy Consumption Data by End Use Sector in FY 2020 (Billion Btu)*, https://ctsedwweb.ee.doe.gov/Annual/Report/AgencyMasterDataByYear.aspx.

42. Each installation consists of one or more sites, which may or may not be located contiguous to the installation. The United States had about 790 bases and smaller "lily pads" in 2015 and about 750 bases in 2021, David Vine, "Lists of U.S. Military Bases Abroad, 1776–2021," American University Digital Research Archive, 2021, https://doi.org/10.17606/7em4-hb13. See also John Conger, "An Overview of the DOD Installations Enterprise," Heritage Foundation, October 4, 2019, https://www.heritage.org/military-strength/topical-essays/overview-the -dod-installations-enterprise.

43. The DOD reports operating over 560,000 buildings and structures at its installations in the United States and oversease. See Office of the Assistant Secretary of Defense for Sustainment, "Installation Energy," https://www.acq.osd.mil/eie/IE/FEP_index.html, accessed February 13, 2022.

44. Department of Defense, "2016 Operational Energy Strategy," 4.

45. Lawrence H. Suid, *The Army's Nuclear Power Program: The Evolution of a Support Agency*. Contributions in Military Studies, No. 98, 1990, https://flibe-energy.com/pdf/ArmyNuclear PowerProgram.pdf.

46. Richard B. Andres and Hanna L. Breetz, "Small Nuclear Reactors for Military Installations: Capabilities, Costs, and Technological Implications," *Strategic Forum*, Institute for

National Strategic Studies, National Defense University, February 2011, https://ndupress .ndu.edu/Portals/68/Documents/stratforum/SF-262.pdf; Aaron Mehta, "Pentagon Awards Contracts to Design Mobile Nuclear Reactor," *Defense News*, March 9, 2020, https://www .defensenews.com/smr/nuclear-arsenal/2020/03/09/pentagon-to-award-mobile-nuclear -reactor-contracts-this-week/, accessed 23 June 2021.

47. While for a time there were other nuclear-powered ships in the navy's surface fleet, those were retired.

48. See Department of the Navy, "United States Naval Nuclear Propulsion Program," September 2017, https://www.energy.gov/sites/prod/files/migrated/nnsa/2018/01/f46/united_states _naval_nuclear_propulsion_program_operating_naval_nuclear_propulsion_plants_and _shipping_rail_naval_spent_fuel_safely_for_over_sixty_years.pdf.

49. The DOD notes that "*Traditionally, the scope of operational energy excludes nuclear energy used for the propulsion of the U.S. Navy's aircraft carriers and submarines, as well as the energy used for military space launch and operations. Operational energy does include the energy needed to operate the carrier's embarked aircraft and helicopters.*" Department of Defense, Office of the Assistant Secretary of Defense for Sustainment, "Operational Energy," https://www.acq.osd.mil/eie/OE /OE_index.html, accessed June 23 2021. Emphasis in the original.

50. A million metric tons of CO2e (MMTCO2e) can also be called a megaton or Mt CO2e. I use MMTCO2 to avoid any confusing associations with the unit megaton (Mt) which is used to describe the explosive power, or yield, of nuclear weapons equivalent to one million tons of conventional TNT explosive.

51. These are Environmental Protection Agency equivalents derived from using their "Greenhouse Gas Equivalencies Calculator," https://www.epa.gov/energy/greenhouse-gas -equivalencies-calculator. For the EPA's explanation of its assumptions, see EPA, "Greenhouse Gases Equivalencies Calculator—Calculations and References," https://www.epa.gov/energy /greenhouse-gases-equivalencies-calculator-calculations-and-references#houseenergy, accessed June 26, 2021. For example, the EPA uses "a common conversion factor of 8,887 grams of CO_2 emissions per gallon of gasoline consumed (Federal Register 2010). For reference, to obtain the number of grams of CO_2 emitted per gallon of gasoline combusted, the heat content of the fuel per gallon can be multiplied by the kg CO_2 per heat content of the fuel."

52. U.S. Energy Information Administration, "Defense Department Energy Use Falls to Lowest Level Since 1975." *Today in Energy*, February 5, 2015, https://www.eia.gov/todayinenergy /detail.php?id=19871, accessed June 26, 2021.

53. Department of Energy, "Comprehensive Annual Data and Sustainability Performance." These were about half of the total federal government scope 3 emissions for fiscal years 2008 and 2010–2016, as measured by the DOE.

54. The greenhouse gas emission factor of diesel fuel and Navy Special Fuel Oil is followed by, in decreasing order, jet fuel, gasoline, aviation gas, and liquid propane gas.

55. Daniel Gouré, "The U.S Army's All-But Forgotten Vehicle Fleet," *Real Clear Defense*, August 22, 2017, https://www.realcleardefense.com/articles/2017/08/22/the_us_armys_all

-but_forgotten_vehicle_fleet_112116.html. The gas-hungry Ford F-150 pickup truck gets seventeen miles per gallon in the city; the hungrier Chevrolet Suburban gets fifteen miles per gallon in the city.

56. Fuel use depends on the flight profile of the aircraft and other factors, such as load. By contrast with military aircraft, a commercial Boeing 747 gets about five gallons per mile and can carry over five hundred passengers.

57. See Bernd Kärcher, "Formation and Radiative Forcing of Contrail Cirrus," *Nature Communications* 9, no. 1824 (May 8, 2018), https://www.nature.com/articles/s41467-018-04068-0.

58. The B-2 carries 167,000 pounds of jet fuel (almost 25,000 gallons) to travel 6,000 nautical miles (c. 6.900 miles) and is capable of mid-air refueling, taking on an additional 99,000 pounds of fuel at each refueling. The F-35A, with a combat radius estimated to be about 584 nautical miles has an internal fuel capacity of 2,761 gallons. The A-10 with an internal fuel capacity of 1,642 gallons of jet fuel has a combat radius of about 250 nautical miles.

59. A 1999 spike in jet fuel purchases corresponds with the seventy-eight-day NATO air war in Kosovo to which the United States contributed more than five hundred aircraft. For statistics on the U.S. role, see https://www.afhistory.af.mil/FAQs/Fact-Sheets/Article/458957/operation-allied-force/.

60. Truly and Alm, *Report of the Defense Science Board on More Capable Warfighting*, 21.

61. David Vine, "Lists of U.S. Military Bases Abroad, 1776–2019," American University Digital Research Archive, 2019, https://doi.org/10.17606/vfyb-nc07; Department of Defense, "DoD Base Realignment and Closure, BRAC Rounds (BRAC 1988, 1991, 1993, 1995 & 2005), Executive Summary," March 2019, https://comptroller.defense.gov/Portals/45/Documents/defbudget/fy2020/budget_justification/pdfs/05_BRAC/BRAC_Exec_Sum_J-Book_FINAL.pdf.

62. See Andrew Feickert and Kathleen McInnis, "Defender Europe 20 Military Exercise, Historical (REFORGER) Exercises, and U.S. Force Posture in Europe," Congressional Research Service, January 14, 2020, https://crsreports.congress.gov/product/pdf/IF/IF11407; Angela O'Mahony et al., *U.S. Presence and the Incidence of Conflict*, RAND Corporation, 2018, 14, https://www.rand.org/content/dam/rand/pubs/research_reports/RR1900/RR1906/RAND_RR1906.pdf.

63. Truly and Alm, "*Report of the Defense Science Board on More Capable Warfighting*, 21.

64. Department of Defense Base Structure Reports for FY2003 and FY2018. See, respectively, for FY2003, https://www.globalsecurity.org/military/library/report/2002/020930_fy03_baseline_dod_bsr.pdf; and FY2018, https://www.acq.osd.mil/eie/Downloads/BSI/Base%20Structure%20Report%20FY18.pdf.

65. McMahon, "The Great White Fleet Sails Today?," 77.

66. McMahon, 87n.

67. McMahon, 77.

68. See, for instance, Department of Energy, "Air Force Achieves Fuel Efficiency Through Industry Best Practices," https://www.energy.gov/sites/prod/files/2013/10/f3/af_fuelefficiency.pdf and Christopher A. Mouton et al., *Fuel Reduction for the Mobility Air Forces*, RAND Corporation, 2015, https://www.rand.org/content/dam/rand/pubs/research_reports/RR700/RR757/RAND_RR757.pdf.

69. Elisabeth Braw, "NATO Needs More Big Exercise, Too," *Defense One*, June 14, 2018. https://www.defenseone.com/ideas/2018/06/nato-needs-more-big-exercises-too/148980/; NATO, "Germany and NATO," Origins: My Country and NATO, https://www.nato.int/cps/fr/natohq/declassified_185912.htm.

70. NATO, "Key NATO and Allied Exercises in 2019," https://www.nato.int/nato_static_fl2014/assets/pdf/pdf_2019_02/1902-factsheet_exercises_en.pdf. Also see NATO, "Key NATO and Allied Exercises in 2018," https://www.nato.int/nato_static_fl2014/assets/pdf/pdf_2018_04/20180425_1804-factsheet_exercises_en.pdf; and NATO, "Key NATO and Allied Exercises in 2021," https://www.nato.int/nato_static_fl2014/assets/pdf/2021/3/pdf/2103-factsheet_exercises.pdf.

71. On military industry see Ann R. Markusen and Sean S. Costigan, eds., *Arming the Future: A Defense Industry for the 21st Century* (New York: Council on Foreign Relations, 1999); Christian Sorensen, *Understanding the War Industry* (Atlanta: Clarity Press, 2020).

72. This can be compared to the $283.4 billion authorized for operations and maintenance and the $162.3 billion authorized for military personnel in FY2021. Office of the Under Secretary of Defense (Comptroller) "Defense Budget Overview, Fiscal Year 2022 Budget Request," Department of Defense, May 2021, A-1, https://comptroller.defense.gov/Portals/45/Documents/defbudget/FY2022/FY2022_Budget_Request_Overview_Book.pdf. See also Office of the Under Secretary of Defense (Comptroller) "Defense Budget Overview, Fiscal Year 2021 Budget Request," Department of Defense, February 2020, revised May 2020, A-1, https://comptroller.defense.gov/Portals/45/Documents/defbudget/fy2021/fy2021_Budget_Request_Overview_Book.pdf.

73. See, for instance, Government Accountability Office, "Overseas Military Construction: Observations on U.S. Contractor Preference," GAO, November 2014, 15–45, https://www.gao.gov/assets/gao-15-45.pdf.

74. Louis Uchitelle, "The U.S. Still Leans on the Military-Industrial Complex," *New York Times*, September 22, 2017, https://www.nytimes.com/2017/09/22/business/economy/military-industrial-complex.html.

75. Environmental Protection Agency, *Inventory of U.S. Greenhouse Gases and Sinks: 1990–2019*, table 4-1. https://www.epa.gov/sites/production/files/2021-04/documents/us-ghg-inventory-2021-main-text.pdf.

76. See Deloitte, "2017 Aerospace and Defense Sector Export and Labor Market Study," 13, https://www2.deloitte.com/content/dam/Deloitte/us/Documents/manufacturing/us-2017-us-A&D-exports-and-labor-market-study.pdf. Manufacturing accounts for most of the industrial sector's greenhouse gas emissions according to the annual EPA inventory; see *Inventory of U.S.*

Greenhouse Gases and Sinks: 1990–2017, https://www.epa.gov/sites/production/files/2019-02/documents/us-ghg-inventory-2019-main-text.pdf. Also see Judith Reppy, "The United States," in *The Structure of the Defense Industry: An International Survey*, ed. Nicole Ball and Milton Leitenberg (New York: St. Martin's Press, 1983), 21–49.

77. Aerospace Industries Association (AIA), "2019 Facts and Figures, U.S. Aerospace and Defense," https://www.aia-aerospace.org/2019-facts-and-figures/.

78. U.S. Bureau of Labor Statistics, "Employment by Major Industry Sector." For an overview, see Allison Thomson, "Defense-related Employment and Spending, 1996–2006," *Monthly Labor Review* (July 2008): 14–33. See table A.10 in the appendix for historical numbers taken from Thomson's analysis.

79. Environmental Protection Agency, *Inventory of U.S. Greenhouse Gases and Sinks: 1990–2019*, table 4-1, https://www.epa.gov/sites/production/files/2021-04/documents/us-ghg-inventory-2021-main-text.pdf.

80. Stuart Parkinson, "The Environmental Impact of the UK Military Sector," Scientists for Global Responsibility, May 2020, 12, https://www.sgr.org.uk/sites/default/files/2020-05/SGR-DUK_UK_Military_Env_Impacts.pdf.

81. U.S. State Department, "U.S. Arms Transfers Increased by 2.8 Percent in FY2020 to $175.08 Billion," January 20, 2021, https://www.state.gov/u-s-arms-transfers-increased-by-2-8-percent-in-fy-2020-to-175-08-billion/.

82. CDP, https://data.cdp.net/.

83. Lockheed Martin, "Products," https://www.lockheedmartin.com/en-us/products.html.

84. See Lockheed Martin, *Climate Change 2020*, CDP report, https://www.lockheedmartin.com/content/dam/lockheed-martin/eo/documents/2020%20CDP%20Report.pdf, accessed July 1, 2021.

85. "Since 2008, we have installed 13 on-site renewable energy systems, including 12 solar systems and one biomass facility for a total of 9.3 MW of capacity. In 2019, Lockheed Martin spent approximately $15 million on completed projects and initiatives specifically related to energy efficiency and conservation. We completed 41 energy-efficiency and GHG emissions reductions projects in 2019, which contributed towards a 22% energy reduction and a 39% reduction in attributed GHG emissions. In 2019, Lockheed Martin avoided approximately $32 million (compared to a 2010 baseline) in energy and water costs through the implementation of similar projects over the years. The implementation of these projects reduces our GHG emissions and strengthen our resiliency to climate-related risks." Also: "In June 2016 Lockheed Martin became the off-taker of 30 MW from a solar power purchase agreement in North Carolina. The Renewable Energy Certificates (REC's) produced from this project contribute to the company's energy mix." Lockheed Martin, *Climate Change* 2020, 24, 40, 48.

86. Lockheed Martin reports that it reduced 250 metric tons of CO_2e through purchases of renewable energy and reduced 48,113 metric tons of CO_2e through initiatives that improved efficiency and reduced consumption, Lockheed Martin, *Climate Change 2020*,

33. Also see Lockheed Martin, *2019 Sustainability Report: The Science of Citizenship,* https://www.lockheedmartin.com/content/dam/lockheed-martin/eo/documents/sustainability/Lockheed_Martin_Sustainability_Report_Full_2019.pdf; and *2020 Sustainability Report: Propelled by Principle,* https://sustainability.lockheedmartin.com/sustainability/content/Lockheed_Martin_2020_Sustainability_Report.pdf, accessed July 1, 2021.

87. Sources: Lockheed Martin, *Climate Change* 2020 and *2020 Sustainability Report*; Northrop Grumman Corp, *Climate Change 2020,* 16, https://www.northropgrumman.com/wp-content/uploads/CDP-2020-Climate-Change-Response.pdf; Northrop Grumman Corp, *2020 Sustainability Report,* https://www.northropgrumman.com/wp-content/uploads/Northrop-Grumman_2020-SustainabilityReport.pdf. In some years Northrup Grumman's reports conflict with each other. See also General Dynamics, *2020 Corporate Sustainability Report,* https://www.gd.com/-/media/GD-Corporate/Responsibility/gd-2020-sustainability-report.ashx.

88. Boeing, "Boeing Reaches Net-Zero Carbon Emissions from Manufacturing and Worksites," December 22, 2020 https://www.boeing.com/features/2020/12/boeing-reaches-net-zero-carbon-emissions-from-manufacturing-and-worksites.page.

89. Boeing reported 1.105 MMTCO2e in scope 1 and 871,000 metric tons in scope 2 emissions in 2019. Boeing, *Global Environmental Report 2020,* 43–44, https://www.boeing.com/resources/boeingdotcom/principles/environment/pdf/2020_environment_report.pdf.

90. Arms sales data from Stockholm International Peace Research Institute (SIPRI), "SIPRI Arms Industry Database," https://www.sipri.org/databases/armsindustry. Corporate emissions from Lockheed Martin, *Climate Change 2020* and *2020 Sustainability Report*; Boeing, *Global Environmental Report, 2020*; Northrop Grumman Corp, *Climate Change 2020.* Also see Northrop Grumman Corp, *2020 Sustainability Report*; Raytheon, *Climate Change 2020,* https://www.rtx.com/social-impact/environment-health-and-safety/environment; L3Harris, *Sustainability Report 2020,* https://www.l3harris.com/sites/default/files/2021-03/L3Harris-SustainabilityReport-2020.pdf; General Dynamics, *2020 Corporate Sustainability Report*; Huntington Ingalls, UMass PERI database, http://grconnect.com/green100/ry2018/index.php?search=yes&database=t1&detail=1&datype=T&reptype=a&company2=&company1=&parent=huntington+ingalls&sectfac=fac&advbasic=bas; Honeywell International, *Climate Change 2020,* https://www.honeywell.com/content/dam/honeywell/files/Honeywell-CDP-Climate-Change-Questionnaire-2020.pdf; Leidos, "Corporate Footprint and Environmental Stewardship," https://www.leidos.com/company/responsibility-and-sustainability/environment/corporate-footprint-and-environmental-stewardship; Booze Allen Hamilton, "Fiscal Year 2020 Greenhouse Gas Emissions Statement and Methodology," https://www.boozallen.com/content/dam/boozallen_site/esg/pdf/publications/fy2020-greenhouse-gas-emissions.pdf, and "Environmental Sustainability," https://www.boozallen.com/e/insight/thought-leadership/commitment-to-sustainability.html; General Electric, "2019 Environmental, Social, and Governance Results," https://www.ge.com/sites/default/files/GE_ESG_Results_RY2019.pdf; Textron, "Textron 2019 Corporate Responsibility Report," https://www.textron.com/assets/CR/2019/TextronAndTheEnvironment.html.

91. Calculated from Lockheed Martin, *Climate Change 2020*, 31–32.

92. Stockholm International Peace Research Institute, *SIPRI Yearbook, 2021* (Oxford: Oxford University Press, 2021); and online at https://www.sipri.org/yearbook/2021/09.

93. They report scope 3 emissions of 370,000 metric tons of CO2e from the production of purchased capital goods and 7,700,000 metric tons of CO2e emitted for purchased goods and services, totaling 8,070,000 metric tons of CO2e. Of these, I attribute about 89 percent of those emissions (7,182,300 metric tons of CO2e) to military-industrial purposes. They calculated emissions from purchased goods and services (7.7 million MTCO2e), capital goods (370,000 MTCO2e), transportation and distribution loss in the delivery of electricity and natural gas to its facilities (105,000 MTCO2e), transportation to and from most of their vendors (60,000 MTCO2e), waste generation (4,500 MTCO2e), business travel (190,000 MTCO2e), employee commuting (215,000 CO2e), and the use of products sold (22 million MTCO2e). Lockheed Martin also suggests that its Scope 3 emissions are enormous when it notes that in 2019 it surveyed 327 of its suppliers, representing c. 53 percent of the money they spent on supplies in 2018. Lockheed Martin, *Climate Change 2020*, 25–27.

94. Thus, the approximate ratio of scope 3 emissions from the goods and services and capital goods in the supply chain to the finished product is about 8.78 to 1.

95. A conservative estimate of U.S. military-industrial greenhouse gas emissions would discount some of the emissions from military industry because a portion of these corporations' workers are located outside the United States For example, 7 percent of Lockheed Martin's workforce was located outside the United States in 2019, working in facilities located in the UK, Poland, Canada, Australia, and Mexico. And in 2020 for example, while Lockheed Martin received 93 percent of its supplies from the United States, the company also had 17,200 "active tier 1 suppliers from 64 countries." Lockheed Martin, *Climate Change 2020*, 8; Lockheed Martin, *2020 Sustainability Report*, 40.

96. Parkinson, "The Environmental Impact of the UK Military Sector."

97. Parkinson and Cottrell, *Under the Radar*.

98. Ho-Chih Lin and Deborah Burton, *Indefensible: The True Cost of the Global Military to Our Climate and Human Security*, Tipping Point North South, October 2020, 10, https://thefivepercentcampaign.files.wordpress.com/2020/12/indefensible-the-true-cost-of-the-global-military-to-our-climate-and-human-security.pdf.

99. Congressional Research Service, "The Environmental Impact of the Gulf War," for the United States Senate Committee on Environment and Public Works Gulf Pollution Task Force, March 1992, 10 and 24. See also Erich R. Gundlach, John C. McCain, and Yusef H. Fadallah, "Distribution of Oil along the Saudi Arabian Coastline (May/June 1991) as a Result of the Gulf War Oil Spills," *Marine Pollution Bulletin* 27 (1993): 93–96; Dagmar Schmidt-Etkin, "Spill Occurrences: A World Overview," in *Oil Spill Science and Technology*, ed. Mervin Fingas (Amsterdam: Elsevier, 2011), 8; and Jacqueline Michel, "1991 Gulf War Oil Spill," in Fingas, *Oil Spill Science and Technology*, 1127–1132.

100. Saif Hameed and Dominic Evans, "Islamic State Torches Oil Field Near Tikrit as Militia Advances," Reuters, March 5, 2015, https://www.reuters.com/article/us-mideast-crisis-iraq-idUSKBN0M10Z420150305.

101. See Matthew Reed, "Blowing up the Islamic State's Oil Company," *Foreign Policy*, October 26, 2016, https://foreignpolicy.com/2016/10/26/blowing-up-the-islamic-states-oil-company-isis-abu-sayyaf/, accessed June 27, 2021.

102. Brian Glyn Williams, "Who Defeated Isis? The Pentagon's War Maps," *Middle East Policy* 27, no. 3 (Fall 2020): 152–193; 164–165.

103. Data compiled from Operation Inherent Resolve, "Airstrike Updates," https://dod.defense.gov/OIR/Airstrikes/, accessed June 27, 2021.

104. Lisa Ferdinando, "OIR Spokesman: Coalition Cripples ISIL Oil Distribution," *DOD News*, November 18, 2015, https://www.defense.gov/Explore/News/Article/Article/630352/oir-spokesman-coalition-cripples-isil-oil-distribution/, accessed June 27, 2021; Joseph Trevithick, "Here Are the Leaflets the United States Dropped on Islamic State," *War Is Boring*, July 31, 2017, https://warisboring.com/we-got-copies-of-leaflets-dropped-on-islamic-state/, accessed June 27, 2021.

105. See Wim Zwijnenburg, "Dying to Keep Warm: Oil Trade and Makeshift Refining in North-West Syria," *Bellingcat*, April 24, 2020, https://www.bellingcat.com/news/2020/04/24/dying-to-keep-warm-oil-trade-and-makeshift-refining-in-north-west-syria/; Kaamil Ahmed, "Makeshift Oil Refineries a Necessary Evil for Locals In North-West Syria," *Guardian*, April 24, 2020, https://www.theguardian.com/global-development/2020/apr/24/makeshift-oil-refineries-a-necessary-evil-for-locals-in-north-east-syria-study-finds.

106. See, for example, Joint Chiefs of Staff, "Joint Targeting Cycle and Collateral Damage Estimation Methodology (CDM)," General Counsel, November 10, 2009, unclassified, https://www.aclu.org/files/dronefoia/dod/drone_dod_ACLU_DRONES_JOINT_STAFF_SLIDES_1-47.pdf.

107. Joint Chiefs of Staff, *Joint Targeting*, JP 3–60, January 31, 2013, A-2, https://www.justsecurity.org/wp-content/uploads/2015/06/Joint_Chiefs-Joint_Targeting_20130131.pdf.

108. Joint Chiefs of Staff, *Joint Targeting*, A-3.

109. Joint Chiefs of Staff, *Joint Targeting*, A-4.

110. Joint Chiefs of Staff, *Joint Targeting*, A-5.

111. Joint Chiefs of Staff, *Joint Targeting*, II-8 and II-9.

112. Zabihullah Ghazi, "Afghanistan's Forest Cover Illegally Stripped Away," *Environment New Service*, August 29, 2013, https://ens-newswire.com/2013/08/29/afghanistans-forest-cover-illegally-stripped-away/; UN Environment, "Salvaging Iraq's Remaining Wilderness," July 10, 2018, https://www.unenvironment.org/news-and-stories/story/salvaging-iraqs-remaining-wilderness.

113. Robert Burns, "Rumsfeld Proposes Defense Cuts," *AP*, June 27 2001, https://apnews.com/article/0891eefe0316f530e3e9660fa650de3b; Thom Shanker, "Rumsfeld Is Facing

a Deadline in Effort to Reshape the Military," *New York Times*, August 9, 2001, https://www.nytimes.com/2001/08/09/us/rumsfeld-is-facing-a-deadline-in-effort-to-reshape-the-military.html.

CHAPTER 5

1. Marine Corps General John Mattis at the Committee on Armed Services, Hearing on National Defense Authorization Act for Fiscal Year 2012, March 3, 2011, https://www.govinfo.gov/content/pkg/CHRG-112hhrg65114/html/CHRG-112hhrg65114.htm.

2. Secretary of Defense Lloyd Austin, Forward to the Department of Defense, Office of the Undersecretary of Defense for Acquisition and Sustainment, *Department of Defense Climate Adaptation Plan. Report Submitted to National Climate Task Force and Federal Chief Sustainability Officer*, September 1, 2021, 1, https://www.sustainability.gov/pdfs/dod-2021-cap.pdf.

3. National Public Radio, "Pakistan Blocks NATO Supply Route To Afghanistan," *National Public Radio*, September 30, 2010, https://www.npr.org/templates/story/story.php?storyId=130234301; Tim Craig and Haq Nawas Khan, "Pakistani Political Leader Says NATO Supply Routes Will Be Cut if U.S. Drone Strikes Continue," *Washington Post*, October 31, 2013, https://www.washingtonpost.com/world/asia_pacific/pakistani-political-leader-says-nato-supply-routes-will-be-cut-if-us-drone-strikes-continue/2013/10/31/3c0e002e-4260-11e3-b028-de922d7a3f47_story.html.

4. Office of the Assistant Secretary of Defense for Sustainment, *Department of Defense Annual Energy Management and Resilience Report* FY2019, 13.

5. Martin Liptrot, "After Hurricane Michael, Tyndall Air Force Base Is Returning Stronger than Ever," *30a.com*, April 26, 2021, https://30a.com/tyndall-air-force-base/.

6. Patrick Martin, "Critical Air Base Flooded in Nebraska," *Washington Post*, March 18, 2019, https://www.washingtonpost.com/national-security/2019/03/18/critical-air-force-base-flooded-nebraska/; AP, "Air Force Estimates $420M Needed to Rebuild Nebraska Base," *Associated Press*, May 1, 2019, https://apnews.com/article/098114673f2445f190a674331ba52278.

7. Stephen Losey, "After Massive Flood, Offutt Looks to Build a Better Base," *Air Force Times*, August 7, 2020, https://www.airforcetimes.com/news/your-air-force/2020/08/07/after-massive-floods-offutt-looks-to-build-a-better-base/.

8. Austin, *Department of Defense Climate Adaptation Plan*, September 1, 2021, 10.

9. Quoted in CNA Analysis and Solutions, *Powering America's Defense: Energy and the Risks to National Security*, May 2009, 14, https://www.cna.org/cna_files/pdf/MAB_2-FINAL.pdf. "Fuel use, according to Army estimates, is about 7 gallons of fuel per mile, with the tank consuming about the same amount of fuel idling as cruising. The Abrams tank is idling about 70 percent of the time in order to run the tank's electrical subsystems." General Accounting Office, *Operation Desert Storm: Early Performance Assessment of Bradley and Abrams*, Washington, DC, 1992, 26.

10. McMahon, "The Great White Fleet Sails Today?," 77, 81.

11. Truly and Alm, *Report of the Defense Science Board on More Capable Warfighting*, 10.

12. Truly and Alm, ES-1 and 13.

13. Truly and Alm, 13.

14. Truly and Alm, 18–20.

15. Truly and Alm, 21.

16. Defense Science Board, *Report of the Defense Science Board Task Force on DoD Energy Strategy "More Fight—Less Fuel,"* Office of the Under Secretary of Defense for Acquisition, Technology, and Logistics, Washington, DC, February 2008, https://dsb.cto.mil/reports/2000s/ADA477619.pdf.

17. Defense Science Board, *"More Fight—Less Fuel,"* 21.

18. Defense Science Board, 22.

19. Steve Siegel, Steve Bell, Scott Dicke, and Peter Arbuckle, *Sustain the Mission Project: Energy and Water Costing Methodology and Decision Support Tool,* Army Environmental Policy Institute, August 2008, vii. https://apps.dtic.mil/sti/pdfs/ADB346027.pdf.

20. Defense Science Board, *"More Fight—Less Fuel,"* 5.

21. Defense Science Board, 30.

22. "The OSD(PA&E) guidance memo for estimating FBCF cautions 'the fully burdened cost of fuel should generally assume peacetime OPTEMPO. Using "worst case" or combat scenarios to evaluate alternatives for the entire life cycle of a platform will skew the burdened rates, making some technologies appear to have a greater return on investment than they are likely to achieve in actual practice during a life cycle exceeding 30 years.' The Task Force does not support this approach. FBCF is a wartime capability planning factor, not a peacetime cost estimate. Because it is scenario dependent, it should be estimated across the range of scenarios and missions envisioned for the system in question. The logistics structure needed to deliver fuel for those scenarios and missions and the operational assets used to protect that fuel during transit are part of the FBCF for that system." Defense Science Board, *"More Fight—Less Fuel,"* 31.

23. Defense Science Board, 15.

24. Energy Independence and Security Act of 2007, Public Law 110–140, Section 526, https://www.govinfo.gov/content/pkg/BILLS-110hr6enr/pdf/BILLS-110hr6enr.pdf/; See the Department of Defense, "Operational Energy Strategy: Implementation Plan," March 2012, https://www.acq.osd.mil/eie/Downloads/OE/20120306_OE_Strategy_Implementation_Plan.pdf. Also see Duncan Hunter National Defense Authorization Act For Fiscal Year 2009, Public Law 110–417, Section 2911, https://www.govinfo.gov/content/pkg/PLAW-110publ417/html/PLAW-110publ417.htm.

25. Duncan Hunter National Defense Authorization Act For Fiscal Year 2009," Public Law 110–417, Section 2911.

26. Deloitte, *"Energy Security: America's Best Defense,"* 2009, 1.

27. Deloitte, 10.

28. Elisabeth Rosenthal, "U.S. Military Orders Less Dependence on Fossil Fuels," *New York Times*, October 4, 2010, https://www.nytimes.com/2010/10/05/science/earth/05fossil.html.

29. Craig Whitlock, "U.S. Turns to Other Routes to Supply Afghan War as Relations with Pakistan Fray," *Washington Post*, July 2, 2011, https://www.washingtonpost.com/world/national-security /us-turns-to-other-routes-to-supply-afghan-war-as-relations-with-pakistan-fray/2011/06/30 /AGfflYvH_story.html.

30. Kathy Gannon, "Low Pay, Big Risks for Fuel Haulers in Afghan War," *NBC News*, February 26, 2011, https://www.nbcnews.com/id/wbna41801027.

31. South Asia Terrorism Portal, "Attacks on NATO," https://www.satp.org/datasheet-terrorist -attack/other-data/pakistan; Faisal Aziz, "Militants Set Fire to NATO Tankers in Pakistan," *Reuters*, October 1, 2010, https://www.reuters.com/article/us-pakistan-nato/militants-set -fire-to-nato-tankers-in-pakistan-idUSTRE69025T20101001; Amy Fallon, "Oil Tank-ers Destroyed in Attack on NATO Supply Route in Pakistan," *Guardian*, October 3, 2010, https://www.theguardian.com/world/2010/oct/03/pakistan-oil-tankers-nato-afghanistan; CBS, "NATO Fuel Tankers Torched in Pakistan Attack," *CBS News*, December 8, 2011, https://www.cbsnews.com/news/nato-fuel-tankers-torched-in-pakistan-attack/. Further, mili-tants in Pakistan have taken to attacking gas pipelines, staging about 350 such attacks from 2008 through 2020. For the number of attacks, see Pak Institute of Peace Studies, Security Reports from 2008 through 2020, https://www.pakpips.com/. For descriptions of gas pipe-line attacks see South Asia Terrorism Portal, https://www.satp.org/datasheet-terrorist-attack /other-data/pakistan, accessed on July 7, 2021.

32. Pak Institute of Peace Studies, Security Reports from 2008 through 2020..

33. Duncan Hunter National Defense Authorization Act For Fiscal Year 2009, Public Law 110–417, Section 2911.

34. Department of Defense, *Base Structure Report, FY 2018*, DOD-15, https://www.acq.osd.mil /eie/Downloads/BSI/Base%20Structure%20Report%20FY18.pdf.

35. Office of the Assistant Secretary of Defense for Sustainment, *Department of Defense Annual Energy Management and Resilience Report* FY2019, appendix P.

36. Department of Defense, *Report on the Effects of a Changing Climate to the Department of Defense*, 5. Thawing permafrost is already occurring at Fort Greeley, AK.

37. See U.S. Navy Department, "Climate Change Roadmap," April 2010, https://www.hsdl.org /?abstract&did=8466.

38. Department of Defense, *Quadrennial Defense Review, 2010*, 85, https://history.defense.gov /Historical-Sources/Quadrennial-Defense-Review/.

39. Office of the Undersecretary of Defense for Acquisition, Technology and Logistics, *Depart-ment of Defense, Climate-Related Risk to DOD Infrastructure Initial Vulnerability Survey (SLVAS) Assessment Report*, January 2018, 1, https://climateandsecurity.files.wordpress.com

/2018/01/tab-b-slvas-report-1-24-2018.pdf. "An installation is defined as a base, camp, post, station, yard, center, homeport facility for any ship, or other activity under the jurisdiction of the DoD, including any leased facility, which is located within any of the States, the District of Columbia, the Commonwealth of Puerto Rico, American Samoa, the Virgin Islands, the Commonwealth of the Northern Mariana Islands, or Guam."

40. Office of the Undersecretary of Defense for Acquisition, Technology and Logistics, *Department of Defense, Climate-Related Risk to DOD Infrastructure Initial Vulnerability Survey (SLVAS) Assessment Report*.

41. Department of Defense, Office of the Undersecretary of Defense for Acquisition and Sustainment, *Report on the Effects of a Changing Climate to the Department of Defense*, January 2019, https://climateandsecurity.files.wordpress.com/2019/01/sec_335_ndaa-report_effects_of_a _changing_climate_to_dod.pdf.

42. McKinley, "Cruise of the Great White Fleet."

43. CNA Military Advisory Board, *National Security and the Accelerating Threat of Climate Change*, May 2014, 25, https://www.cna.org/cna_files/pdf/MAB_5-8-14.pdf; Maria McCollester, Michelle E. Miro, and Kristin Van Abel, *Building Resilience Together: Military and Local Government Collaboration for Climate Adaptation*, RAND Corporation, 2020, 7, https://www .rand.org/pubs/research_reports/RR3014.html.

44. Department of Defense, *Report on the Effects of a Changing Climate to the Department of Defense*, 8.

45. See Walter C. Ladwig III, Andrew S. Erickson, and Justin Mikolay, "Diego Garcia and American Security in the Indian Ocean," in *Rebalancing U.S. Forces: Basing Forward Presence in the Asia Pacific*, ed. Carnes Lord and Andrew S. Erickson (Annapolis: Naval Institute Press, 2014), 130–179, http://www.andrewerickson.com/wp-content/uploads/2019/02/Lord-Erickson _Rebalancing-US-Forces_NIP_2014_Ladwig-Erickson-Mikolay_Diego-Garcia_SINGLE -FILE.pdf.

46. David Vine, *Island of Shame: The Secret History of the U.S. Military Base on Diego Garcia* (Princeton, NJ: Princeton University Press, 2009), 10.

47. David Vine and Laura Jeffery, "'Give Us Back Diego Garcia': Unity and Division among Activists in the Indian Ocean," in *The Bases of Empire: The Global Struggle against U.S. Military Posts*, ed. Catherine Lutz (New York: New York University Press, 2009), 181–217.

48. United Nations, "General Assembly Welcomes International Court of Justice Opinion on Chagos Archipelago, Adopts Text Calling for Mauritius' Complete Decolonization," May 22, 2019, https://www.un.org/press/en/2019/ga12146.doc.htm.

49. Marwaan Macan-Markar, "Mauritius Makes Play for Future with US Base on Diego Garcia," *Nikkei Asia*, November 18, 2020, https://asia.nikkei.com/Editor-s-Picks/Interview /Mauritius-makes-play-for-future-with-US-base-on-Diego-Garcia.

50. Curt Storlazzi et al., "Most Atolls Will Be Uninhabitable by the Mid-21st Century Because of Sea-Level Rise Exacerbating Wave-Driven Flooding," *Science Advances* 4, no. 4 (April 2018), https://advances.sciencemag.org/content/4/4/eaap9741.

51. Haunani Kane and Charles Fletcher, "Rethinking Reef Island Stability in Relation to Anthropogenic Sea Level Rise," *Earth's Future* 8, no. 10 (October 2020): 1–14.

52. Jeff D. Colgan, "Climate Change and the Politics of Military Bases," *Global Environmental Politics* 18, no. 1 (February 2018): 33–51; 36.

53. The United States had facilities in Greenland between 1953 and 1967. See Colgan, "Climate Change and the Politics of Military Bases."

54. National Research Council Committee, *National Security Implications of Climate Change for U. S. Naval Forces* (Washington, DC: The National Academies Press, 2011).

55. Department of Defense, "Climate Change Adaptation Roadmap," 2014, https://www.acq .osd.mil/eie/Downloads/CCARprint_wForward_e.pdf. The DOD takes up the challenge of reducing its emissions in a separate document.

56. Memorandum for the Heads of Executive Departments and Agencies, Climate Change and National Security, September 21, 2016, https://www.justice.gov/opa/file/895016/download.

57. Klare, *All Hell Breaking Loose.*

58. Kidd quoted in Patrick Tucker, "Climate Change Is Already Disrupting the Military. It Will Get Worse, Officials Say," *Defense One*, August 10, 2021, https://www.defenseone.com /technology/2021/08/climate-change-already-disrupting-military-it-will-get-worse-officials -say/184416/.

59. Hicks interview with Noel King, "Climate Change Is a Risk to National Security, the Pentagon Says," *National Public Radio*, October 26, 2021, https://www.npr.org/2021/10/26 /1049222045/the-pentagon-says-climate-change-is-having-a-negative-impact-on-national -securit.

60. Department of the Army, Office of the Assistant Secretary of the Army for Installations, Energy and Environment, *United States Army Climate Strategy* (Washington, DC: February 2022), 5, https://www.army.mil/e2/downloads/rv7/about/2022_army_climate_strategy .pdf.

61. U.S. Army, *Army Climate Resilience Handbook*, August 2020, 98, https://www.asaie.army.mil /Public/ES/doc/Army_Climate_Resilience_Handbook_Change_1.pdf.

62. David Hasemyer, "U.S. Soldiers Falling Ill, Dying in the Heat as Climate Warms," *Inside Climate News*, July 23, 2019, https://insideclimatenews.org/news/23072019/military-heat -death-illness-climate-change-risk-security-global-warming-benning-bragg-chaffee/.

63. Patricia Kime, "After Hitting Record High, Military Heatstroke Cases May Be on the Decline," *Military.com*, May 13, 2020, https://www.military.com/daily-news/2020/05/12 /after-hitting-record-high-military-heatstroke-cases-may-be-decline.html; Armed Forces Health Surveillance Branch, "Update: Heat Illness, Active Component, U.S. Armed Forces, 2020," *Health.mil*, April 1, 2021, https://www.health.mil/News/Articles/2021/04/01 /Update-Heat-MSMR-2021?type=Fact+Sheets.

64. Kelley, *Global Climate Change Implications for the United States Navy.*

65. United States Arctic Research Commission, "The Arctic Ocean and Climate Change," 16.

66. National Research Council, *National Security Implications of Climate Change for U.S. Naval Forces*, 165.

67. George C. Wilson, "Weather Trouble Reported before Launching Rescue Mission," *Washington Post*, May 10, 1980, https://www.washingtonpost.com/archive/politics/1980/05/10/weather-trouble-reported-before-launching-rescue-mission/177cae4e-90b0-47c4-8135-9573b0accba8/.

68. Marine Corps General John Mattis at the Committee on Armed Services, Hearing on National Defense Authorization Act for Fiscal Year 2012.

CHAPTER 6

1. Department of Homeland Security, "DHS Strategic Framework for Addressing Climate Change," October 21, 2021, 9, https://www.dhs.gov/sites/default/files/publications/dhs_strategic_framework_10.20.21_final_508.pdf.

2. Department of Defense *Quadrennial Defense Review, 2010*, iv.

3. Department of Defense, Office of the Undersecretary for Policy (Strategy, Plans, and Capabilities), *Department of Defense Climate Risk Analysis*, report to the National Security Council," October 2021, https://media.defense.gov/2021/Oct/21/2002877353/-1/-1/0/DOD-CLIMATE-RISK-ANALYSIS-FINAL.PDF.

4. Bender, "Chief of US Pacific Forces Calls Climate Biggest Worry."

5. Department of Defense, *Quadrennial Defense Review, 2014*, 8.

6. For instance, see Campbell et al., *The Age of Consequences*; CNA Corporation, *National Security and the Threat of Climate Change*, Center for Naval Analysis, 2007, https://www.cna.org/cna_files/pdf/national%20security%20and%20the%20threat%20of%20climate%20change.pdf.

7. Dyer, *Climate Wars*; Welzer, *Climate Wars*; Todd Miller, *Storming the Wall: Climate Change, Migration, and Homeland Security* (New York: City Lights, 2017); Anatol Lieven, *Climate Change and the Nation State: The Case For Nationalism in a Warming World* (New York: Oxford University Press, 2020); David Wallace-Wells, *The Uninhabitable Earth: Life after Warming* (New York: Tim Duggan, 2019); Klare, *All Hell Breaking Loose*.

8. Carol Polsgrove, "War Is Hell—So Is Climate Change: The US Military Readies for Battle in a Harsh New Theater," *Sierra*, February 23, 2020, https://www.sierraclub.org/sierra/war-hell-so-climate-change-michael-klare-all-hell-breaking-loose.

9. Matthew Smith, "Most People Expect to Feel the Effects of Climate Change, and Many Think It Will Make Us Extinct," *YouGovAmerica*, September 16, 2019, https://today.yougov.com/topics/science/articles-reports/2019/09/16/global-climate-change-poll.

10. See Franziskus von Lucke, "The Securitisation of Climate Change in the United States: The Integration of Climate Threats into the Security Sector," University of Tübingen, *Climasec Working Paper #10* (June 2015); and Michael Durant Thomas, *The Securitization of Climate Change:*

Australian and United States Military Responses (2003–2013) (Cham: Springer, 2017); Rita Floyd, *Security and the Environment: Securitization Theory and U.S. Environmental Security Policy* (Cambridge: Cambridge University Press, 2010); Matt McDonald, *Ecological Security: Climate Change and the Construction of Security* (Cambridge: Cambridge University Press, 2021).

11. Jon Barnett, "The Geopolitics of Climate Change," *Geography Compass* 1, no. 7 (2007): 1361–1375; 1365.

12. President George H. W. Bush, *National Security Strategy of the United States*, The White House, Washington, DC, 1991, 2, https://history.defense.gov/Portals/70/Documents/nss/nss1991.pdf?ver=3sIpLiQwmknO-RplyPeAHw%3d%3d.

13. President Bush, *National Security Strategy of the United States*, August 1991, 22.

14. Parties to the treaty are to "develop, periodically update, publish and make available to the Conference of the Parties . . . national inventories of anthropogenic emissions by sources and removals by sinks of all greenhouse gases not controlled by the Montreal Protocol, using comparable methodologies to be agreed upon by the Conference of the Parties." United Nations Framework Convention on Climate Change, Article 4, paragraph 1a, https://unfccc.int/files/essential_background/background_publications_htmlpdf/application/pdf/conveng.pdf.

15. Shannon Casey, "Around the Pier: Gore: Roger Revelle Was My Inspiration," Scripps Institution of Oceanography, June 1, 2007, https://www.nytimes.com/2010/12/22/science/earth/22carbon.html; Justin Gillis, "A Scientist, His Work, and a Climate Reckoning," *New York Times*, December 21, 2010, https://www.nytimes.com/2010/12/22/science/earth/22carbon.html, accessed May 31, 2021.

16. Secretary of Defense Les Aspin, *Report on the Bottom Up Review*, Department of Defense, October 1993, iii, https://apps.dtic.mil/sti/pdfs/ADA359953.pdf.

17. Aspin, *Report on the Bottom Up Review*, 8.

18. Jonathan Schell, *The Unconquerable World: Power, Nonviolence, and the Will of the People* (New York: Metropolitan Books, 2003), 5.

19. Aspin, *Report on the Bottom Up Review*, 99.

20. Aspin, 99.

21. Aspin, 7.

22. U.S. Navy Task Force on Climate Change, "The United States Navy Arctic Roadmap, 2014–2030," February 2014, http://navysustainability.dodlive.mil/files/2014/02/USN-Arctic-Roadmap-2014.pdf.

23. President Bill Clinton, *A National Security Strategy of Engagement and Enlargement*, July 1994, White House, Washington, DC, 15, https://nssarchive.us/wp-content/uploads/2020/04/1994.pdf.

24. President William Jefferson Clinton, State of the Union Address, January 23, 1996, https://clintonwhitehouse4.archives.gov/WH/New/other/sotu.html.

25. Sherri W. Goodman, "The Environment and National Security," Remarks to the National Defense University, August 8, 1996, https://evergreen.loyola.edu/khula/www/strategic -intelligence/intel/goodman.html.

26. Gerald Kutney, *Carbon Politics and the Failure of the Kyoto Protocol* (New York: Routledge, 2014), 139–151.

27. Peter Schwartz and Doug Randall, *An Abrupt Climate Change Scenario and Its Implications for United States National Security*, prepared for Global Business Network, October 2003; National Research Council Committee on Abrupt Climate Change, *Abrupt Climate Change: Inevitable Surprises* (Washington, DC: National Academy Press, 2002).

28. Schwartz and Randall, *An Abrupt Climate Change Scenario*, 1.

29. Schwartz and Randall, 2.

30. Schwartz and Randall, 2.

31. For a discussion of the Schwartz and Randall report, see Allan Shearer, "Whether the Weather: Comments on 'An Abrupt Climate Change Scenario and Its Implications for United States National Security,'" *Futures* 37 (2005): 445–463.

32. Schwartz and Randall, *An Abrupt Climate Change Scenario*, 2–3.

33. Thomas, *The Securitization of Climate Change*, 14.

34. Vice Admiral Lee Gunn, "National Security and the Accelerating Risk of Climate Change," *Elementa* 5 (2017): 30–37l 31.

35. Gunn, "National Security and the Accelerating Risk of Climate Change," 34.

36. CNA Military Advisory Board, *National Security and the Threat of Climate Change*, 2007, https://www.cna.org/CNA_files/pdf/National%20Security%20and%20the%20Threat%20 of%20Climate%20Change.pdf.

37. CNA Military Advisory Board, *National Security and the Accelerating Risks of Climate Change*, May 2014, 25.

38. Norwegian Nobel Committee, "The Nobel Peace Prize for 2007," http://nobelpeaceprize.org /en_GB/laureates/laureates-2007/announce-2007/.

39. Campbell et al., *The Age of Consequences*.

40. Campbell et al., 8.

41. Campbell, et al., 7.

42. Joshua W. Busby, *Climate Change and National Security: An Agenda for Action*, Council on Foreign Relations, Council Special Report, CSR 32, November 2007, https://cdn.cfr.org /sites/default/files/report_pdf/ClimateChange_CSR32%20%281%29.pdf. Also see Joshua W. Busby, "Beyond Internal Conflict: The Emergent Practice of Climate Security," *Journal of Peace Research* 58 no. 1 (2021): 186–194.

43. Busby, "Climate Change and National Security," 5.

44. Busby, 11.

45. Department of Defense, *Quadrennial Defense Review, 2010*, 84–85.

46. Department of Defense, 89.

47. Defense Science Board, *Trends and Implications of Climate Change for National and International Security*, October 2011, x, https://fas.org/irp/agency/dod/dsb/climate.pdf.

48. Defense Science Board, *Trends and Implications of Climate Change for National and International Security*, xi.

49. Director of National Intelligence James Clapper, "Statement for the Record Worldwide Threat Assessment of the US Intelligence Community Senate Select Committee on Intelligence," January 29, 2014, https://www.odni.gov/files/documents/Intelligence%20Reports/2014%20WWTA%20%20SFR_SSCI_29_Jan.pdf.

50. Department of Defense, "Climate Change Adaptation Roadmap," 2014, 4.

51. President Barack Obama, *National Security Strategy*, The White House, Washington, DC, February 2015, 12, https://obamawhitehouse.archives.gov/sites/default/files/docs/2015_national_security_strategy_2.pdf.

52. White House, "Findings from Select Federal Reports: The National Security Implications of Climate Change," May 2015, 3.

53. For instance, see the National Intelligence Council, *Implications for U.S. National Security of Anticipated Climate Change*, September 21, 2016, https://www.dni.gov/files/documents/Newsroom/Reports%20and%20Pubs/Implications_for_US_National_Security_of_Anticipated_Climate_Change.pdf.

54. National Intelligence Council, *Implications for U.S. National Security of Anticipated Climate Change*.

55. The letter is found at https://langevin.house.gov/sites/langevin.house.gov/files/documents/01-11-18_Langevin_Stefanik_Letter_to_POTUS_Climate_Change_National_Security_Strategy.pdf.

56. James Stavridis, "America's Most Pressing Threat? Climate Change," *Bloomberg Opinion*, January 11, 2018, https://www.bloomberg.com/opinion/articles/2018-01-11/america-s-no-1-enemy-climate-change.

57. Daniel R. Coats, "Worldwide Threat Assessment of the US Intelligence Community, Statement for the Record," Senate Select Committee on Intelligence," January 29, 2019, 23, https://www.dni.gov/files/ODNI/documents/2019-ATA-SFR---SSCI.pdf.

58. International Military Council on Climate and Security, *World Climate and Security Report*, June 2021, https://imccs.org/the-world-climate-and-security-report-2021/.

59. Secretary of Defense Lloyd Austin, Memorandum for Senior Pentagon Leadership, Commanders of the Combatant Commands, Defense Agency and DOD Field Activities Directors, March 9, 2021, https://media.defense.gov/2021/Mar/10/2002597518/-1/-1/0/ESTABLISHMENT-OF-THE-CLIMATE-WORKING-GROUP.PDF.

60. Glenn Thrush and Julian E. Barnes, "Biden's Intelligence Director Vows to Put Climate at 'Center' of Foreign Policy," *New York Times*, May 10, 2021, https://www.nytimes.com/live /2021/04/22/us/biden-earth-day-climate-summit.

61. Austin, *Department of Defense Climate Adaptation Plan*, September 1, 2021.

62. Department of Defense, *Department of Defense Climate Risk Analysis,* https://media.defense.gov /2021/Oct/21/2002877353/-1/-1/0/DOD-CLIMATE-RISK-ANALYSIS-FINAL.PDF; The White House, *Report on the Impact of Climate Change on Migration*, October 2021, https://www .whitehouse.gov/wp-content/uploads/2021/10/Report-on-the-Impact-of-Climate-Change-on -Migration.pdf; Department of Homeland Security, "DHS Strategic Framework for Addressing Climate Change," https://www.dhs.gov/sites/default/files/publications/dhs_strategic_frame- work_10.20.21_final_508.pdf; National Intelligence Council, *National Intelligence Estimate: Climate Change And International Responses Facing Challenges to US National Security through 2040*, NIC-NIE-2021-10030-A, October 2021, https://www.dni.gov/files/ODNI/documents /assessments/NIE_Climate_Change_and_National_Security.pdf.

63. Hicks interview with Noel King, "Climate Change Is a Risk to National Security, the Penta- gon Says."

64. See Government Accountability Office, *Southwest Border Security: Actions Are Needed to Address the Cost of Readiness,* GAO-21–356, February 2021, https://www.gao.gov/assets /gao-21-356.pdf; Steve Beynon, "Military's Border Mission Will Continue for at Least Another Year," *Military.Com*, July 7, 2021, https://www.military.com/daily-news/2021/07 /07/militarys-border-mission-will-continue-least-another-year.html.

65. Christian Parenti, *Tropic of Chaos: Climate Change and the New Geography of Violence* (New York: Bold Type Books, 2011), 11.

66. Parenti, *Tropic of Chaos*, 207–224.

67. Department of Defense, *Department of Defense Climate Risk Analysis*, 8.

68. National Intelligence Council, *National Intelligence Estimate: Climate Change and Interna- tional Responses Facing Challenges to US National Security through 2040*, 11.

69. Department of Defense, *Department of Defense Climate Risk Analysis*, 8.

70. Department of the Army, Office of the Assistant Secretary of the Army for Installations, Energy and Environment, *United States Army Climate Strategy* (Washington, DC: February 2022), 4. https://www.army.mil/e2/downloads/rv7/about/2022_army_climate_strategy.pdf.

71. Department of the Army, *Climate Strategy*, 4–5.

72. Thomas F. Homer-Dixon, Jeffrey H. Boutwell, and George W. Rathjens, "Environmen- tal Scarcity and Violent Conflict," *Scientific American* 268, no. 2 (February 1993): 38–45; Thomas F. Homer-Dixon, "On the Threshold: Environmental Changes as Causes of Acute Conflict," *International Security* 16, no. 2 (Fall 1991): 76–116; Thomas F. Homer-Dixon, "Environmental Scarcities and Violent Conflict: Evidence from Cases," *International Security* 19, no. 1 (1994): 5–40; Thomas F. Homer-Dixon, "Strategies for Studying Causation in

Complex Ecological-Political Systems," *Journal of Environment and Development* 5, no. 2 (1996): 132–148; Thomas F. Homer-Dixon, *Environment, Scarcity and Violence* (Princeton, NJ: Princeton University Press, 1999).

73. Michael T. Klare, "Global Warming Battlefields: How Climate Change Threatens Security," *Current History*, 106, no. 703 (2007): 355–361.

74. See, for example, Peter H. Gleick, "Water, Drought, Climate Change, and Conflict in Syria," *Weather, Climate, and Security* 6, no. 2 (2014).

75. For a review of some of these cases, see Francis Galgano, ed., *The Environment-Conflict Nexus: Climate Change and the Emergent National Security Landscape* (Cham: Springer, 2019).

76. See Parenti, *Tropic of Chaos*.

77. For a review, see Daniel Moran, ed., *Climate Change and National Security: A Country-Level Analysis* (Washington, DC: Georgetown University Press, 2011).

78. Scholars questioning whether environmental stress inexorably cause conflict include Marc A. Levy, "Is the Environment a National Security Issue?" *International Security* 20, no. 2 (Fall 1995): 35–62, and Colin H. Kahl, *States, Scarcity and Civil Strife in the Developing World* (Princeton: Princeton University Press, 2006).

79. Nils Petter Gleditsch, "This Time Is Different! Or Is It? NeoMalthusians and Environmental Optimists in the Age of Climate Change," *Journal of Peace Research* 58, no. 1 (2021): 177–185. Also see Nina von Uexkull and Halvard Buhang, "Security Implications of Climate Change: A Decade of Scientific Progress," *Journal of Peace Research* 58, no. 1 (2021): 3–17.

80. Marwa Daoudy, *The Origins of the Syrian Conflict: Climate Change and Human Security* (Cambridge: Cambridge University Press, 2020).

81. Andrew M. Linke and Brett Reuther, "Weather, Wheat, and War: Security Implications of Climate Variability for Conflict in Syria," *Journal of Peace Research* 58 no. 1 (2021): 114–131.

82. See Katharine J. Mach et al., "Climate as a Risk Factor for Armed Conflict," *Nature* 571, July 11, 2019, 193–197; Kathrine J. Mach and Caroline M. Kraan, "Science–Policy Dimensions of Research on Climate Change and Conflict," *Journal of Peace Research* 58, no. 1 (2021): 168–176.

83. Alex Michaelowa and Tobias Koch have argued that responding to climate change "could be instrumental in reducing growth of international conflict potential due to climate change." They argue, "National security . . . will in any case be enhanced through lower dependence on fossil fuels which means less need to intervene in producer countries"; Michaelowa and Koch, "Military Emissions, Armed Conflicts," 393–394.

84. Jacob D. Petersen-Perlman, Jennifer C. Veilleux, and Aaron T. Wolf, "International Water Conflict and Cooperation: Challenges and Opportunities," *Water International* (January 2017): 105–120, https://www.tandfonline.com/doi/full/10.1080/02508060.2017.1276041. Also see National Research Council Committee on Abrupt Climate Change, *Abrupt Climate Change*, 146.

85. National Research Council, *Climate and Social Stress: Implications for Security*, eds., John D. Steinbruner, Paul C. Stern, and Jo L. Husbands, *Analysis* (Washington: The National Academies Press, 2013), 100.

86. See articles in Ashok Swain and Joakim Öjendal, *Routledge Handbook of Environmental Conflict and Peacebuilding* (New York: 2018); James R. Lee, *Environmental Conflict and Cooperation: Premise, Purpose, Persuasion and Promise* (New York: Routledge, 2020).

87. Tamar Meshel and Moin A. Yahya, "International Water Law and Fresh Water Dispute Resolution: A Cosean Perspective," *University of Colorado Law Review* 82, no. 2 (March 18, 2021), https://lawreview.colorado.edu/printed/international-water-law-and-fresh-water -dispute-resolution-a-cosean-perspective/.

88. Lee, *Environmental Conflict and Cooperation*: 157.

89. Von Lucke, "The Securitisation of Climate Change in the United States, 29.

CHAPTER 7

1. Mattis quoted in "Breaking the Tether of Fuel," *Marine Corps Gazette* 90, no. 8 (August 2006): 49–52; 49.

2. Magnuson quoted in Camille von Kaenel, "Energy Security Drives U.S. Military Toward Renewables," *Scientific American*, March 16, 2016, https://www.scientificamerican.com /article/energy-security-drives-u-s-military-to-renewables/.

3. Nonfood waste biofuels are blended specifically for this purpose from nonfood feed stocks such as carinata seed (also known as Ethiopian mustard), algae, and even wood.

4. Department of Defense *Quadrennial Defense Review, 2010*, 87.

5. Ray Mabus quoted in David Alexander, "'Great Green Fleet' Using Biofuels Deployed by U.S. Navy," *Reuters*, January 20, 2016, https://www.reuters.com/article/us-usa-defense-greenfleet -idUSKCN0UY2U4.

6. Jim Lane, "Launch of the Great Green Fleet," *The Digest*, January 20, 2016, https://www .biofuelsdigest.com/bdigest/2016/01/20/launch-of-the-great-green-fleet/.

7. Diane Ley, "The Great Green Fleet Makes History!," U.S. Department of Agriculture, February 21, 2017, https://www.usda.gov/media/blog/2012/08/15/great-green-fleet-makes-history.

8. Vilsack quoted in "U.S. Navy Deploys 'Great Green Fleet' in Bid to Reduce Military Energy Use," Environmental and Energy Study Institute, January 22, 2016, https://www.eesi.org /articles/view/u.s.-navy-deploys-great-green-fleet-in-bid-to-reduce-military-energy-use.

9. Department of the Army, Office of the Assistant Secretary of the Army for Installations, Energy and Environment, *United States Army Climate Strategy* (Washington, DC: February 2022), 4. https://www.army.mil/e2/downloads/rv7/about/2022_army_climate_strategy.pdf.

10. Michael Klare, "A Military Perspective on Climate Change Could Bridge the Gap Between Believers and Doubters," *The Conversation*, February 18, 2020, https://theconversation.com

/a-military-perspective-on-climate-change-could-bridge-the-gap-between-believers-and-doubters-128609.

11. Office of the Assistant Secretary of Defense for Sustainment, *Department of Defense Annual Energy Management and Resilience Report* FY1999, 13.

12. Marine Corps General John Mattis at the Committee on Armed Services, Hearing on National Defense Authorization Act for Fiscal Year 2012.

13. Department of Defense *Quadrennial Defense Review, 2010*, 87.

14. Heather L. Greenley, "Department of Defense Energy Management: Background and Issues for Congress," Congressional Research Service, R45832, July 25, 2019, https://fas.org/sgp/crs/natsec/R45832.pdf.

15. Office of the Assistant Secretary of Defense for Sustainment, *Department of Defense Annual Energy Management and Resilience Report FY*2019, 13.

16. Department of Defense, *Quadrennial Defense Review, 2010*, 87.

17. Jon Powers and Michael Wu, "A Clean Energy Agenda for the U.S. Department of Defense," Atlantic Council, January 14, 2021, https://www.atlanticcouncil.org/blogs/energysource/a-clean-energy-agenda-for-the-us-department-of-defense/.

18. Von Kaenel, "Energy Security Drives U.S. Military Toward Renewables."

19. On the other hand, a 2017 GAO report suggested that the U.S. military had not consistently taken the likely budgetary impacts of climate change into account. Government Accountability Office, *Climate Change Adaptation: DOD Needs to Better Incorporate Adaptation into Its Planning and Collaboration at Overseas Installations*, November 2017, https://www.gao.gov/assets/690/688323.pdf.

20. Jeremy Rosenberg, "U.S. Navy Bracing for Climate Change," *NASA Global Climate Change*, March 21, 2012, https://climate.nasa.gov/news/699/us-navy-bracing-for-climate-change/.

21. Department of Defense, *Quadrennial Defense Review, 2010*, 87.

22. See Marine Corps, *United States Marine Corps Energy Expeditionary Strategy and Implementation Plan: Bases to Battlefields*, 2010, https://www.hqmc.marines.mil/Portals/160/Docs/USMC%20Expeditionary%20Energy%20Strategy%20%20Implementation%20Planning%20Guidance.pdf.

23. Suzanne Goldenberg, "US Marines in Afghanistan Launch First Energy Efficiency Audit in War Zone," *Guardian*, August 13, 2009, https://www.theguardian.com/environment/2009/aug/13/us-marines-afghanistan-fuel-efficiency.

24. Spencer Ackerman, "Afghanistan's Green Marines Cut Fuel Use by 90 Percent," *Wired*, January 13, 2011, https://www.wired.com/2011/01/afghanistans-green-marines-cut-fuel-use-by-90-percent/.

25. Lauren Hepler, "Energy Meets Security: Can the Military Scale Green Power?" June 22, 2016, https://www.greenbiz.com/article/energy-meets-security-can-military-scale-clean-power.

26. Department of Defense, *Quadrennial Defense Review, 2010*, 87.

27. Department of Defense, *Energy for the Warfighter: The Operational Energy Strategy*, DOD Report to Congress, May 2011, https://www.acq.osd.mil/eie/Downloads/OE/Operational%20 Energy%20Strategy,%20Jun%2011.pdf; Department of Defense, "Operational Energy Strategy: Implementation Plan," March 2012, https://www.acq.osd.mil/eie/Downloads/OE /20120306_OE_Strategy_Implementation_Plan.pdf.

28. Department of Defense, *Energy for the Warfighter*, 1.

29. Department of Defense, 2.

30. Associated Press, "The Military Is Getting Greener, but That Clashes with Trump's Promises," *Fortune*, January 14, 2017, http://fortune.com/2017/01/14/military-oil-trump-green-power/.

31. Timothy Gardner, "U.S. Military Marches on Toward Green Energy, Despite Trump," Reuters, March 1, 2017, https://www.reuters.com/article/us-usa-military-green-energy -insight/u-s-military-marches-forward-on-green-energy-despite-trump-idUSKBN1683BL.

32. Executive Order 13693 of March 19, 2015, "Planning for Federal Sustainability in the Next Decade" was revoked by President Trump with Executive Order 13834 on May 17, 2018. See https://www.fedcenter.gov/programs/eo13834/ and https://www.fedcenter.gov /programs/eo13693/.

33. Gardner, "U.S. Military Marches on Toward Green Energy, Despite Trump."

34. Department of the Army, *Climate Strategy*, 5.

35. Department of the Army, *Climate Strategy*, 8.

36. Department of the Army, *Climate Strategy*, 8.

37. See Reuters, "Military Getting Greener," http://fingfx.thomsonreuters.com/gfx/rngs/USA -TRUMP-ENERGY-MILITARY/0100400G00X/index.html; Gardner, "U.S. Military Marches on Toward Green Energy, Despite Trump."

38. Wilson Rickerson, Michael Wu, and Meredith Pringle, "Beyond the Fence Line: Strengthening Military Capabilities Through Energy Resilience Partnerships," Association of Defense Communities, November 2018, https://static1.squarespace.com/static/58c0207d15d5db7d6b968444 /t/5ea73aa1f943bc783bc072bb/1588017862496/Beyond+The+Fence+Line.pdf.

39. See the Defense Logistics Agency fact books for FY2014–FY2020, https://www.dla.mil /Energy/About/Library/.

40. Department of Defense, *Sustainability Report& Implementation Plan, 2020*: 14.

41. Office of the Assistant Secretary of Defense for Sustainment, *Department of Defense Annual Energy Management and Resilience Report* FY2019, 5, 13, 24, 77, and 74.

42. Department of Defense, *Sustainability Report & Implementation Plan 2020*, 14, https://www .sustainability.gov/pdfs/dod-2020-sustainability-plan.pdf.

43. Office of the Assistant Secretary of Defense for Sustainment, *Department of Defense Annual Energy Management and Resilience Report* FY2016, 35; Department of Defense, *Fiscal Year*

2020 Annual Energy Management Report, https://www.acq.osd.mil/eie/Downloads/IE/FY%202002%20AEMR.pdf.

44. Carolyn Fortuna, "Otis Microgrid: Cape Cod Military Base to Run Fully on Renewable Energy," *Clean Technica*, September 10, 2018, https://cleantechnica.com/2018/09/10/otis-microgrid-cape-cod-military-base-to-run-fully-on-renewable-energy/; NS Energy Staff Writer, "Ecoult Ultrabattery System Installed in Otis Air National Guard Base Microgrid, USA," *NS Energy*, October 17, 2019; Elisa Wood, "Ameresco to Break Ground This Week on Solar and Storage Microgrid at Fort Hunter Liggett," *Microgrid Knowledge*, May 26, 2021, https://microgridknowledge.com/military-microgrids-liggett-ameresco/.

45. Jet-biofuel mixtures are called hydrotreated renewable jet fuels or HRJs. 95th Air Base Wing Public Affairs, "F-22 Raptor Flown on Synthetic Biofuel," Wright Patterson AFB, March 21, 2011, https://www.wpafb.af.mil/News/Article-Display/Article/399906/f-22-raptor-flown-on-synthetic-biofuel/#:~:text=photo%2FKevin%20North)-,EDWARDS%20AIR%20FORCE%20BASE%2C%20Calif.,plant%20not%20used%20for%20food.

46. Office of the Assistant Secretary of Defense for Sustainment, *Department of Defense Annual Energy Management and Resilience Report*, FY2016, 27.

47. Nabila A. Huq et al., "Toward Net-Zero Sustainable Aviation Fuel with Wet Waste-Derived Volatile Fatty Acids," *PNAS Proceedings of the National Academy of Sciences* 118, no. 13 (March 30, 2021), https://www.pnas.org/content/118/13/e2023008118.

48. See Office of the Under Secretary of Defense for Acquisition and Sustainment, *Operational Energy Annual Report, Fiscal Year 2019*, appendix B, https://www.acq.osd.mil/eie/Downloads/OE/FY19%20OE%20Annual%20Report.pdf.

49. This will change when the sustainable aviation fuel biorefinery planned near Trenton, North Dakota, is finished in 2022. Ron Kotrba, "100mgy SAF Project in North Dakota Moves Closer to Realization," *Biobased Diesel Daily*, June 7, 2021, https://www.biobased-diesel.com/post/100-mgy-saf-project-in-north-dakota-moves-closer-to-realization?utm_campaign=4726a444-e656-4979-9dae-16d0669cfdfe&utm_source=so&utm_medium=mail&cid=54d0cb5d-90d2-4af8-8fe4-918af6c4d31e.

50. Ian Duncan, "Biden Administration Sets Goal of Replacing all Jet Fuel with Sustainable Alternatives by 2050," *Washington Post*, September 9, 2021, https://www.washingtonpost.com/transportation/2021/09/09/jets-sustainable-aviation-fuel-goal/.

51. Department of Defense, "Climate Change Adaptation Roadmap," 2014, 1.

52. Department of Defense, "DOD Directive 4715.21, Climate Change Adaptation and Resilience," January 14, 2016, 10–11, https://dod.defense.gov/Portals/1/Documents/pubs/471521p.pdf.

53. Department of Defense, Office of the Undersecretary of Defense (Acquisition and Sustainment), *Department of Defense Climate Adaptation Plan*, September 1, 2021.

54. Rachel S. Cohen, "USAF Fully Funded for Tyndall, Offutt Rebuilds," *Air Force Magazine*, February 18, 2020, https://www.airforcemag.com/usaf-fully-funded-for-tyndall-offutt-rebuilds/;

Erika Orstad, "Tyndall Air Force Base Begins 'Base of the Future' Construction," October 13, 2020, *Mypanhandle.com*, https://www.mypanhandle.com/news/tyndall-afb/tyndall-air-force -base-begins-base-of-the-future-construction/.

55. Martin, "Critical Air Base Flooded in Nebraska; AP, "Air Force Estimates $420M Needed to Rebuild Nebraska Base."

56. David Hasemeyer, "U.S. Military Knew the Flood Risks at Nebraska's Offutt Air Force Base, but Didn't Act in Time," *ABC News*, March 21, 2019, https://www.nbcnews.com/news/us -news/u-s-military-knew-flood-risks-nebraska-s-offutt-air-n985926.

57. Losey, "After Massive Flood, Offutt Looks to Build a Better Base."

58. Kiran S. Jivnani and Inkoo Kang, "Building Smarter Military Bases for Climate Resilient Communities," *Atlantic Council*, October 1, 2021, https://www.atlanticcouncil.org/blogs /geotech-cues/building-smarter-military-bases-for-climate-resilient-communities/.

59. Cohen, "USAF Fully Funded for Tyndall, Offutt Rebuilds."

60. Austin quoted in David Vergun, "Action Team Leads DOD Efforts to Adapt to Climate Change Effects," *DOD News*, April 22, 2021, https://www.defense.gov/Explore/News/Article/Article /2577354/action-team-leads-dod-efforts-to-adapt-to-climate-change-effects/.

61. Kidd quoted in David Vergun, "DOD Working to Mitigate Climate Change Effects on Installations," *DOD News*, April 22, 2021, https://www.defense.gov/Explore/News/Article /Article/2572665/dod-working-to-mitigate-climate-change-effects-on-installations/.

62. The White House, "Fact Sheet: President Biden's Leaders Summit on Climate," April 23, 2021, https://www.whitehouse.gov/briefing-room/statements-releases/2021/04/23/fact -sheet-president-bidens-leaders-summit-on-climate/.

63. The tool is described in Department of Defense, "DoD Climate Assessment Tool," April 2021, https://media.defense.gov/2021/Apr/05/2002614579/-1/-1/0/DOD-CLIMATE -ASSESSMENT-TOOL.PDF.

64. The White House, "Fact Sheet: President Biden's Leaders Summit on Climate."

65. General Services Administration, "Alongside the EPA, DOD, and DHS, GSA Announces New Actions to Reduce Emissions of Super-Polluting Hydrofluorocarbons," September 23, 2021, https://www.gsa.gov/about-us/newsroom/news-releases/alongside-the -epa-dod-and-dhs-gsa-announces-new-actions-to-reduce-emissions-of-superpolluting -hydrofluorocarbons%C2%A0-09232021.

66. Surash quoted in Karlie Goldenberg, "Army Grapples with Very Serious Climate Change Threat," *Association of the United States Army*, October 13, 2021, https://www.ausa.org/news /army-grapples-very-serious-climate-change-threat.

67. National Defense Authorization Act for Fiscal Year 2022, Public Law 117-81, Section 323, passed and signed December 27, 2021, https://www.congress.gov/bill/117th-congress/senate -bill/1605/text.

68. Joseph Biden, "Executive Order Catalyzing America's Clean Energy Economy through Federal Sustainability," December 8, 2021, https://www.whitehouse.gov/briefing-room/presidential-actions/2021/12/08/executive-order-on-catalyzing-clean-energy-industries-and-jobs-through-federal-sustainability/. The White House did not mention the exemption in the fact sheet.

69. "Executive Order Catalyzing America's Clean Energy Economy Through Federal Sustainability," Section 602.

70. Military spending on mitigation would be an important indicator of whether mitigation is receiving an appropriate level of attention at the DOD. But through the FY2022 DOD Budget, it has been difficult to determine how much the Pentagon is spending on adaptation and mitigation. Deputy Defense Secretary Kathleen Hicks said this would change with the FY2023 budget. Sebastian Sprenger, "Next Pentagon Budget Will Detail Climate Change Spending," *Defense News*, September 8, 2021, https://www.defensenews.com/smr/defense-news-conference/2021/09/08/next-pentagon-budget-will-detail-climate-change-spending/.

CHAPTER 8

1. For a discussion of different conceptions of security, see Rita Floyd and Richard A. Matthew, eds., *Environmental Security: Approaches and Issues* (New York: Routledge, 2013); and McDonald, *Ecological Security*.

2. Department of Defense, "DOD Background Paper on a National Security Provision for the Proposed Climate Change Protocol," September 5, 1997. Reprinted in *Inside the Pentagon* 13, no. 41 (October 9, 1997) 1, 10–13; 12.

3. Department of Defense, "DOD Background Paper on a National Security Provision," 13.

4. Department of Defense, 13.

5. Congressman Henry Waxman, May 20, 1998, in *Congressional Record* 144, part 7, Proceedings and Debates of the 105th Congress, 9983.

6. National Intelligence Council, *National Intelligence Estimate: Climate Change and International Responses Increasing Challenges to US National Security Through 2040.*

7. International Institute for Strategic Studies, *The Military Balance, 2021* (London: Routledge, 2021), 23.

8. Stockholm International Peace Research Institute, SIPRI, "Military Expenditure Database," https://www.sipri.org/databases/milex, accessed June 25, 2021.

9. Stockholm International Peace Research Institute, *SIPRI Yearbook 2021*, chap. 10, "World Nuclear Forces," https://sipri.org/sites/default/files/2021-06/yb21_10_wnf_210613.pdf, accessed June 25, 2021.

10. American Security Project, "Powering the Department of Dense: Initiatives to Increase Resiliency and Energy," September 2017, https://www.americansecurityproject.org/wp-content/uploads/2017/09/Ref-0204-Powering-the-DoD.pdf.

11. Benjamin Chiacchia, "What the Great Green Fleet Could Still Be," *Proceedings* 146, no. 6 (June 2020), https://www.usni.org/magazines/proceedings/2020/june/what-great-green -fleet-could-still-be, accessed July 7, 2021.

12. Chiacchia, "What the Great Green Fleet Could Still Be."

13. See Glaser and Kelanic, eds., *Crude Strategy*; John Glaser, "Does the U.S. Military Actually Protect Middle East Oil?," Cato Institute, January 9, 2017, https://www.cato.org/publications /commentary/does-us-military-actually-protect-middle-east-oil; Emma Ashford, "Unbal- anced: Rethinking America's Commitment to the Middle East," *Security Studies Quarterly* 12, no. 1 (Spring 2018): 127–148. Also see Milton R. Copulos, *America's Achilles Heel: The Hidden Cost of Imported Oil*, The National Defense Council Foundation, Washington, DC, October 2003, http://citeseerx.ist.psu.edu/viewdoc/download;jsessionid=DD3F77E8166 A096D9F1BB3B615199125?doi=10.1.1.186.7523&rep=rep1&type=pdf; Joshua Rovner, "After America: The Flow of Persian Gulf Oil in the Absence of US Military Force," in Glaser and Kelanic, *Crude Strategy*, 141–165; Eugene Gholz, "Nothing Much to Do: Why America Can Bring All Troops Home from the Middle East," *Quincy Paper No. 7*, Quincy Institute for Responsible Statecraft, June 24, 2021, https://quincyinst.org/report/nothing-much-to-do -why-america-can-bring-all-troops-home-from-the-middle-east/; Barry R. Posen, *Restraint: A New Foundation for U.S. Grand Strategy* (Ithaca, NY: Cornell University Press, 2014), 106–113 and 132.

14. Greenley, "The Strategic Petroleum Reserve," 6.

15. See Chapman, "Gulf Oil and International Security," 86.

16. See Moran and Russell, "Introduction"; Klare, "Petroleum Anxiety."

17. See Rovner, "After America"; Gholz, "Nothing Much to Do"; Posen, *Restraint*.

18. Rovner, "After America," 160.

19. Glaser, "Withdrawing from Overseas Bases," 2.

20. Andrew Gustafson, "Strategy in the Gulf: It's No Longer 1979," *Proceedings* 146, no. 6 (June 2020), https://www.usni.org/magazines/proceedings/2020/june/strategy-gulf-its-no-longer -1979.

21. Feis, "Petroleum and American Foreign Policy," 18–19.

22. Feis, 20.

23. Truly and Alm, "*Report of the Defense Science Board on More Capable Warfighting*, 1.

24. Truly and Alm, 7.

25. Source: U.S. Energy Information Administration, https://www.eia.gov/dnav/pet/hist /LeafHandler.ashx?n=PET&s=MCRIMUSPG2&f=A. The total level of U.S. oil importa- tion has also declined in recent years.

26. U.S. Energy Information Administration data, Primary Energy Consumption Estimates by Source, https://www.eia.gov/totalenergy/data/monthly/pdf/sec1_7.pdf, accessed July 4, 2021.

27. Johan Schot and Laur Kanger, "Deep Transitions: Emergence, Acceleration, Stabilization and Directionality," *Research Policy* 47 (2018): 1045–1059; 1045.

28. Gregory Hooks and Chad L. Smith, "Treadmills of Production and Destruction: Threats to the Environment Posed by Militarism," *Organization & Environment* 18, no. 1 (March 2005): 19–37.

29. Gregory Hooks, Michael Lengefeld, and Chad L. Smith, "Recasting the Treadmills of Production and Destruction," *Sociology of Development* 7, no. 1 (2021): 52–76; 53.

30. I am influenced here by theories of emergence and complexity.

31. Johnstone and McLeish, "World Wars and the Age of Oil," 101732–101744.

32. Johnstone and McLeish, 101740.

33. Johnstone and McLeish, 101742.

34. Johnstone and McLeish, 101742.

35. See Bildirici, "The Causal Link."

36. Federal Energy Administration, *Project Independence: A Summary* (Washington: U.S. Government Printing Office, November 1974), 4.

37. Corrie E. Clark, "Renewable Energy R&D Funding History: A Comparison with Funding for Nuclear Energy, Fossil Energy, Energy Efficiency, and Electric Systems R&D," Congressional Research Service, RS22858, June 18, 2018, 4, https://fas.org/sgp/crs/misc/RS22858.pdf.

38. Jorgenson, Brett, and Jeffrey Kentor, "Militarization and the Environment," 22. Also see Bildirici, "The Causal Link."

39. John Hamilton Bradford and Alexander M. Stoner, "The Treadmill of Destruction in Comparative Perspective: A Panel Study of Military Spending and Carbon Emissions, 1960–2014," *Journal of World-Systems Research*, 23, no. 2 (2017): 298–325; 320. Also see Bildirici, "The Causal Link."

40. Parkinson, "The Environmental Impact of the UK Military Sector," 12.

41. See Clark, Jorgenson, and Kentor, "Militarization and Energy Consumption," 35.

42. Clark, Jorgenson, and Kentor, 38.

43. I have not counted the emissions resulting from research, development, and construction of nuclear weapons here.

44. Thomas K. Duncan and Christopher J. Coyne, "The Origins of the Permanent War Economy," *The Independent Review* 18, no. 2 (Fall 2013): 219–240.

45. Rebecca U. Thorpe, *The American Warfare State: The Domestic Politics of Military Spending* (Chicago: University of Chicago Press, 2014), 165. Also see Linda Bilmes, *The Ghost Budget*, forthcoming.

46. For a summary and discussion see John Duffield, *Over a Barrel: The Costs of U.S. Foreign Oil Dependence* (Stanford: Stanford University Press, 2008), 157–182; and Mark A. Delucchi and James J. Murphy, "US Military Expenditures to Protect the Use of Persian Gulf Oil for Motor Vehicles," *Energy Policy* 36 (2008): 2243–2264.

47. Emphasis added. U.S. General Accounting Office (GAO), *Southwest Asia, Costs of Protecting U.S. Interests*, GAO NSIAD-91–240, Washington, DC, August 1991, 1, https://www.gao .gov/assets/nsiad-91-250.pdf.

48. GAO, *Southwest Asia, Costs of Protecting U.S. Interests*, 1–2.

49. GAO, 2.

50. Duffield, *Over a Barrel*, 169–182.

51. Delucchi and Murphy, "US Military Expenditures to Protect the Use of Persian Gulf Oil for Motor Vehicles," 2261.

52. Eugene Gholz, "U.S. Spending on Its Military Commitments to the Persian Gulf," in Glaser and Kelanic, *Crude Strategy,* 167–195.

53. Securing America's Future Energy, "The Military Cost of Defending the Global Oil Supply," September 21, 2018, http://secureenergy.org/wp-content/uploads/2020/03/Military -Cost-of-Defending-the-Global-Oil-Supply.-Sep.-18.-2018.pdfhttp://secureenergy.org/wp -content/uploads/2020/03/Military-Cost-of-Defending-the-Global-Oil-Supply.-Sep.-18. -2018.pdf.

54. Kai-Hua Wang et al., "Whether Crude Oil Dependence and CO2 Emissions Influence Military Expenditures in Net Importing Countries," *Energy Policy* 153 (June 2021): 112281–112290; Chia-Yi Lee, "China's Energy Diplomacy: Does Chinese Foreign Policy Favor Oil-Producing Countries?," *Foreign Policy Analysis* 15, no. 4 (October 2019): 570–588.

55. David Styan, "China's Maritime Silk Road and Small States: Lessons from the Case of Djibouti," *Journal of Contemporary China* 29 no. 122 (2020): 191–206.

56. McKenzie, Jr., "Posture Statement of General Kenneth F. McKenzie, Jr."

57. Aaron L. Friedberg, "What's at Stake in the Indo-Pacific," U.S. Naval Institute *Proceedings* 147, no. 10 (October 2021), https://www.usni.org/magazines/proceedings/2021/october /whats-stake-indo-pacific.

58. Stockholm International Peace Research Institute, "SIPRI Arms Industry Database." q

59. Harvey Sapolsky and Eugene Gholz, "Private Arsenals: America's Post-Cold War Burden," in Markusen and Costigan, *Arming the Future,* 191–206; 201. See also Aerospace Industries Association (AIA), "2020 Facts and Figures, U.S. Aerospace and Defense," 3, http://aiafactstg .wpengine.com/.

60. Hannah Arendt, *The Origins of Totalitarianism* (New York: Harcourt, 1968), xviii.

61. Department of the Army, *Climate Strategy*, 13.

62. Department of Defense, *Department of Defense Climate Risk Analysis*, 6. Also see the National Intelligence Council, "National Intelligence: Climate Change and International Responses Increasing Challenges to US National Security Through 2040."

63. See Department of the Army, *Climate Strategy*, 11.

64. I thank Alexander Thompson for raising this point.

65. Further, while the rest of the U.S. government does not use coal, the U.S. military still has coal power generation at some bases. Those coal plants should be a high priority for closure or conversion to another fuel.

66. Christopher T. Mann, *Base Closure and Realignment (BRAC): Background Issues for Congress*, Congressional Research Service, April 25, 2019, 7, https://www.everycrsreport.com/files /20190425_R45705_9e300ef394d6f4dabc78a7ef8fbbc33ef9bd01e7.pdf.

67. Department of Defense, *Department of Defense Infrastructure Capacity, October 2017*, 2, https://man.fas.org/eprint/infrastructure.pdf; Department of Defense, *Department of Defense Infrastructure Capacity, March 2016*, interim report. Mann notes that "In April 2016, DOD submitted to the House Armed Services Committee an *Infrastructure Capacity Report* (interim version) that assessed 22% of the Department's base infrastructure excess to its needs." Mann, *Base Closure and Realignment (BRAC)*, 9.

68. Leo Shane, "Plans for a New Base Closing Round May Be Running out of Time: Report," *Military Times* 15 (August 2019), https://www.militarytimes.com/news/pentagon-congress /2019/08/15/plans-for-a-new-base-closing-round-may-be-running-out-of-time-report/.

69. For ideas on base closures, see Glaser, "Withdrawing from Overseas Bases," 4; Michael J. Lostumbo et al., *Overseas Basing: An Assessment of Relative Costs and Strategic Benefits*, RAND Corporation, 2013, https://www.rand.org/content/dam/rand/pubs/research_reports/RR200 /RR201/RAND_RR201.pdf.

70. See, for instance, Michael Touchton and Amanda J. Ashley, *Salvaging Community: How American Cities Rebuild Closed Military Bases* (Ithaca: Cornell University Press 2019).

APPENDIX

1. Thomas, *The Securitization of Climate Change*, 31.

2. See, for instance, Nikki Reisch and Steve Kretzman, "A Climate of War: The War in Iraq and Global Warming," *Oil Change International* (March 2008), http://priceofoil.org/content /uploads/2008/03/A%20Climate%20of%20War%20FINAL%20(March%2017%20 2008).pdf.

3. Fiscal years for the U.S. government start on October 1 and end on September 30. Prior to 1976, the federal fiscal year began on July 1 and ended on June 30.

4. Lin and Burton, *Indefensible,* 18.

5. Lin and Burton, 5.

6. Belcher et al., "Hidden Carbon Cost of the 'Everywhere War.'"

7. Liska and Perrin, "Securing Foreign Oil," 11.

8. The Department of Energy, Comprehensive Annual Energy Data and Sustainability Performance, https://ctsedwweb.ee.doe.gov/Annual/Report/Report.aspx, accessed June 24, 2021.

9. For instance, see Defense Logistics Agency Energy, *Fiscal Year 2020 Fact Book*.

10. The Pentagon calculates fuel consumption for internal planning purposes, but the DOD has explicitly withheld this information in its reporting to Congress. The DOD's OP-26A form "POL Consumption and Costs" explicitly states that fuel consumption data is not to be shared with Congress: "The OP-26A exhibit will not be included in justification material forwarded to Congress." Emphasis in the original. Department of Defense, Comptroller, DOD Financial Management Regulation, Chapter 3, pp. 3–108, https://comptroller.defense .gov/Portals/45/documents/fmr/archive/02aarch/02a_03old.pdf.

11. See Department of Energy, "Federal Government Energy and Water Use in 2020," Comprehensive Annual Energy Data and Sustainability Performance, https://ctsedwweb.ee.doe.gov /Annual/Report/Report.aspx.

12. Department of Energy, Comprehensive Annual Energy Data and Sustainability Performance, http://ctsedwweb.ee.doe.gov/Annual/Report/ComprehensiveGreenhouseGasGH GInventoriesByAgencyAndFiscalYear.aspx, data as of June 1, 2018; Department of Energy, "Government-wide Energy Use (BBtu)," Comprehensive Annual Energy Data and Sustainability Performance, http://ctsedwweb.ee.doe.gov/Annual/Report/HistoricalFederalEnergy ConsumptionDataByAgencyAndEnergyTypeFY1975ToPresent.aspx.

13. Carbone dioxide, methane, and nitrous oxide emission factors for each fuel are from Office of Energy Efficiency & Renewable Energy, *Federal Comprehensive Annual Energy and Reporting Requirements*, Federal Energy Management Program, https://www.energy.gov/eere/femp /federal-facility-consolidated-annual-reporting-requirements.

14. GWP emissions coefficients, https://www.eia.gov/environment/emissions/co2_vol_mass .php.

15. Because the heat content and greenhouse gas emission of the various products of crude oil (e.g., diesel and jet fuel) are different, calculations must use the specific heat content and emissions profiles for each fuel. The average heat content of crude oil is 5.80 MMBTU per barrel. The average carbon coefficient of crude oil is 20.31 kg carbon per MMBTU. The fraction oxidized is 100 percent. 5.80 MMBTU/barrel $\times$ 20.31 kg C/MMBTU $\times$ 44 kg CO_2/12 kg C $\times$ 1 metric ton/1,000 kg = 0.43 metric tons CO_2/barrel. Sources for emissions factors, Environmental Protection Agency, "Emissions Factors for Greenhouse Gas Inventories," March 2018, https://www.epa.gov/sites/default/files/2018-03/documents/emission -factors_mar_2018_0.pdf.

16. PFCs, HFCs, NF3, and SF6 have global warming potentials that range from 7,390 to 22,800. While the global warming effects of methane, nitrous oxide, and water vapor are well understood, when they are emitted during jet fuel combustion at high altitudes the effects are not as well understood as the effects of CO2. See the Environmental Protection Agency, "Emissions of Flourinated Gases," https://www.epa.gov/ghgemissions/overview-greenhouse -gases#f-gases; and "Understanding Global Warming Potentials," https://www.epa.gov /ghgemissions/understanding-global-warming-potentials. The Department of Energy uses the EPA GWP factors. See their *Energy Management Data Report*, https://www.energy.gov

/eere/femp/downloads/annual-energy-management-data-report; and Council on Environmental Quality, *Federal Greenhouse Gas Accounting and Reporting Guidance*, January 17, 2016, 4, https://www.sustainability.gov/pdfs/federal_ghg%20accounting_reporting-guidance.pdf.

17. IPCC Second Assessment Report. See the IPCC Fifth Assessment Report, *Climate Change 2014: Synthesis Report. Contribution of Working Groups I, II and III to the Fifth Assessment Report of the Intergovernmental Panel on Climate Change,* ed. R. K. Pachauri and L. A. Meyer (Geneva: IPCC, 2014). https://www.ipcc.ch/site/assets/uploads/2018/02/SYR_AR5 _FINAL_full.pdf.

18. European Environment Agency, European Union Aviation Safety Agency, and Eurocontrol, *European Aviation Environment Report, 2019*, 88, https://www.easa.europa.eu/eaer/system /files/usr_uploaded/219473_EASA_EAER_2019_WEB_LOW-RES.pdf. Also see Martin Cames, Jakob Graichen, Anne Siemons, and Vanessa Cook, *Emission Reduction Targets for International Aviation and Shipping*, Policy Department A: Economic and Scientific Policy, European Union, November 2015, 13–14, http://www.europarl.europa.eu/RegData/etudes /STUD/2015/569964/IPOL_STU(2015)569964_EN.pdf.

19. See Ulrike Burkhardt and Bernd Kärcher, "Global Radiative Forcing from Contrail Cirrus," *Nature Climate Change* (March 2011): 54–58; and Ulrike Burkhardt, Lisa Bock, and Andreas Bier, "Mitigating the Contrail Cirrus Climate Impact by Reducing Aircraft Soot Number Emissions," *Climate and Atmospheric Science* (October 2018), https://www.nature .com/articles/s41612-018-0046-4.

20. Note for table 2-13 Transportation Related Greenhouse Gas Emissions, Environmental Protection Agency, *Inventory of U.S. Greenhouse Gases and Sinks: 1990–2019*, 2–38.

21. The EPA notes that "Civil aviation comprises aircraft used for the commercial transport of passengers and freight, military aviation comprises aircraft under the control of national armed forces, and general aviation applies to recreational and small corporate aircraft." *Environmental Protection Agency, Inventory of U.S. Greenhouse Gases and Sinks: 1990–2019*, 3–117.

22. Environmental Protection Agency, *Inventory of U.S. Greenhouse Gases and Sinks: 1990–2019*, 3–26.

23. Note for table 2-13, "Transportation Related Greenhouse Gas Emissions," in Environmental Protection Agency, *Inventory of U.S. Greenhouse Gases and Sinks: 1990–2019*, 2-38–2-39.

24. The IPCC method is not decisive. "The 2006 IPCC Guidelines do not provide a distinct method for calculating military water-borne emissions. Emissions from military water-borne fuel use can be estimated using the equation 3.5.1 and the same calculation approach is recommended for non-military shipping. Due to the special characteristics of the operations, situations, and technologies (e.g., aircraft carriers, very large auxiliary power plants, and unusual engine types) associated with military water-borne navigation, a more detailed method of data analysis is encouraged when data are available. Inventory compilers should therefore consult military experts to determine the most appropriate emission factors for the country's military water-borne navigation." IPCC, *2006 IPCC Guidelines for National Greenhouse Gas Inventories*, vol. 2, 3.53.

25. Environmental Protection Agency, *Inventory of U.S. Greenhouse Gases and Sinks: 1990–2019*, 3–120.

26. Environmental Protection Agency, 3–120.

27. Environmental Protection Agency, 3–37.

28. "Additionally, there are uncertainties in historical aircraft operations and training activity data. Estimates for the quantity of fuel actually used in Navy and Air Force flying activities reported as bunker fuel emissions had to be estimated based on a combination of available data and expert judgment. Estimates of marine bunker fuel emissions were based on Navy vessel steaming hour data, which reports fuel used while underway and fuel used while not underway. This approach does not capture some voyages that would be classified as domestic for a commercial vessel. Conversely, emissions from fuel used while not underway preceding an international voyage are reported as domestic rather than international as would be done for a commercial vessel. There is uncertainty associated with ground fuel estimates for 1997 through 2019, including estimates for the quantity of jet fuel allocated to ground transportation. Small fuel quantities may have been used in vehicles or equipment other than that which was assumed for each fuel type.

There are also uncertainties in fuel end-uses by fuel type, emissions factors, fuel densities, diesel fuel sulfur content, aircraft and vessel engine characteristics and fuel efficiencies, and the methodology used to back-calculate the data set to 1990 using the original set from 1995. The data were adjusted for trends in fuel use based on a closely correlating, but not matching, data set. All assumptions used to develop the estimate were based on process knowledge, DoD data, and expert judgments. The magnitude of the potential errors related to the various uncertainties has not been calculated but is believed to be small. The uncertainties associated with future military bunker fuel emission estimates could be reduced through revalidation of assumptions based on data regarding current equipment and operational tempo, however, it is doubtful data with more fidelity exist at this time." *Environmental Protection Agency, Inventory of U.S. Greenhouse Gases and Sinks: 1990–2019*, 3-120–121.

Index

Page numbers in italics indicate figures or tables.

preserving and sequestration, 206, 227, 239

use during Civil War, 34–35

Forrestal, James, 68

Fossil fuel use. *See* Coal; Oil

Framework Convention on Climate Change, United Nations (UNFCCC). *See* United Nations Friedberg, Aaron, 280–281

Fuel energy content, 105

Fulcrum Sierra Biofuels, 244

Fulton, Robert, 29

Fulton (steamer), 32

Garfield, Harry, 50

General Dynamics, 167, 169

General Electric, 168, 169

Geothermal energy, growth in consumption of, 265, 266

Germany

Luftwaffe, 54

nonaggression pact with Soviet Union, 54

World War I, 49–50

World War II oil strategy, 53–55

Gholz, Eugene, 262, 279

Glaser, John, 263

Glasgow Conference of the Parties, 218

Gleditsch, Nils Petter, 225

Global Business Network, 209–210

Global warming. *See* Climate change

Global War on Terrorism (GWOT), 89

Goodman, Sherri W., 208–209, 211, 217

Gorbachev, Mikhail, 82, 85

Gore, Al, 207, 213

Göring, Hermann, 54

Graham, Lindsey, 93

Grand Ethiopian Renaissance Dam (GERD), 226–227

Gray, Steven, 46

Great Green Fleet, 231–232, 243, 258–259

Great White Fleet, 43–45, 132, 159, 184, 194

emissions, 134–135, 337n

Greenhouse effect, 103

Greenhouse gas (GHG) emissions

calculating, 2–4, 103–104, 302 (*see also* Military emissions)

reference point for, 104

omission of military emissions, 104

Greenhouse gases, 4

Greening Government Initiative, 248, 249

Greenspan, Alan, 90

Guam, 80

climate-related vulnerabilities, *193*, 194–195, 197

coaling station on, 41

Guantanamo Bay, Cuba, 43

Gulf War (1991), 83–85, 90, 92, 130, 260, 278

fuel use, 184

military emissions during, 150, 152, 154, 172

oil field burned during, 172, 260

Operation Desert Shield, 83–84, 278

Operation Desert Storm, 84, 158

Gunn, Lee, 211–212

Gustafson, Andrew, 263

Hagel, Chuck, 114, 116, 118, 211

Hague Center for Strategic Risks, 217–218

Haig, Alexander, 76, 117

Haines, Avril, 218

Haiti, deforestation in, 209

Hamilton, Lee, 74, 77

Hampton Roads (Virginia), 33, 40, 43–44, 193–194

Hansen, James, 109

Harrison, Benjamin, 38

Hawaii

climate-related vulnerabilities, *193*, 194–195

Pearl Harbor, 38, 40, 43, 56, 69, 194

and Tyler Doctrine, 30

U.S. annexation of, 41

Heat exhaustion and heat stroke, 200

Heath, Edward, 76